# Controversies
# in the Earth Sciences

# Controversies in the Earth Sciences

## A Reader

Richard Cowen
Jere H. Lipps
**University of California, Davis**

**WEST PUBLISHING CO.**  St. Paul • New York • Boston
Los Angeles • San Francisco

Library of Congress Cataloging in Publication Data

   Cowen, Richard, 1940–      comp.
      Controversies in the earth sciences.
       1.  Earth sciences—Addresses, essays, lectures.
I.  Lipps, Jere H., 1939 -      joint comp.  II.  Title.
QE35.C68     550'.8     75–1395
**ISBN** 0–8299–0044–6

# Preface

For a long time geology has been predominantly a descriptive science. The earth is indeed vastly variable and complex, and much description was and is still necessary. Yet there is a much more exciting aspect of geology: one that presents information about the earth in the context of dynamic processes. Here we move from description to inference and even speculation, and immediately it seems to us that the science takes on a vigor and excitement that can be truly thrilling. In our courses in beginning geology we have found that students respond and become involved to a higher degree when we use this dynamic approach. One of the most successful parts of this approach, even for nonmajors, is to focus attention on the current controversies in earth science. Almost any lecture material and textbook can be supplemented or enhanced by this technique, and we have found that readings from the original scientific literature best serve this purpose.

This approach serves a second valuable educational goal. After having been exposed to the original works on both sides of a scientific controversy, students rapidly become aware that science is a creative process, that it is not all dogma simply to be learned, and that they themselves could learn to practise this kind of science. They seem to absorb much more geology, and to enjoy absorbing it.

We have therefore assembled readings centered on several of the major current controversies in earth science. The readings are necessarily brief, but they are chosen with the intention of giving a concise and up-to-date overview of each problem. We intend that each controversy shall be fully comprehensible to a nonmajor student provided that he has read and understood the relevant part of any introductory text in geology.

Although the reader has been compiled for use by students in introductory courses, the controversies we have chosen are so topical and so important to modern geology that we expect the reader will serve as seed material for upper division and even graduate level general courses, discussion groups and seminars.

We have written brief introductions which place the readings in perspective, and further readings are suggested where we think they would be appropriate. No lists of questions are included, because each controversy is in itself a large question, often with multiple smaller questions wrapped up in it.

Because the readings were written by scientists for scientists, there will inevitably be occasions when student readers will be puzzled by a word or phrase. We assure them that they are not alone; the remedy is to look up the textbook, ask a professor, or worry at it until it is clear. Working scientists use the same methods -- and very often a new idea comes out of the effort.

We would like to express our thanks to the individuals and bodies who gave us permission to reproduce articles in this reader. They are acknowledged with each reading. Marilee Kindelt typed the manuscript.

Richard Cowen
Jere H. Lipps

February 1, 1975

# Table of Contents

## SECTION 2: PLATE TECTONICS AND HOT SPOTS

## SECTION 3:   PLATE TECTONICS AND THE FOSSIL RECORD

# Controversies
# in the Earth Sciences

# Origin of the Earth and Moon
## Section 1.

Mercury, as seen by Mariner 10 spacecraft cameras on March 29, 1974 from a distance of 200,000 kilometers. The surface looks uncannily like that of the Moon.   Jet Propulsion Laboratory/NASA photograph.

FORMATION OF THE EARTH AND MOON

It is really not possible to separate the question of the origin of the Earth and Moon from the formation of the entire Solar System, because the Sun and planets probably formed in one interconnected process between 5 and 4.5 billion years ago.  But we can best test theories about planetary formation on the basis of data from Earth and the Moon, because we know so much more about them than we do about the other planets.

Stars form from clouds of inter- stellar gas and dust which collapse by gravitational attraction into a central mass.  But are planetary systems an inevitable result of star formation? If our system is any guide, it seems that the planets contain very little of the mass, but most of the angular momentum, of the solar system as a whole, and this has to be explained.

1

## FORMATION FROM A DISC-SHAPED NEBULA

Most calculations of planetary formation assume that the cloud of dust and gas (the nebula) was originally disc-shaped. A. G. W. Cameron and others assume that a star and its planets are formed at the same time, with angular momentum being transferred outwards from young star to planets in rather complex processes.

The star forms by gravitational collapse in the center of the nebula. Meanwhile, dust grains in the disc, moving randomly, collide and stick together, quickly forming into larger and larger particles. Planetesimals and finally protoplanets are formed, and they attract other particles to them more by gravitational attraction than by random collision. More violent impacts heat up the protoplanets, and primitive atmospheres are "boiled off" their surfaces.

The young star has by this time reached a stage called the "T Tauri" stage, where it is emitting great quantities of gas and energy in violent solar flares. A devastating "solar wind" of charged particles blows away the primitive atmospheres from the protoplanets in the inner part of the system, and light elements are swept outwards to be gathered by the outer protoplanets into giant planets formed largely from gas molecules. The inner planets are left as dense rocky bodies with low quantities of light and volatile substances.

In the late stages of formation of the planetary system, the T Tauri solar wind dies down, and some of the remaining protoplanets may collide, disturbing their orbit or their rotation, and sometimes leaving great scars on their surfaces as giant craters. Smaller pieces of debris may never be swept up, and they remain in orbit within the planetary system as meteorites, asteroids and comets.

This model, or different versions of it, explains the major features of our planetary system. But many details of, say, the pressure, temperature and chemistry of the nebula, and the speed at which planets form from it, may cause spectacular differences in the theories of different investigators. We shall look at different versions of planetary formation, all of them based on this sort of model.

## CONDITIONS IN THE NEBULA

As the Sun formed, the nebula probably reached a temperature near $2000^\circ$ K., and it would then have been wholly gaseous. On cooling, the gases would gradually have condensed into mineral grains of different chemistry, each with its own temperature of formation.

If planetesimals, protoplanets and planets did not begin to form until the nebula was cool, they would have formed from a vast mixture of mineral grains, and so would have been chemically very well mixed--overall they would have been homogeneous. But Earth is a layered planet, with crust, mantle and core of different chemistry. If Earth formed from a cold nebula, it would have to differentiate into crust, mantle and core at a later time.

At present it is fashionable to suppose that the nebula was still hot, but cooling, at the time the planets formed. Mineral grains of different chemistry would have been gradually condensing as planetesimals formed. It now becomes important to estimate how fast planetesimals and protoplanets can form, compared to the cooling of the nebula. If protoplanets formed quickly, there would be time for one generation of condensing minerals to aggregate into protoplanets before the next generation of grains condensed. The protoplanets would each then accrete a "skin" of successive generations of dust particles. Rapid accretion would thus encourage the formation of layered planets. If accretion was slow, then the planetesimals themselves would be formed from more than one generation of mineral grains, and the planets would be fairly homogeneous in their final chemistry.

## EARTH AND MOON

The planets all seem to have different densities, and therefore different overall chemical composition. This suggests that they were formed from different sets of minerals, and therefore that the different regions of the nebula were at different temperatures when the planets formed. Some people believe that the planets formed at different times in the nebula, beginning from the innermost regions; other people suggest a temperature gradient in the nebula to allow planets of different chemistry to form at the same time; and still others wonder what would have happened in the third dimension, away from the main plane of the disc of the nebula.

In terms of forming the Earth and Moon, some special conditions must be postulated, because they now share the same orbit round the Sun. Yet they are chemically different as well as different in size (the Moon is only 1/81 of the mass of the Earth). The Moon is less dense than the Earth, and it is probably made almost entirely of silicate minerals, without an appreciable iron core. The lunar highlands have a high proportion of high-temperature refractory minerals (which should have condensed early from the nebula), while the lunar basins are covered with basaltic lavas which are quite normal (like Earth's). The Moon is low in light volatile elements and compounds, compared with Earth. These differences must be explained.

There seem to be only two approaches to the problem. One is to form the Earth and Moon more or less where they are now, close to one another. Some mechanism must be found which would distribute the chemical components unequally between the two bodies, to account for their differences in size and composition. The second approach is to suggest that the Moon formed somewhere else in the Solar System, and thus it accumulated a different amount of material of a different chemical mixture than the Earth; the problem here is to account for the fact that Earth and Moon then became associated later in their history.

THE READINGS

As we have seen, ideas about the formation of the Earth and Moon are difficult to separate. We begin the readings with three papers on the Moon, published in 1970. The first Apollo results (Apollo 11) had just come in, and the American Geophysical Union sponsored a discussion of their effects on theories of formation of the Moon: three of these papers were published in the AGU journal EOS, and they show very well the state of the science at that time. CAMERON favored a hypothesis which had been put forward by the Australian geologist Ringwood, that the Earth and Moon had formed at the same time in the same orbit, and that the chemical differences between the two bodies were the result of unequal sharing during formation. Although the presentation is convincing, it is only fair to say that Cameron no longer believes this idea: new calculations have produced a better model, in his opinion (see the list of further readings).

O'KEEFE had previously suggested that the Moon formed from the early Earth by a process of fission, and he found support for this idea from the Apollo results. SINGER, on the other hand, had already suggested that the Moon had been captured by Earth (that is, attracted into orbit round the Earth after they had both formed independently). Singer did not find any Apollo evidence to contradict his ideas. So three leading scientists can all work with the same data, and come in all honesty to quite different conclusions.

In the late 1960's and early 1970's, ideas on inhomogeneous accretion were becoming increasingly fashionable--they would allow the layering of the planets to be an original feature as different minerals were swept up from a cooling nebula. In addition, data on planetary interiors, which can be examined indirectly at least, can be used to test theories of the events occurring in the early solar nebula, which can only be studied by making mathematical models.

We reproduce a paper by TUREKIAN and CLARK (1969), not because it was the first to suggest inhomogeneous accretion (that was in 1944), but because it brought together the new ideas into a concise and testable argument.

By 1972, inhomogeneous accretion models had become well enough polished to be used to account for features of Earth and Moon. On June 16, 1972, the journal Nature published an introduction by P. J. S., who summarized work so far. Earth's core is in two layers at least, a solid inner core and a molten outer core. Melting requires energy, and the source of energy for the outer core was disputed. The "cold accretion" theory of the formation of the Earth tried to explain melting of the outer core after Earth formed. But P. J. S. then introduced a paper by ANDERSON and HANKS (1972a) in the same issue of Nature: they pointed out that some of the early condensing materials from a cooling nebula are rich in radioactive elements. These would provide enough energy to melt the outer core, and there might be enough energy left over to produce "Hot Spots" (see a

later section of this book).

Later that year (September 29, 1972), W. H. McC. introduced another twist to the theory in Nature. He introduced a paper in which ANDERSON extended the idea of inhomogeneous accretion to the Moon as well as the Earth. He made the original suggestion that the Moon might have begun to condense in an area of space that was about at its present distance from the Sun, but off the median plane of the disc of the nebula. Making some simple assumptions about temperatures and pressures there, Anderson was able to make a model explaining most of the strange features of the Moon as it compares with Earth.

Just as the core of the Earth is important in theories of its origin, so the deep interior of the Moon is important. Anderson's idea for the formation of the Moon more or less demanded that its interior should be hot, although a significant number of scientists were arguing that the Moon was cold. ANDERSON and HANKS re-examined the problem, and on 22 December 1972 they published a paper in Science in which they said that the evidence for a cold Moon was not convincing (fortunately for Anderson's ideas). Since then, seismic evidence from recording stations set up on the Moon by Apollo crew members to measure moonquakes suggests that the Moon has a layered structure including a molten core. On this point Anderson and Hanks have been supported by new data.

At present the idea is running into controversy for chemical reasons (see Grossman and Larimer 1974, listed below), and the argument is sure to continue over the next few years.

KAULA and HARRIS (1973) have taken a different approach, concentrating on the physical rather than the chemical aspects of the formation of the Moon. They say that the capture of the Moon as a single body is a very difficult process, even though this is demanded by Anderson's model, and also by Cameron's present ideas of its formation (Cameron 1973, Cameron and Pine 1973). Kaula and Harris say that we should seriously consider a model in which the Moon is captured in the form of large and small chunks which collide and coalesce with a cloud of pre-existing particles orbiting Earth.

So the problem of the formation of the Earth and Moon, and of the other planets, is nowhere near solved. The discussions look like continuing with equal excitement into the future, especially as we begin to learn more about other planets (from the Pioneer missions to Mercury and the outer planets, and the Viking landers on Mars). As a foretaste of this, we include a report (METZ 1974) on the state of our ideas about Mars. The intriguing and unusual features which suggest the presence of running water on the planet at some time are the most exciting find in planetary astronomy for years. Here again, there is controversy--see Schumm's paper listed below!

FURTHER READINGS

CAMERON, A. G. W. 1973. Accumulation processes in the primitive solar nebula. Icarus 18, 407-450.

CAMERON, A. G. W. and PINE, M. R. 1973. Numerical models of the primitive solar nebula. Icarus 18, 377-406.

GROSSMAN, L. and LARIMER, J. W. 1974. Early chemical history of the solar system. Rev. Geophys. Space Phys. 12, 71-101.

LEWIS, J. S. 1974. The temperature gradient in the solar nebula. Science 186, 440-443.

---- 1974. The chemistry of the solar system. Sci. Amer. 230, 50-65.

SCHUMM, S. A. 1974. Structural origin of large Martian channels. Icarus 22, 371-384.

WEIDENSCHILLING, S. J. 1974. A model for accretion of the terrestrial planets. Icarus 22, 426-435.

# Formation of the Earth-Moon system

## A.G.W. Cameron (1970)

It is often stated that there are three types of theory concerning the origin of the earth-moon system. I feel that there are serious objections to each of them. They are: (1) Formation of the moon by fission of the earth. This idea has been attractive in the past because of the low density of the moon, which suggests that the moon might be made out of the terrestrial mantle material. However, we now know that the moon is also very much more depleted in the more volatile mantle elements than is the earth, and in my view this would be difficult to reconcile with such a mechanism. In addition, to trigger rotational instability by reduction of the moment of inertia of the earth through formation of the iron core, requires that the system already be on the verge of instability, which is a situation of rather low plausibility. The angular momentum of such a system would be very much greater than the angular momentum of the present earth-moon system. (2) Independent formation of the moon in orbit close to the earth. It is very difficult to see why a body formed independently out of the same kinds of materials that are forming the earth should differ in bulk chemical composition in such a radical way from the earth, being low both in iron and in the more volatile materials. (3) Formation of the moon elsewhere in the solar system followed by capture by the earth. The same compositional objection given immediately above would apply in this case also. In addition, those who wish the moon to be captured by the earth have shown that the process is a very unlikely, if possible, one.

A fourth type of theory has been suggested by Ringwood (1966). This theory states that the moon should be formed by a condensation from a hot extended silicate atmosphere of the primitive earth. At first glance this idea might seem to be the most implausible of all, simply because it is a strange idea, but I shall argue in this article that the conditions that seem to me most probable in the formation of the solar system lead naturally to this picture.

## THE PRIMORDIAL SOLAR NEBULA

The story should properly start with a discussion of the formation of stars in our galaxy. Space does not permit any details of this discussion to be given here. I have given some brief accounts of it elsewhere (Cameron, 1961, 1969). The physical processes involved are the compression of an interstellar gas cloud to a density sufficient to induce gravitational collapse, followed by the fragmentation of the cloud into many smaller pieces during the collapse process. Angular momentum must be conserved during these processes, and as a result the ultimate fragments, which will have masses comparable to that of the sun, will flatten into disk structures.

A typical model that I have investigated is shown in Figure 1. It has a total of 2 solar masses.

In order to construct the model shown in Figure 1, the radial centrifugal equilibrium of the disk has been decoupled from the vertical structure. The equilibrium in the radial direction is achieved by finding a balance between gravitational forces toward

the center of the disk and the centrifugal forces of each part of the disk arising from the local angular momentum of that part. The techniques for adjusting the distribution of mass in the radial direction to achieve this balance everywhere in the disk are adopted from methods used to find the structure of spiral galaxies. In the vertical direction, the gas is everywhere in hydrostatic equilibrium. Figure 1 shows only the integrated mass in each square centimeter of the disk, the surface density. The vertical structure is not indicated by this plot.

One important feature should be noticed at once: there is no central body in hydrostatic equilibrium in the disk. Hence the formation of the disk does not immediately lead to the

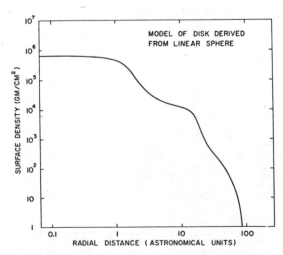

Fig. 1. The surface density of a possible model of the primordial solar nebula as a function of radius. This particular model was obtained by flattening a uniformly spinning sphere with a linear variation of density from the center to zero at the surface.

formation of a central star. The sun would have to form from the gases in the disk by a dissipation process.

## DISSIPATION OF THE PRIMORDIAL NEBULA

The disk model shown in Figure 1 is subject to differential rotation. Very roughly speaking, the angular velocity in the disk is inversely proportional to the radial distance. Hence there is a shear between adjacent layers in the radial direction in the disk, and the dissipation of the nebula must result from frictional forces between the adjacent layers. Such frictional forces would tend to accelerate the outer layer and to decelerate the inner layer, thus leading to an outward transport of angular momentum and an inward transport of mass. The amount of the friction depends upon the viscosity of the gas.

If the only source of viscosity were to be molecular viscosity, dissipation would take an impossibly long time, and the disk would not be appreciably altered after a period of some billions of years. A much more rapid dissipation is possible if there is turbulent viscosity in the disk, since mass transport by turbulence can mix elements of the material over much larger radial distances than is possible in the case of molecular viscosity. Thus the primary criterion for a rapid dissipation must be the presence of turbulence in the disk.

Von Weizsacker (1944) introduced the idea of turbulence in large astronomical gaseous systems during the 1940's. He claimed that wherever such a system had a very large Reynolds number, it must be turbulent. However, there is a flaw in this argument. Laboratory systems with high Reynolds number are turbulent only because of the presence of rigid boundaries, whose motion provides an energy input into the largest scale eddies in the system. However, in the disk under consideration, there are no rigid boundaries, and hence a high Reynolds number is not by itself a sufficient condition for the existence of turbulence.

Turbulence will exist, however, providing there is convection present that transports heat from the central regions of the disk to the surfaces where the heat can be radiated away into space. The presence or absence of convection is determined through the solution of the vertical structure equations for the hydrostatic equilibrium of the disk perpendicular to the plane. In addition to the equations for ordinary hydrostatic equilibrium, it is necessary to calculate the temperature gradient perpendicular to the disk which provides a continuous flow of energy toward the upper and lower boundaries. If the temperature gradient is not very great, then the necessary energy flow can occur by radiative energy transfer, and the structure is not convective. However, if the temperature gradient is great enough so to exceed the adiabatic temperature gradient of the medium, then the medium is unstable against convection and turbulence will be present. The temperature gradient needed to transport the energy by radiative transfer depends upon the opacity of the medium. If the opacity is sufficiently great, a large temperature gradient would be required for radiative transport, and hence convection is more likely to exist.

The sources of opacity in the medium consist of the $H^-$ ion at higher temperatures, molecular opacity, opacity due to small iron grains, and opacity due to ice at low temperatures. There are many other contributors to the opacity, but these appear to be the principal ones. The disk is formed initially at rather high temperature because of the compression of the gas during its collapse from an inter-stellar cloud. Temperatures of the order of $10^4$ °K are likely to exist initially in the central regions, and these temperatures will decrease out-ward in the radial direction.

The calculation of the opacity in the gas, over the complete range of temperatures and densities expected to be of interest, is currently being carried out by one of my graduate students, Milton Pine. However, some older calculations, carried out with iron grain opacities only, indicate the disk will be unstable against convection for surface densities at least as low as $10^5$ gm/cm$^2$, and this instability may possibly exist as low as $10^{4.5}$ gm/cm$^2$. It may be seen from Figure 1 that these surface densities are exceeded out to radial distances of several astronomical units. Thus the inner part of the solar nebula is unstable against convection, and there should be a high turbulent viscosity everywhere. The dissipation times for a turbulent primordial solar nebula were estimated roughly by Von Weizsacker (1944) as about $10^3$ years and by ter Haar (1948; unpublished data, 1950) as about $10^2$ years. It is hoped that an evolutionary calculation that follows the changing structure of the disk with time will refine these dissipation times. However, the essence of the matter is that the dissipation of the disk should not require very many orders of magnitude more in time than the orbital revolution time of the disk. The very short dissipation time of this large gaseous system is one of the key conclusions from which the further discussion of the development of the planets within the gaseous disk depends.

ACCUMULATION PROCESSES

It is virtually impossible for small chemically condensed particles to accumulate into planetary sized bodies in a vacuum. If there is to be any appreciable velocity dispersion of these particles, then the collisions between them will tend to be disruptive rather than accumulative. We see such disruptive processes going on at the present time in the asteroid belt, where collisions among asteroidal bodies appear to be continually breaking these bodies into smaller pieces. For this reason the accumulation of the planets must be constrained to occur in a time shorter than the dissipation time of the inner parts of the primitive solar nebula. It is important to note in this regard that the accumulation of the inner terrestrial planets appears to have been a very inefficient process. The present masses of the terrestrial planets amount to only about 1% of the chemically condensed material that would be associated with the gas in the inner part of the solar nebula shown in Figure 1. There are only very small gravitational potential gradients present in the disk, and consequently very large solid bodies are easily transported by the bulk motion of the gas. Thus most chemically condensed

material will accompany the net flow of the gas inward to form the sun. Only by growing rapidly to large size can a planet become sufficiently resistive to the motion of the gas so that it does not follow the gas into the sun.

The presence of the gas appears likely to promote the accumulation of chemically condensed material into larger sized bodies in a variety of ways. There is actually a pressure gradient of the gas in the radial direction that has been neglected in the construction of Figure 1. This means that solid bodies will follow Keplerian orbits about the center of the disk, but the gas will rotate somewhat more slowly than the Keplerian velocity owing to its partial support by the radial pressure gradient. Hence there is a continuing friction between the gas and the solid particles that will result in a net inward drift of the particles.

Everywhere in the inner solar nebula the shear motions associated with the presence of turbulence will cause particles on slightly different streamlines to drift together. However, this process of accumulation is probably not very efficient.

There are radioactivities such as $K^{40}$ present in the primordial solar nebula, and there may also be cosmic rays present. The charged particles that are produced continually create ion pairs in the gas. The shearing motion of the turbulence will thus tend to separate charges from one another, and lead to the generation of electric fields in the gas. One might expect an approach toward an equipartition of energy in which the electric field energy density would approach the energy density of the turbulence. However, electrical breakdowns in the gas will occur long before such an equality could be achieved. The small particles in the gas will acquire charges of either sign, and hence the particles will undergo acceleration in the electric fields that are present. This acceleration is probably a much more efficient mechanism for promoting the accumulation of small particles, since the electric fields will move the particles toward one another with a much greater efficiency than the shearing motions alone.

We know very little about the surface chemistry of particles in the solar nebula under these conditions, and it is not easy to say what the efficiency for sticking together will be after a couple of particles collide. Lightning flashes in the solar nebula may assist in welding the surfaces of particles together.

When the growing bodies have reached larger sizes, inertial effects in their motions with respect to the gas will become important. In a fully turbulent medium, the pattern of turbulent flows will be continually changing from one point to another. If a larger body is following such a flow and the character of the flow changes in a relatively short length interval, then the inertial motion of the body will carry it some distance beyond the flow before its motion can be redirected. In this way the motion of the larger particle will cross the motions of many smaller particles, leading to enhanced accumulation of the small particles onto larger ones.

HIGH TEMPERATURE PLANETS

A great deal of gravitational potential energy is released when a planet the size of the earth is accumulated. If the accumulation is a very rapid one, it must be expected that the temperature within the earth is very high. Let us make a rough calculation in which we neglect the energy that is stored by a planet as it grows, and set the surface temperature of the planet high enough to radiate away the released gravitational potential energy. For this purpose, consider a planetary body that grows as a result of a steady accumulation of very much smaller bodies. The surface temperature condition is then

$$4\pi R^2 \sigma T^4 \approx GM/R \; dM/dt$$

Here $M$ is the mass inside the radius $R$, $\sigma$ is the radiation constant, and $T$ is the radiation temperature.

Let us see what radiating temperatures are implied for the earth, if we set $M$ equal to the mass of the earth and $R$ equal to the present radius of the earth. Then if $dM/dt$ is constant during the accumulation time of the earth, $T = 4500\ ^{o}K$ for an accumulation time of $10^3$ years, and $T = 8000\ ^{o}K$ for an accumulation time of $10^2$ years. On the other hand, if we assume $dR/dt$ is constant during the accumulation time, then $T = 5920\ ^{o}K$ for an accumulation time of $10^3$ years, and $T = 10520\ ^{o}K$ for an accumulation time of $10^2$ years. Clearly these temperatures are fictitious in the sense that the ordinary rocky material of the earth would be vaporized at these temperatures, and the accumulating radius would actually be considerably greater than the one assumed. Thus the surface temperature would be lower but the matter would be heated on compression.

Now let us consider the opposite extreme of the accumulation process. The accumulation will result in the presence at any time of a hierarchy of body sizes, extending from very large bodies, approaching planetary size, down to dust grains. The terminal phase of accumulation of a planet may involve collisions between bodies having comparable mass. If two bodies, each containing half the mass of the earth, collide and fuse together, then the released gravitational potential energy is sufficient to raise the internal temperature of the combined body by about 33,000 $^{o}K$. This neglects the breaking of chemical bonds, and if one were to take that into account, the rise in temperature would be closer to 20,000 $^{o}K$. In the case where the final accumulation involves a few bodies of comparable mass, then the resulting planet is likely to be given a fairly rapid spin, simply from the large statistical fluctuations that can arise in the impact parameters of the collision. However, if the body grows by accumulation of many much smaller bodies, there will be a much smaller statistical fluctuation in the average impact parameter of the colliding bodies, and only a small net spin is to be expected for the planet.

It may thus be seen that the accumulation of the earth is likely to produce initial temperatures in the planet of the order of $10^4\ ^{o}K$. Such a primitive planet will in no respect resemble the planetary body on which we live today; the high temperatures would result in the formation of a planet having a radius of 5 to 10 times the present radius of the earth, with the interior consisting of high temperature gases: gaseous iron and gaseous magnesium silicates together with their high temperature decomposition products, in the main. The structure of such a high temperature planet must be calculated by techniques similar to those used in constructing a a model of a star, and calculations directed toward this end are presently being performed by one of my graduate students.

## THE FORMATION OF THE MOON

This concept of a high temperature planet with an extended gaseous silicate atmosphere is precisely the structure required by Ringwood (1966) in his suggested origin of the moon. There are a number of interesting numbers that emerge when we consider some features of the earth-moon system that would have to be present at such a time. If the present angular momentum of the earth-moon system were concentrated in such an extended planet, then the planet would be rotationally unstable beyond about three earth radii from the center of the equatorial plane. This happens to be approximately the Roche limit for the earth, so that solid bodies can accumulate and stick together against the effects of the disrupting tidal forces of the earth at distances beyond three earth radii. Therefore it makes a great

deal of sense to consider that the high temperature initial earth was in fact formed as a result of a collision between two bodies of comparable mass, which produced the very large extended object under discussion with a sufficiently rapid spin to have the angular momentum of the current earth-moon system. The outer part of the silicate envelope in the equatorial plane of the planet would then be rotationally unstable, and it would flatten down into a short disk in the equatorial plane. The moon may rather easily condense from such a disk. As the disk cools, solid condensations will occur within it, and these are capable of collecting together to form the moon. However, it is entirely possible that the dynamics of the condensation may favor the growth of the moon during the phase of strong interaction between the gases and their solid condensates, leading to a very rapid accumulation time indeed. If the moon accumulates in a time of about a century, then interior temperatures in the moon are likely to be in the vicinity of 1500 $^{\circ}$K, so that the interior would be extensively molten.

In the high temperature gaseous planet under consideration, iron would condense at a higher temperature than magnesium silicates. Consequently, it is to be expected that liquid droplets of iron would rain out through the large gaseous atmosphere and tend to collect toward the center of gravity of the combined body, even if they had not already collected toward the centers of the fairly large bodies that collided to form the earth. In this way it is a fairly straightforward matter to understand why the moon should have a low content of iron.

The outer radius of the extended planetary **atmosphere** will to a certain degree be defined by the place at which the silicates condense to form clouds. Since the silicates will form the main bulk of the atmosphere, a major change in atmospheric properties would occur at the silicate cloud tops. Beyond this point, a gaseous atmosphere of the more volatile elements is likely to exist. However, this gaseous atmosphere may well be removed from the planet by turbulent mixing with the surrounding primitive solar nebula. The turbulent velocities in the solar nebula would be a few kilometers per second, which is comparable to the escape velocity from the earth at the large radii under consideration. Therefore, such volatile elements as are brought into the earth by the accumulation of cold condensed smaller bodies can be lost from the atmosphere during the accumulation phase. This process requires additional study,

because it has not as yet been analyzed formally.

A related question is the degree to which the more volatile of the elements found in the earth's mantle actually condense on solid bodies during the accumulation stage of the earth. The calculations reported here have not yet produced a detailed temperature history of the solar nebula at the distance at which the earth will accumulate. However, it is likely that the temperature will continually decline at that distance during the accumulation of the earth, and hence it is possible that the more volatile elements are of low abundance in the solid bodies that participate in the early accumulation stages of the earth, but they may be present to a considerably greater degree in the smaller bodies that accumulate later on. I shall return to the question of the volatiles later.

THE T TAURI SOLAR WIND

Newly formed stars in dense regions of gas and dust in the galaxy are usually observed to be emitting gas at a prodigious rate. This stage of rapid mass loss in a young star is called the T Tauri phase. The rates of mass loss appear to lie in the range of $10^6$ to $10^7$ times the present rate of

mass loss in the solar wind.  The
mechanism by which the mass is lost is
not at present understood, but it is
probably a variation of the mechanism
of the hydrodynamic expansion of the
solar corona that constitutes the
present solar wind.

Once the sun has formed by the
dissipation of the inner parts of the
solar nebula, it must be expected that
the T Tauri solar wind will be promptly
established.  This rapid mass loss must
have had some extremely important
effects on the primitive solar system.
It should result in an extremely rapid
removal of the remaining gas in the
primitive solar nebula to large
distances away from the sun, leaving a
variety of solid material 'stranded' in
space.  It should remove any primordial
atmospheres consisting of gases
captured from the primordial solar
nebula onto the inner terrestrial
planets.  It must therefore terminate
the rapid rate of accumulation of
material to form the terrestrial
planets.  The time scale for the
'sweep-up' process by which the remaining material in space between the
planets will either collide with the
planets or be ejected from the solar
system is likely to be of the order of
$10^7$ years, judging from meteorite
accumulation times estimated by the
Monte Carlo method by Wetherill (1969)
and references cited therein.

## THE RESIDUAL ACCUMULATION

This final accumulation of small
bodies onto the surfaces of the earth
and the moon will add an unknown amount
of material.  If the earth is to
acquire all of its more volatile
elements in this way, then estimates by
Turekian (personal communication)
indicate that a final addition of about
1/10 of the present mass of the earth
in the form of a mixture of ordinary
chondritic material and carbonaceous
chondritic material would suffice to
bring in the necessary volatiles,
including the extreme volatiles such
as water and the rare gases.  This does
not seem to be an unreasonable figure
for the amount of mass that could be
swept up by the earth from its
vicinity.

It might be expected that this
process would result in the enrichment
of the moon in the volatile materials
relative to the earth, since the moon,
being a smaller body, has a greater
amount of surface area per unit mass.
However, what counts is not the geo-
metric cross section of the bodies but
the gravitational capture cross
section, which depends upon the mass of
the bodies as well as upon the radius.
The slower the relative velocity of the
bodies to the earth-moon system, the
greater the probability that such

bodies would be captured by the earth rather than the moon. Thus if most of the bodies that are swept up by the earth in this final accumulation stage are moving in nearly circular orbits near the orbit of the earth, the earth would receive a greater enrichment in the volatiles than would the moon. The bodies that have initial orbits farther away from the orbit of the earth would be travelling with considerable elliptical motion by the time that they are perturbed enough to cross the orbit of the earth, and hence there is a greater probability that their motions would be perturbed to cross the orbit of Jupiter, from which they would be lost to the inner solar system.

There is one other important feature of the sweep-up of small solid bodies by the earth-moon system. In the process for formation of the moon described above, the moon would collect in the equatorial plane of the primitive earth. Its subsequent recession from the earth would maintain it in the equatorial plane. However, the moon does not lie in the equatorial plane of the earth, and celestial mechanics calculations that carry the motion of the moon back in time indicate that the angle of inclination of the moon's orbit to the equatorial plane of the earth was greater in the past. This might at first sight seem to be an insuperable difficulty with

the above mechanism for origin. However, the addition of the residual material to the earth during the sweep-up process would result in a random bombardment of the surface of the earth by the swept-up particles. This process would thus add an additional angular momentum vector to the spin of the earth, and this additional angular momentum vector may very well tilt the axis of the earth away from the pole of the equatorial plane in which the moon was formed. Particularly if one of the residual swept-up bodies is considerably larger than most of the others, a very large change in the tilt of the axis could occur.

It may thus be seen that Ringwood's (1966) suggested mechanism for the origin of the moon appears to be a natural consequence of a rapid accumulation of the earth, and that such a rapid accumulation seems required by any process of turbulent dissipation of a primordial nebular disk out of which the solar system was formed. Detailed theories for many of the processes described above have not yet been worked out. However, the issue of the hot extended earth seems a very fundamental one that would hardly be affected by more detailed considerations than those given here, and that is the principal point that should be established at this time.

REFERENCES

Cameron, A. G. W., The formation of the sun and planets, Icarus, 1, 13, 1961.

Cameron, A. G. W., The pre-Hayashi phase of stellar evolution, in Low Luminosity Stars, edited by S. S. Kumar, Gordon and Breach Science Publishers, New York, 1969.

Ringwood, A. E., and D. H. Green, An experimental investigation of the Gabbro to Eclotite transformation and its petrological applications, Geochim. Cosmochim. Acta., (30) 5, 767-834, 1967.

ter Haar, G. L., Studies on the origin of the solar system, Kgl. Danske Videnskip. Selskab., Mat.-Phys. Medd., 25, 3, 1948.

Von Weizsacker, C. F., Uber Die Entstehung Des Planetensystems, Z. Astrophys., 22, 319-355, 1943.

Wetherill, G. W., Relationships between orbits and sources of chondritic meteorites, in Meteorite Research, edited by P. M. Millman, D. Reidel Publishing Co., Dordrecht, Holland, 1969.

ACKNOWLEDGMENT

This research has been supported in part by the National Science Foundation and by the National Aeronautics and Space Administration.

# Apollo 11: implications for the early history of the solar system

## J.A. O'Keefe (1970)

In this paper, I shall attempt to show that the Apollo 11 data support the idea that the moon was formed by the breakup of the earth, and that they suggest that after the breakup, the moon went through a heating episode that boiled away most of its mass. I will go on to discuss the possibility that the planets of the solar system might have been formed in a similar way, that is by breakdown of Jupiter-sized objects rather than by buildup from smaller ones.

From the cosmological standpoint, the most interesting results from the Apollo 11 samples were the chemical measurements. To appreciate these, it is necessary to have some idea as to what the chemical measurements should have shown. In other words, if the moon had formed from the primeval materials of the system, what would a chemical analysis have been like? The

answer to this question is called the cosmic abundance scheme. In this scheme, hydrogen forms 90% of the material, helium 10%, and the other elements, including the nonvolatile elements, constitute less than 1%. The element abundances have a fixed ratio to one another; that is, there is a definite answer to the question: What was the initial ratio of silicon to iron? A systematic listing of these ratios was first done by Henry Norris Russell in the 1930's. The most recent and useful table is by Cameron (1968).

In examining a lunar sample we do not really expect to find the gases still present. No object that you could hold in your hand at ordinary temperatures could consist mostly of hydrogen and helium. In any solid object that you could hold in your hand, we would have lost nearly all of the elements that would ordinarily be in the gaseous form--the so-called atmophile elements. These include hydrogen, nitrogen, and the rare gases.

When we study the lunar sample we find that these elements are indeed missing. Their abundances are in fact even lower than in the earth.

A second important class of elements covers a large part of the righthand side of the periodic table. These elements are often associated with odors; they include things like sulfur, phosphorus, iodine, chlorine. At the present time, it is considered that the most important fact about these elements in governing their behavior is the fact that they or their oxides are volatile. The analyses of the lunar samples show that these elements are deficient in the moon to an extent that is greater than their deficiencies on the earth. The indication is that the moon has been through a period of severe heating.

On the left side of the periodic table is a group of elements that are called the lithophile elements. These elements normally go into the formation of rocks; that is what their name means. They are ordinarily present in rocks in the form of their oxides, and for this reason they are sometimes called oxyphile elements. In the lunar sample most of the lithophile elements are present at a level above their cosmic abundance in somewhat the same proportions as on the earth. There are two interesting differences: first, the alkali elements, which are more volatile than the average, are somewhat below their terrestrial abundances although they are above their cosmic abundances; second, a group of very refractory elements including titanium, zirconium, and hafnium are present in abundances considerably higher than their cosmic abundances or their abundance in the earth.

The meaning of these two peculiarities of the lithophile elements is the same as that which we have mentioned previously. The alkalies are missing because there has been strong heating; the refractories are enhanced because, apparently, the heating has been so intense as to cause the loss of a major fraction of the original mass of the moon and so to bring about a concentration of the elements that are hardest to get rid of.

In the middle of the periodical table is the most interesting group of elements called the siderophile elements. Some of these, like gold and osmium, are relatively volatile; others, such as platinum or iridium, are very refractory. All have the character that their energy of oxidation, per oxygen atom, is less than that of iron. As a result oxides of nickel, for example, will not exist in the presence of molten free iron. The iron will rob the nickel of its oxygen.

In many mixes of the nonvolatile constituents of the world, it is found that the iron is partly oxidized and partly in the form of the free metal. The same thing could happen in principle with other metals, but it happens most often with iron because of its abundance. Examples are: a blast furnace with pig iron and slag; a chondritic meteorite with silicate and iron phases; the earth with its mantle and core. In all such cases it is found that the siderophile elements are strongly concentrated in the metal phase. Evidently they will always be metal because they cannot hold the oxygen; and it develops that the free metal dissolves better in metallic iron than in the slaggy silicate liquid.

It is believed to be for this reason that the siderophile elements, which are not particularly rare in the universe as a whole, are quite rare in the crust of the earth. Nickel, for example, constitutes about 1% of the mass of a typical chondritic meteorite, but only a few parts per million of the abundance in the crust. Some of this is due to the crust-mantle differenti-ation; but a large part is also apparently owing to the concentration of the siderophiles in the core of the earth.

With these considerations in mind, it is extremely interesting to see that

the siderophile elements are depleted in the Apollo lunar samples by an amount that is, in many elements, comparable with the depletion in the earth. There is a strong suggestion that the material of the moon has likewise been through a process of leaching by contact with metallic iron. It is difficult to explain the pattern of siderophile deficiency which includes some quite refractory elements in any other way.

Where then did the moon's nickel go? It is apparently not simply a deficiency of nickel in the outermost layers of the moon, because the lavalike rocks of the lunar crust at the Apollo 11 site contain some bits of metallic iron that have only a fraction of a per cent of nickel. If these had come from regions where the nickel content approached the cosmic composition relative to a silicon, then, because there is so little free iron, its scavenging action should have led to very high nickel content-- perhaps 50% or more. It therefore cannot be concentrated in the moon at levels from which the lavas come.

Neither can it be concentrated in a lunar core of any size. The moment of inertia of the moon according to Michael (1970) at Langley, is 0.4015 $^{+0.0048}_{-0.0015}$ $Mr^2$, where $M$ is the mass of the moon and $r$ is the radius. The moment of inertia of a homogeneous sphere is

exactly 0.4. With a little arithmetic
it can be shown that even if heating
effects are allowed, the maximum core
size is about 1% of the moon by weight
if the core is assumed to be essentially iron-nickel. This would be barely
sufficient to accommodate the cosmic
abundance of nickel even if the mass
were pure nickel, which would be
extraordinary. It is essentially
certain that the moon's nickel is not
concentrated, like the earth's, in an
iron-nickel core.

In fact there can be at most a
very small amount of free iron (whose
density is about 7.8) in the moon. It
has been found that the mean density
of the moon is approximately 3.34 g
$cm^{-3}$; this is almost the density of the
earth's mantle. This density is
actually slightly <u>less</u> than the density
of the crustal rocks at the Apollo 11
site. Any significant mixture of
metallic iron would give the moon a
higher over-all density than is
observed.

<u>The logical conclusion is then
that the moon's nickel is in the
earth's core and that the moon formed
by fission of the earth after the core-
mantle separation had taken place.</u>

The idea of the formation of the
moon by fission has been discussed for
many years. <u>Darwin</u> (1898) (son of the
biologist, Sir Charles Darwin) drew
attention to the fact that the action
of the tides throughout the history of
the earth has always been such as to
transfer angular momentum from the
earth's rotation to the moon's orbital
motion. Following backward in time, it
can be deduced that a few billion years
ago the moon must have been at a distance of about 2.8 earth radii and
revolving around it in a period of
about four hours. The earth at that
time was turning in a period of about
four hours. The problem is to pass
from a rapidly rotating earth with an
iron core to this kind of earth-moon
system.

The primary difficulty is that if
we imagine the angular momentum and
mass of the earth and moon (neither of
which is changed by the course of tidal
evolution as we have described it)
concentrated in a single mass, then
that mass will rotate in a period of
about four hours and will not fission.
An attempt was made in the early part
of the 20th century to explain the
fission of this body as a result of a
resonance effect due to solar tide.
However, <u>Jeffreys</u> (1930) showed that
this idea will not work.

A more hopeful idea, which seems
to solve several difficulties at once,
is the suggestion that after the
separation of the earth and moon, the
strong tidal interaction between the
two bodies heated both, but especially
the moon, and caused the loss of large

amounts of mass and angular momentum from the moon. Interactions of this kind, with the escape of a large amount of mass and angular momentum, are actually observed in close binary stars. Of course, there are many differences in the circumstances. Nevertheless, the resemblance may be more than just suggestive.

If we suppose that over half of the moon's mass was boiled away at this point, then we explain simultaneously the deficiency of volatile elements in the moon, the enhancement of the refractory elements, the loss of angular momentum, and a great loss of mass which Lyttleton (1953) has shown is required if we are to understand the formation of the moon by fission.

Theories of this kind in which the fission of the earth is followed by an episode of very strong heating in the moon were produced by Wise (1969) and O'Keefe (1969) before the Apollo 11 results became available. Thus although the results have been presented here as leading toward a theory of the origin of the moon, the historical sequence is that the facts appeared afterward as a verification for theories already in existence. It is natural to feel somewhat encouraged by this circumstance. Perhaps there is some reality in these theories after all.

In the early days of the space age it was very often said that the moon was the key to the solar system. Let us, therefore, put that key in the lock and see if it turns. Is it possible, in other words, that the planets were produced by processes like the moon?

The processes that we see acting inside the solar system are more often processes of breakup than of building. Meteors appear to be cometary fragments; comets are seen to break in two; asteroids belong to families that come from some initial breakup; meteorites always bear the clear marks of a formation from larger bodies.

On the planetary scale, it is widely believed that Pluto is an escaped satellite of Neptune. If the moon had escaped, as it almost did, it would certainly be regarded as a planet.

Since we have among the objects of the solar system this hierarchy of destruction, it is strange that our theories always contemplate a hierarchy of aggregation. In our theories we begin with gases that condense to small particles. These in our theories accrete to larger and larger bodies until finally the great planets are formed. How did we get started on this rather strange path?

The basic impulse came from some work by celestial mechanicians of the late eighteenth century whose work showed that the major features of the planetary orbits do not change with time. A planetary orbit may precess, or the long axis may swing around in space, or the planet may be a little late or a little early at arriving at its perihelion, but the size and the shape of the orbits do not change. If they do not, then the planetary system must always have been as it is now--a thing as thin and flat as a pancake. This is why theories were devised for spreading gas out in a thin layer and then condensing it in place.

In the twentieth century, however, these arguments became obsolete. Brown (1933) pointed out that when it is a question of billions of years instead of the millions of years assumed in the eighteenth century then these theorems can no longer be relied on for guidance. Hagihara (1961) in a sweeping review of the problem from all aspects finally came to the same conclusion as Brown; namely that we do not know whether the present configuration of the solar system is the same as the initial configuration. This is true even apart from the possible influence of nongravitational effects.

It seems to me, therefore, that we are free to approach the problem from a more natural standpoint. There is immense difficulty in explaining how small particles got built up step by step into bodies of the size of the moon or Mercury; but this difficulty is greatly reduced if we suppose that the first bodies that were formed were the size of Jupiter. In this case the gravitational binding energy is such as to make aggregation entirely plausible. The immense difficulty of explaining why smoke particles should clump together to form pebbles and then boulders and then asteroids is done away with.

The principal difficulty with this new idea about the formation of the planets is that if our system began with large bodies we must explain what caused them to spin more and more rapidly until they burst. Wise (1969) has suggested that it is a process of sedimentation in which the core materials sink to the bottom. He has applied this idea to the earth-moon system. However, there are difficulties in using this idea if we intend, for example, to regard meteorites as fractions of a broken planet. The chondritic meteorites have not been sorted out this way. One could imagine, following McCrea (1969), that perhaps the sorting took place in an atmosphere full of volatiles that has since become lost.

It might also be possible to follow the lead of Marsden (1970) who has been drawing attention to the important role played in the evolution of comets by the escape of volatiles.

These produce nongravitational forces that change the cometary orbits. Might they not also bring about rotation in a comet? Is it not imaginable that in some similar way the escape of volatiles leads to a spinup of a great planet and so to its fission?

It used to be thought that the density of the planets formed an orderly scheme with the densest near the sun and the lightest further out. Recently, however, at the Naval Observatory Duncombe et al. (1968) have shown that Pluto has a density greater than that of the earth and very likely as great as iron. With this fact the chance of getting some order out of the densities of the planets seems to be lost. The fact seems to be that there is no order; that they are composed of different kinds of material; that in short they were differentiated when they were formed.

To sum up, therefore, it appears that we should regard the accretion process with a good deal of skepticism, and that we ought to consider the possibility that Jupiter or another planet like it (similar to those that van de Kamp (1969) has found around other stars) is the parent of the planets of the solar system.

## REFERENCES

Brown, E. W., Observation and gravitational theory in the solar system, Pub. Astron. Soc. Pacific, 44, 21-40, 1932.

Cameron, A. G. W., A new table of the abundances of the elements in the solar system, in Origin and Distribution of the Elements, edited by L. H. Ahrens, Pergamon Press, Oxford and New York, 125-243, 1968.

Darwin, George Howard, The Tides, 1898, reprinted by W. H. Freeman and Company, 1962.

Duncombe, R. L., W. J. Klepczynski, and P. K. Seidelman, Orbit of Neptune and the mass of Pluto, Astron. J. 73(9), 830-835, 1968.

Hagihara, Yusuke, The stability of the solar system, in Planets and Satellites, Chapter 4, 95-158, edited by G. P. Kuiper and B. M. Middlehurst, Volume 3 of The Solar System, University of Chicago Press, 1961.

Jeffrey, H., The resonance theory of the origin of the moon, 2, Mon. Notices Roy. Astron. Soc., 91(1), 169-173, 1930.

Lyttleton, R. A., The Stability of Rotating Liquid Masses, 150 pp., Cambridge (England) University Press, 1953.

Marsden, B. G., On the relation between comets and mirror planets, Astron. J. 75, 206-217, 1970.

McCrea, W. H., Densities of the terrestrial planets, Nature, 224(28), 1969.

Michael, W. H., Jr., Moments of inertia of the moon, preprint for 1970 Conference on the Origin and Evolution of the Planets, Pasadena, California, 1970.

O'Keefe, J. A., Origin of the moon, J. Geophys. Res., 74(10), 2758-2767, 1969.

van de Kamp, P., Parallax, proper motion, acceleration, and orbital motion of Barnard's Star, Astron. J., 74, 238-240, 1969.

Wise, D. U., Origin of the moon from the earth: some new mechanisms and comparisons, J. Geophys. Res., 74(25), 6034-6045, 1969.

ADDITIONAL NOTE

Dr. O'Keefe added the following note on 1 October 1974.

In 1971, Seidelman, Klepczynski, Duncombe, and Jackson, Astronomical Journal, vol. 76, pp. 488-492, announced that an error was found in the above mentioned mass of Pluto, which makes its mass only about half of that earlier reported. It is now between 1/2500000 and 1/3500000 of the mass of the sun. The density turns out to be around $5 \text{ g cm}^{-3}$, with a mean error around 17%.

# Origin of the Moon by capture and its consequences

## S.F. Singer (1970)

I am going to discuss **four** topics, ranging from celestial mechanics to speculative propositions regarding the early history of the solar system: (1) A new calculation of the orbit evolution of the moon that suggests the capture of the moon as a separate body in tidal interactions with the earth; (2) The consequences of this capture on the earth itself and on its early history; (3) The consequences of the capture on the moon, its thermal history, and its surface features; and (4) Some consequences to the early history of the solar system: the existence of many moon-like bodies in the inner part of the solar system; the angular momentum of Venus; the origin of Phobos and Deimos.

Of the many proposed modes of origin of the moon, some violate physical laws; many are in conflict with observations; all are improbable.

Reprinted from E∂S 51, 637-641 by permission of the author and publisher. Copyright 1970 American Geophysical Union.

Perhaps the least improbable--based on new tidal theory calculations and on the interpretation of lunar surface material data--is capture of the moon as it passed near the earth in a <u>direct</u> (prograde) orbit, shortly after the formation of moon and earth, about 4.5 billion years ago. (Capture of the moon from an initially <u>retrograde</u> orbit that had been proposed some years ago, leads to physically unacceptable consequences.) The effects of **capture** on the earth would have been cataclysmic, leading to intensive heating of its interior, to volcanism, and to the immediate formation of an atmosphere and hydrosphere. Thus capture of a moon may have given rise to the unique properties of the earth (in the solar system) and to the early evolution of life, about 3.5 billion years ago.

Because of the strong tidal interactions, the moon itself would be fragmented to some extent and probably heated during the early stages of the capture phase.  It is likely that the surface features of the moon also bear witness to such an event.  Observations on the moon could establish with more certainty whether simultaneous heating of the earth and moon did occur.

Some more speculative consequences of the new calculation for the history of the solar system are:  a number of moons may have formed in the inner solar system, all but one having dis-appeared by impact with planets; that the planet Venus was de-spun following the retrograde capture of one of these moons; that Phobos and Deimos may have accumulated at the time of the formation of Mars rather than being the remnants of asteroids that were later captured.

## ORBITAL EVOLUTION OF THE MOON

Beginning with Darwin, various authors have investigated the effect of the earth's tides on the moon's orbit and have concluded that the moon was at one time much closer to the earth.  Gerstenkorn (1955) has carried the calculation further and found that after the orbit shrinks to about 2.8 earth radii it would suddenly be trans-formed into an escape orbit of high inclination.  Taking this result at face value, he proposed that the moon approached the earth about 2 billion years ago along a retrograde orbit;

i.e., moving in a direction opposite to that of the earth's spin.  In the capture process, the orbit was flipped over the earth's pole and changed into a prograde (direct) orbit corresponding to the present sense of motion of the moon.  Basically the same result was derived by MacDonald (1964) who has given the most recent and extensive treatment of the problem.

In the last few months I have undertaken a new calculation that gives a fundamentally different result; namely, that the moon was captured from a prograde orbit.  The calculation is based on a frequency-dependent tidal perturbation.[1]  Rather than assume a constant phase angle $\delta$ between the tidal bulge and the earth-moon line, I

[1] It is important to avoid con-fusion between two entirely different definitions of frequency.  Even if dissipation, i.e., the Q of the earth, is frequency-independent (MacDonald, 1969) (over periods ranging from the Chandler wobble, i.e., 14 months to seismic periods), I would claim that the tidal perturbation depends on a frequency that is defined in terms of the relative angular velocity of earth and moon; i.e., $(\Omega - f)$ where $f$ is the moon's true anomaly.  In an eccentric orbit, $f$ is not constant.  When the moon moves within the synchronous orbit, $f > \Omega$.

find then that the angle varies in magnitude, in phase, and in sign during a single orbit of the moon, particularly when the moon approaches a synchronous orbit, where its mean motion is close to the earth's angular velocity $\Omega$.

Details of Calculation. In carrying out the calculation it is essential to use proper averaging. There are three time scales; fortunately, they are widely separated, which greatly simplifies the calculation: (1) the time scale of the orbital period; (2) the time scale of motion of perigee; and (3) the time scale of secular variation of the orbit elements a, e, and i. It is not permissible to average over the orbital time scale as has been done in the past calculations.

The calculation involves four additional features: (1) the effects of solar torques are considered in the manner of Goldreich (1966); (2) the higher harmonics of the tidal potential are investigated, especially when the moon comes close to the earth; (3) we allow for the development of a core of the earth during the evolution of the lunar orbit; this affects the moment of inertia and spin of the earth and therefore the position of the synchronous orbit limit; and (4) the time scale, which is inversely related to $\delta$, enters in a nonlinear manner; i.e., the time scale affects the manner

in which the orbit evolves. This point has been investigated.

Results: The basic result is shown in Figure 1. The lunar orbit, which is now at 60 earth radii and has an eccentricity of 0.055, approaches the earth to a minimum distance of 2.6 earth radii at which point the orbit is very nearly circular; then suddenly the eccentricity increases as the orbit

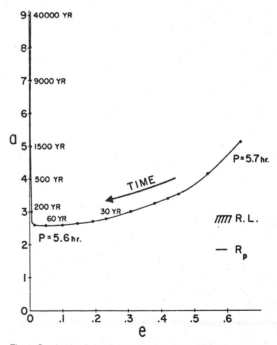

Fig. 1. Semimajor Axis (a) (in units of earth radii) vs. Eccentricity (e) of lunar orbit; the evolution of the orbit of the moon under the influence of earth tides, using a frequency-dependent tidal theory. The progression of time is indicated by the arrow, although the calculation starts with the present orbit and proceeds backwards in time. Time measured from capture (i.e., transition from hyperbolic to elliptic orbit) is indicated alongside the curve. Note that the lunar orbit shrinks to a small circle in about 100 years, while the spin period of the earth changes only slightly (from 5.7 to 5.6 hours). At this point, intensive energy dissipation sets in in the earth's interior. The conventional Roche limit (R.L.) and initial perigee distance (Rp) are shown. Note that the inclination of the lunar orbit remains close to 10° during and after capture (for about 50,000 years); i.e., the capture is from a prograde orbit. This is in marked contrast to previous calculations using frequency-independent tidal theory, in which capture took place from a retrograde orbit.

grows in size, eventually reaching a parabolic orbit. There is little change of orbit inclination during the rapid capture phase which occupies only a few thousand years.

This basic result is essentially the same as that from my two-dimensional calculation (Singer, 1968), which at that time suggested capture from a prograde orbit. Interestingly, the angular momentum of the moon-- proportional to

$$|\underline{a}\ (1 = \underline{e}^2)|^{1/2}$$

--is very nearly constant during the entire period of about a hundred years when the moon changes from a parabolic to a circular orbit; consequently, the spin period of the earth changes only slightly from an initial 5.7 hours to 5.6 hours at minimum distance.

This result should be contrasted to the calculations using a frequency-independent phase angle that gave an initial retrograde orbit of the moon and an initial spin period for the earth of less than 1 hour (MacDonald, 1964). This is a physically unrealistic result. Equally unrealistic would be the energy dissipation that must occur within the earth; for this reason MacDonald did not accept Gerstenkorn's proposal of tidal capture of the moon.

The fundamental difference introduced by the frequency-dependent tidal theory can be illustrated in the case where the lunar inclination is 0°. The frequency-dependent calculation (Singer, 1968) gave basically the same result as Figure 1. The frequency-independent calculation would not give capture at all, but a moon spiralling into the earth.

Time Scale: The time scale for lunar orbit evolution is about 2 billion years (Gerstenkorn, 1955; MacDonald, 1964) if the present elastic parameters and dissipation constants are assumed to be valid throughout the earth's geological history. Some authors have taken this time scale more seriously than others. I have rejected any time scale derived from celestial mechanics calculations and have chosen one of about 4.5 billion years on the basis of two arguments (Singer, 1968): (1) in order to make capture probable, the moon would have to be captured from a heliocentric orbit similar to that of the earth's; this capture must occur very shortly after its formation; and (2) the capture of the moon initiates --I assume--the formation of a core in the earth, and this formation took place during the first hundred million years of the earth's existence according to geological and geochronological data. Now we can add a third argument; namely, the dating of lunar surface material.

It seems difficult in any case to maintain a constant dissipation parameter since we know that the dissipation takes place mainly in the oceans and therefore depends crucially on the details of ocean-land boundaries. Under some theories of continental drift, the continental masses would have been originally amalgamated into two major masses, and therefore the ocean tide dissipation would have been substantially less in earlier times.

EFFECTS OF CAPTURE ON THE EARTH

The major effect of the huge internal and surface tides that must have existed following capture is to produce energy dissipation within the earth by solid friction (Kaula, 1964). In the first phase of capture, as the moon changes from a parabolic to a circular orbit, the energy dissipated in the earth averages only $2 \times 10^9$ erg/g, corresponding to the kinetic energy loss of the moon. This is a maximum value and assumes that no energy dissipation takes place within the moon. It corresponds to 10% of the energy that is required to induce melting. During this phase of capture, the spin angular momentum of the earth hardly changes so that practically none of the Earth's spin kinetic energy is dissipated. However, during the second phase, when the moon moves from its minimum-distance circular orbit out to its present orbit, a much larger amount of energy can be dissipated. It is during this phase that the lunar orbital angular momentum increases greatly--roughly as $\underline{a}^{1/2}$; therefore, the spin angular momentum of the earth, $\underline{C}\Omega$, must decrease correspondingly. This means that the kinetic energy of rotation, $\frac{1}{2}C\Omega^2$, of the earth decreases, with the energy dissipated at the surface or within the earth, depending on where the major dissipation areas are located.

If we hypothesize, with Urey, Harrison Brown, and Don Anderson, that the earth assembled in a reasonably cold form, then at the time of lunar capture, which might have occurred during the last stages of assembly or shortly after, the earth did not have an atmosphere or ocean. Hence all tidal dissipation took place internally, at least until surface water was formed. This internal dissipation averages about $1.5 \times 10^{10}$ erg/g, which is quite adequate to produce melting in 'hot spots' and to initiate the formation of a core. Once the core begins to assemble, the gravitational energy thereby released accelerates the heating and melting process, as described by Elsasser. And, as described by Rubey, Cloud, and others, the resultant volcanism leads to the very early formation of atmosphere and oceans on the earth. This hypothesis for core formation should be contrasted with the usual one, in which the earth assembled cold, and heating is mainly due to radio-activity.  This process would take a longer time and give a much later date of formation for atmosphere and oceans on the earth.  It should also be contrasted with the hypothesis under which the earth formed hot, either because of the high temperature of the solar gas or because of kinetic energy dissipated during the accretion process.  In the latter case, the core would form immediately, even if the moon were not captured.

An attractive feature of the present hypothesis of core formation by lunar capture is that it combines the uniqueness of the moon with other unique features of the earth, such as its atmosphere and liquid water oceans, and the existence of life.

EFFECTS OF CAPTURE ON THE MOON

The most probable method of cap-turing the moon, short of introducing ad hoc assumptions concerning extra bodies or additional friction, is to combine tidal dissipation, produced by a close approach of the moon, with the formalism of the restricted three-body problem.  If the moon is formed in an orbit similar to that of the earth, then it will make close encounters with the earth every few years.  After a

certain number of near misses, the moon may experience a so-called Jacobi capture, where it has an opportunity to make a number of successive close approaches to the earth (Singer, 1968). (1) We can show that (impulsive) tidal energy losses can change the zero-velocity surfaces in the restricted three-body problem so as to cut off the escape route of the moon, which then becomes permanently captured.  (2) This process can be further aided if the capture occurs during a favorable phase on the earth's (eccentric) orbit.  (3) Finally, the moon will approach the earth well within the so-called Roche limit.  Depending on its elastic properties, it is quite possible for the (original) moon to be split into pieces.[2]  The resultant reaction can further aid in capturing the moon (or its remaining portion).

[2]Some of the fragments may be immediately re-accreted by the moon as it swings out towards apogee.  Other fragments may escape, only to be swept up by the earth some years later.  As fragments are pulled off the moon within the Roche limit, the time scale of orbital evolution slows down; as pointed out by Opik (1969), it is basically determined by the mass of the largest object orbiting the earth. All these points bear further detailed investigation.

Any initial spin of the moon is immediately dissipated by tides but does not produce much heating. A typical average value, corresponding to a four-hour spin period of the moon, is $5 \times 10^8$ erg/gm, which is only 3% of the energy required to produce melting. However, part of the gravitational energy that is lost by the moon in the capture process can appear within the moon; the rest appears within the earth. The division depends on the elastic parameters of both moon and earth at the time of capture.

Once the moon's spin has been removed, the actual dissipation takes place by radial 'pumping' as the moon moves around the earth in an eccentric orbit. This process, first described by Urey, depends very much on the elastic parameters of the moon that determine not only the tidal distention but also the extent to which the moon relaxes to its original spherical shape as it moves away from the earth.

If all of the energy of capture were to be dissipated within the moon, it would come to about $1.6 \times 10^{11}$ erg/gm, or nearly ten times the amount necessary to produce melting of the moon. Pending a more detailed investigation, therefore, it seems reasonable to assume widespread heating and at least some melting of the moon during the capture phase.

Strong circumstantial evidence for the moon having been extremely close to the earth about 4 billion years ago comes from the asymmetric distribution of ringed maria (and perhaps also of 'mascons') that appear to be all on the earth-facing side.

## CONSEQUENCES FOR THE SOLAR SYSTEM

The moon is not a very typical satellite, in the sense that it has an extremely high mass-ratio with respect to its parent planet. (There are many massive satellites in the outer solar system. They appear to have been formed in place rather than captured, as judged by their orbits.) The capture of the moon may be a unique event in the solar system; is the formation of a moon-like object also unusual? If so, then the existence of our moon would not be very probable.

## MULTI-MOON HYPOTHESIS AND THE ANGULAR MOMENTUM OF VENUS

It seems entirely possible that many moon-like objects were assembled in the inner part of the solar system, as was earlier suggested by Urey (1952). Any of these near earth-orbit would have disappeared by impact on the earth, preferentially to being captured. We are now investigating the probability of capture versus impact by extensions to the restricted three-body problem.

One can apply these ideas to explain the anomalous angular momentum of the planet Venus. We speculate that Venus encountered one or more moons that underwent retrograde capture. Under the frequency-independent tidal calculations, this moon would change into a prograde orbit and remain captured. Under the new calculation it would impact on Venus but first transmit its large (negative) angular momentum to the planet. A sample calculation, using a moon with mass double that of our moon, gives an angular momentum loss of 100%.

## PHOBOS AND DEIMOS

Based on a calculation of orbit evolution by tidal forces (Singer, 1968), I had suggested that Phobos and Deimos may be asteroids or asteroidal fragments that have been captured by Mars. But the times involved for orbit evolution were tens of billions of years and therefore much too long (reflecting the small masses of Phobos and Deimos--about $10^{-9}$ that of Mars). In view of the more recent three-dimensional calculations, it seems even more difficult to suggest capture by tidal effects. Their very nearly equatorial orbits can be explained only on the basis that the satellites had equatorial orbits when captured.

It seems more likely, therefore, that Phobos and Deimos were formed in the vicinity of Mars as the planet was assembled, and that they have undergone only a small degree of orbit evolution based on tidal forces. If this is the case, then Phobos and Deimos should allow us the very exciting opportunity to sample primordial planetesimal stuff stuff, perhaps corresponding to the raw material out of which the planets were assembled. They are both so small that they should not have undergone any internal heating or modification and therefore should have preserved their original chemical composition and petrographic structure.

## SUMMARY

Celestial mechanics calculations based on tidal theory suggest that the moon was captured from an initially prograde orbit. While the inherent probability of capture is low, its relative probability in relation to other theories of origin is respectably high; the main advantage of the capture hypothesis is that it does not require ad hoc assumptions concerning other bodies that have now disappeared.

Strong support for the capture theory comes from the need for a heat source early in the history of the earth. Even stronger support would come from a finding that the earth and the moon experienced simultaneous heating, i.e., within a few hundred years' interval about 4.5 billion years ago.

Finally, an interpretation of the earth-facing ringed maria on the lunar surface would provide evidence for the moon having been very close to the earth during the early part of its history.

REFERENCES

Gerstenkorn, H., Uber Gezeitenreibung beim Zweikorperproblem, Z. Astrophys. 36, 245, 1955.

Goldreich, P., History of the lunar orbit, Rev. Geophys., 4, 411, 1966.

Kaula, W. M., Tidal dissipation by solid friction and the resulting orbital evolution, Rev. Geophys., 2, 661-1964.

MacDonald, G. J. F., Tidal friction, Rev. Geophys., 2, 467, 1964.

Opik, E. J., The moon's surface, Ann. Rev. Astronomy Astrophys., 7, 473, 1969.

Singer, S. F., Capture and orbit evolution of the moon through tidal forces, Geophys. J. Roy. Astron. Soc., 15, 205, 1968.

Urey, H. C., The Planets: Their Origin and Development, 245 pp., Yale University Press, New Haven, Conn., 1952.

*

# Inhomogeneous accumulation of the Earth from the primitive solar nebula

## K.K. Turekian and S.P. Clark (1969)

It has been a common assumption in many if not most theories of the origin of the Earth that the Earth was accumulated in such a fashion as to approach initial homogeneity of composition. The present zonation of the Earth into a core, mantle, and crust is then presumed to be the result of planetary segregation processes occurring some time after the beginning of accumulation. Ringwood (1) has argued effectively that if we choose such a model there are certain constraints that must be met.

(1) The metallic iron formed by reduction of iron oxides or silicates in the Earth and coalescing in the core would almost quantitatively extract nickel (and a few other siderophile elements) from the silicate or oxide phases. The high concentration of nickel in ultramafic rocks (2000 ppm) and basalts (200 ppm) seems to require, on the other hand, a reasonably high

Reprinted from Earth and Planetary Sciences Letters 6, 346-348 by permission of the authors and publisher. Copyright 1969 Elsevier Publishing Co.

concentration of this element in the upper mantle.

(2) The oxidation state of the upper mantle as reflected in the $Fe^{+2}/Fe^{+3}$ ratio of ultramafic rocks and basalts appears to be too high to be compatible with intimately mixed metallic iron.

(3) If the metallic iron in the core were truly made by carbon reduction of iron in oxides or silicates the release and subsequent escape of a mass of CO or $CO_2$ equal to about ½ of the mass of the core is required.

Rather than evaluating Ringwood's attempts to avoid the difficulties posed by this model of accumulation and subsequent planetary processing of homogeneous material from the solar nebula, we consider the case of the accretion of the Earth and planets in a non-homogeneous manner. Eucken (2) suggested such a model to accumulate the Earth's iron core from the primitive nebula. Wood (3) suggested that chondrules represented condensations of pre-planetary material from the primitive nebula. The extension of this postulate to the zoned accumulation of the planets is inspired by some more recent papers, Larimer's (4) on the condensation temperatures of various elements and compounds from the solar nebula in which Eucken's calculations have been redone and expanded, Cameron's (5) on the temperature and cooling rate of the early solar nebula. Larimer and Anders (6) have used these results to explain differences between the different meteoritic types and to propose that the Earth may be a mixture of two end members of which most chondritic meteorites are presumed composed. Anders (7) in a subsequent paper shows that if the Earth had the composition of ordinary chondrites the crustal concentrations of the volatile elements Kr, Ar, Pb, Bi, In, Tl and $H_2O$ would indicate total concentration of these elements in the crust. As an alternate he suggests the Earth may have been veneered with low temperature material accumulated late in the history of accretion. We propose that a combination of parts of these earlier models can be made to provide a unified model for the accumulation and struct- ure of the Earth and other planets.

Starting at a temperature greater than $2000^{\circ}K$ the primitive solar nebula with approximately solar composition and about $10^{-3}$ to 1 atmosphere pressure cools to its present temperature in

less than $10^5$ years. During the time of cooling, elements and compounds condense in the order of increasing vapor pressure. The order of condensa- tion, calculated by Larimer assuming solar composition and a total pressure between $10^{-3}$ and 1 atmosphere, is: iron and nickel; magnesium and iron silicates; alkali silicates; metals such as Ag, Ga, Cu, etc.; iron sulfide; and finally metals such as Hg, Tl, Pb, In, and Bi. Organic compounds and rare gases would also condense late in the process. When the gas cools below about $400^{\circ}K$, conditions change from reducing to oxidizing (8). Any metallic iron still exposed would be converted to $Fe_3O_4$ at this stage, and conversion of iron-magnesium silicates to hydrous minerals could also occur in the latter stages of cooling.

Thus the order of condensation coincides grossly with the stratifica- tion observed in the Earth and inferred in other planets. Such stratification is usually attributed to the settling of the densest material towards the planetary center. But high density turns out to be associated with low volatility, enabling planets to accrete in a way that is automatically stable gravitationally. The iron body that is now the Earth's core formed by accumulation of the condensed iron- nickel in the vicinity of its orbit. It then served as the nucleus upon which the silicate mantle was deposited, and the mantle in turn shielded the core from subsequent reaction with $H_2S$ and $H_2O$ to form sulfides and oxides. The last accumulates would be FeS, $Fe_3O_4$, the volatile trace elements, organic compounds, hydrated silicates, and rare gases.

This model of planets stratified initially due to inhomogeneous accumulation of the elements has the potential of resolving a variety of long-standing problems concerning the origin of the solar system.

(1)  The iron core of the Earth by this model is condensed first and is not produced by carbon reduction in the Earth after accumulation; there is thus no need to provide a mechanism for the escape of an enormous quantity of $CO_2$ or CO.

(2)  The outer layers of the Earth are more oxidized than the inner layers because of the changing nature of the nebular gas during cooling, and possibly also because of the loss of hydrogen from the solar system. The relatively oxidized near-surface material was never intimately mixed with the reduced core, subsequent to condensation.

(3)  The trace element composition of the crust and upper mantle can be high, especially for the volatile elements such as In, Hg, Tl, Pb and Bi because of the low temperature of accumulation of the outer portion. This is compatible with Urey's (8) early observations about the properties of mercury and its terrestrial abundance.

(4)  The source of $H_2O$, $CO_2$ and other volatiles as well as Urey's "solubles", the halide salts (9) can be derived from a surface layer (the crust and upper mantle) and need not involve "degassing" of the whole Earth.

(5)  Since the iron incorporated in the outer layers of the Earth is the product of reactions with particles of iron dust left in the nebula, the outer layers can be rich in metals associated with the early high temperature condensation of the iron. This explains, in particular, the high abundance of nickel in the upper mantle.

(6)  Extension to other planets accounts for numerous features of the solar system. The terrestrial planets owe their high densities to the fact that they were in more direct competition with the sun in the later stages of accumulation. (The orbital radius of Mars is 1.5 A.U.; that of Jupiter is 5.2 A.U.) Mercury's apparent high density results from its being closest to the sun. (Ironically, of all the planets Mercury may be most deficient in mercury.) The low density of the moon, the Galilean satellites of Jupiter, Titan and Triton and their small sizes would result if their

accumulation did not begin until the planets had already swept up most of the condensed iron-nickel. The density of Mars can be explained in a similar way, but with accretion starting somewhat earlier so that a small core was formed.

The model implies that there is no reason why bodies in the solar system should have closely similar overall chemical composition. In particular, there is no reason why any class of meteorite should closely approximate the bulk composition of the Earth.

We hope in the future to put this model on a firmer, more quantitative basis.

This research was supported by NASA under Grant NAS-9-8032 and the National Science Foundation under grant GA 1354.

REFERENCES

(1) A. E. Ringwood, Chemical evolution of the terrestrial planets, Geochim. Cosmochim. Acta 30 (1966) 41-104.

(2) A. Eucken, Physikalisch-chemische Betrachtungen über die früheste Entwicklungsgeschichte der Erde: Nachr. d. Akad. d. Wiss. in Göttingen, Math.--Phys. Kl., Heft 1 (1944) 1-25.

(3) J. A. Wood, Chondrules and the origin of the terrestrial planets, Nature 194 (1962) 127-130.

(4) J. W. Larimer, Chemical fractionations in meteorites-I. Condensation of the elements, Geochim. Cosmochim. Acta 31 (1967) 1215-1238.

(5) A. G. W. Cameron, The formation of the Sun and planets, Icarus I (1962) 13-69.

(6) J. W. Larimer and E. Anders, Chemical fractionations in meteorites-II. Abundance patterns and their interpretation, Geochim. Cosmochim. Acta 31 (1967) 1239-1270.

(7) E. Anders, Chemical processes in the early solar system as inferred from meteorites, Accts. Chem. Res. 1 (1968) 289-298.

(8) H. C. Urey, The Planets (Their Origin and Development) (Yale University Press, New Haven, Conn., 1952).

(9) H. C. Urey, On the concentration of certain elements at the Earth's surface, Proc. Roy. Soc. A219 (1953) 281-292.

# Inhomogeneous Earth

## P.J.S. (1972)

Ideas about the formation of the Earth's core--the latest of which are discussed by Anderson and Hanks on page 387 of this issue of Nature--have proliferated so much in recent years that the average reader, perhaps not intimately concerned with the problems involved, may forgiven for wondering who now to believe. The essential difficulty is, of course, that the formation of the core is so far away in both time and distance that Earth scientists must be content with intelligent speculation and only the minimum of fact, the idea being to try to formulate the simplest, most plausible hypotheses which do not actually conflict with the few hard bits of information which exist. The result is a welter of different hypotheses, each of which is based on geochemical and geophysical evidence which is often quite detailed but circumstantial rather than observational (for example, the results of meteorite or laboratory studies rather

than investigations of Earth materials in situ and none of which is entirely satisfactory.

But in spite of the apparent confusion, it is possible to discuss some important trends in the evolution of ideas concerning the Earth's core, ideas which must inevitably be bound up intimately with the wider problem of the origin of the whole Earth and the solar system. It is now many years since the belief that the Earth formed in high temperature conditions gave way to the opposing theory involving low temperature accretion from particles which on average probably represented a roughly chondritic composition. For such accretion to have taken place at low temperature, however, the speed of the process would have had to have been slow enough to enable the gravitational potential energy to be radiated away, a condition which is frequently interpreted in terms of an accretionary period of about 100 million years. The cold accretion theory then usually goes on to say that heating of the Earth by long-lived radioactive isotopes took place later, giving rise to the differentiation which resulted in the formation of the core. In short, core formation may well have taken place some considerable time after accretion (that is, after the formation of the Earth itself) and may even be continuing today.

This is what may be regarded as the basic conventional view of core formation; but there is an important variation on it which, ironically, marks a return to a "hot origin" theory, though in quite a different sense than that intended before the acceptance of accretion. The idea here is that accretion took place too rapidly for the gravitational potential energy to be radiated away effectively, that the accretion temperature was thus high, and that the conditions necessary for core formation obtained more or less at the time of accretion. The first proponent of this view was probably Ringwood (Geochim. Cosmochim. Acta, 20, 241; 1960) who supported it on geochemical grounds, although it was later taken up by Hanks and Anderson (Phys. Earth Planet. Interiors, 2, 19; 1969) and, more recently, received experimental support from Oversby and Ringwood (Nature, 234, 463; 1971).

But whether the core formed during, or later than, accretion, there has been little suggestion so far that the mixture of iron and silicates from which the Earth accreted was anything but homogeneous; and it is this assumption of homogeneity which effectively defines what is now recognized as the conventional view of core formation. In this connexion the work of Elsasser (Earth Science and Meteorites, North-Holland, Amsterdam; 1963) may be regarded as representative.

In considering the early history of the Earth, Elsasser started with an Earth which accreted cold and in which the material was uniformly distributed. The principal feature of such a model is that the melting point curve of the silicates rises much more steeply with depth than the actual temperature, which implies an appreciable rise in the viscosity of the silicates with depth. As the Earth is later heated by radioactive decay, the outer layers are the first to become soft enough to allow iron to move towards the centre. At greater depths, however, the sinking of the iron is slowed down by the increased viscosity, the iron forms a coherent layer which is gravitationally unstable, and the large drops which result fall rapidly to the centre forming a core. Elsasser's estimate of the time taken to form a core slightly smaller than the present one was about $10^8$ years. Another, equally well known, model beginning with a homogeneous Earth was that of Birch (J. Geophys. Res., 70, 6217; 1965).

With Orowan (Nature, 222, 867; 1969), however, a new idea emerged (albeit within the framework of a model which was perhaps less in keeping with mainstream thinking); namely, that the Earth may have accreted inhomoeneously right from the start. Orowan's point was that, even at low temperatures, iron is plastic-ductile as long as its carbon content does not vastly exceed that of meteorites. As a result, colliding metal particles may be expected to cohere because they can absorb kinetic energy by plastic deformation and can thus combine by cold or hot welding. Silicates, on the other hand, being brittle, break up on collision except at near melting point temperatures. The idea is then that the Earth's accretion began with metallic particles. Once sufficiently large, the growing Earth would then collect non-metallic particles by embedding them in ductile material and later by gravitational attraction. In other words, according to this view, the Earth formed with a metal core already partially differentiated, core formation and Earth accretion were thus simultaneous, and later core formation by differentiation was unnecessary.

The same year, Turekian and Clark (Earth Planet. Sci. Lett., 6, 346; 1969) also proposed an inhomogeneous model, but based on rather different principles. Their idea was that as the primitive solar nebula cooled, elements and compounds would condense in the order of increasing vapour pressure, an order defined by Larimer (Geochim. Cosmochim. Acta, 31, 1215; 1967) as iron and nickel, magnesium and iron silicates, alkali silicates, metals such as Ag, Ga and Cu, iron sulphide, and metals such as Hg, Tl, Pb, In and Bi. In general terms such an order agrees with the stratification in the Earth usually attributed to differentiation. Thus, according to this model, condensed iron-nickel would first form the core which, as in Orowan's model, would become the nucleus for the deposition of silicate mantle.

There is little doubt that such inhomogeneous models need to be tested, if only because, as Ringwood (Geochim. Cosmochim. Acta, 30, 41; 1966) pointed out some years ago, there are a number of difficulties with the conventional model. On the other hand, as Anderson and Hanks point out in this issue of Nature, the Turekian-Clark model has also hitherto been stuck with a problem--namely, that there is apparently not enough energy available to melt the outer core either during or after accretion. By eliminating this problem, Anderson and Hanks have now made the idea of initial inhomogeneity that much more attractive.

# Formation of the Earth's core

## D.L. Anderson and T.C. Hanks (1972)

In the conventional view of core formation the Earth accreted from a homogeneous mixture of iron and silicates, and the core formed when radioactive heating melted the iron which subsequently drained toward the centre[1], the change in gravitational potential energy increasing the average temperature of the Earth by some 2,000 K. In these circumstances core formation would be the major event in the thermal evolution of the Earth[2].

Assuming the Earth accreted uniformly with respect to composition 4.5 x 10[4] yr ago, we have concluded that core formation could have been simultaneous with the accretion process[3]. The initial temperatures necessary to lead to core formation, together with the temperature rise on core formation, would generate temperatures greater than 3,000 K in most of the planet, lending further credibility to the "hot origin" theory. Earlier, Ringwood[4] had inferred a "hot origin" for the Earth,

as well as core formation on accretion, by geochemical arguments.

The conventional view of core formation within an originally homogeneous Earth, however, is not so much the most attractive hypothesis as merely the simplest possible assumption. Turekian and Clark[5] and Clark et al.[6] have offered an alternative in which the Earth accreted inhomogeneously, material accumulating on the proto-Earth in the sequence from which it condenses out of the solar nebula. The core formed immediately from the high temperature condensate, iron, and was essentially intact before any important amount of silicates condensed. The difficulty with this model, however, is that there is insufficient energy, either gravitational or compressional, to melt the outer core on or after accretion; the core would remain solid because of the depletion, based on chemical and meteoritic evidence, of the heat producing radioactive elements in iron[7]. But the inhomogeneous accretion model is attractive because it helps to explain the chemical differences among

Reprinted from Nature 237, 387-388 by permission of the authors and publisher. Copyright 1972 Macmillan Journals Ltd.

the meteorites and among the terrestrial planets, including the Moon. We propose that the inhomogeneous accretion hypothesis, when looked at in more detail, can explain early melting of the core.

The condensation sequence of elements and compounds from a cooling cloud of solar composition has been considered in detail[8,9]. The first compounds to condense are the Ca and Al rich oxides, silicates and titanates, which include $MgAl_2O_4$, $CaTiO_3$, $Ca_2SiO_4$, $Al_2SiO_5$, $CaSiO_3$, $CaMgSi_2O_6$, $Ca_2Al_2SiO_7$, $Ca_3MgSi_2O_8$ and $Ca_2MgSi_2O_7$, all of which condense at temperatures above the inhomogeneous condensation of iron[9]. $CaTiO_3$, $MgAl_2O_4$, $Al_2SiO_5$ and $CaAl_2Si_2O_8$ condense before Fe even under equilibrium conditions. Uranium, thorium, and the rare earth elements (REE) are also refractory and can be expected to condense early. For example, Ca-rich achondrites are high in Ca, Al, Ti, U, Th, and REE relative to the chondrites, and it has been proposed that these meteorites represent high temperature condensates or a differentiate of such a condensate[8,10]. They are low in Fe and volatile fractions and presumably accreted before significant iron had condensed in their vicinity. The Moon also has these chemical characteristics and an origin in terms of a high temperature condensation process is appropriate for that body as well[11].

Such an origin for the Moon explains the Ca-Al enrichment at the lunar surface and interior[11], the U, Th, and REE enrichment of the surface and the deficiency in volatiles.

In our model the nucleus of the proto-Earth consists of the early high temperature condensates and is rich in Ca, Al, Ti, Th, U, and REE. The Moon and the calcium-rich achondrites may be taken as prototypes. The uranium content of these bodies is much higher than chondritic abundances. In particular, the eucrites, a class of Ca-rich achondrites, have U concentrations an order of magnitude larger than carbonaceous chondrites. The Ca-rich inclusions in the Type III carbonaceous chondrites[12] may represent an even more primitive condensate and differentiation of such material may be the source of eucrites and lunar basalts and anorthosites. These are also deficient in Fe and enriched in rare earths and, by analogy, in U. Iron condenses and accretes after the formation of the highly radioactive nucleus; the planet Mercury presumably accreted to this stage. Important mantle minerals such as $MgSiO_2$ and $Mg_2SiO_4$ condense and accrete later. Potassium, another important heat producing element, becomes significant at the later stages of accretion, as do the volatiles such as $H_2O$ and FeS.

The presence of a radioactive

nucleus provides a mechanism for melting the iron core. For illustration we have taken a very simple model for the primordial layering in the Earth. We assume that the refractory nucleus has the composition of eucrites and accreted before iron. The Ca-rich inclusions of Type III carbonaceous chondrites may be a better analogue of the primitive condensate, but their U and Th contents have not been measured. The size is somewhat arbitrary. It is presumably at least as large as the Moon and its size may be related to the present size of the solid inner core. We assume that a mixture of iron and eucritic material accretes on the nucleus to form the protocore and that the radio-activity decreases linearly from eucritic values at the top of the nucleus to zero at the top of the iron-rich shell. The mantle accretes later and has the radioactive abundances of chondrites. An accretion time of 5 x $10^4$ yr is assumed with the Hanks-Anderson[3] accretion rate. The initial temperature profile is calculated from Larimer and Anders's condensation temperatures, the assumption of radiative equilibrium during accretion and adiabatic compression in the interior (Fig. 1). We note that the initial temperatures are well below the melting curve for iron in the iron regions of the Earth. Faster accretion can increase the initial temperatures, of

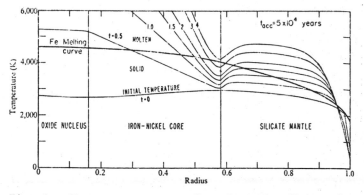

Fig. 1. Temperatures as a function of radius and time for an Earth model that accretes in 50,000 yr during condensation. The radius coordinate is fractions of the Earth's radius. Temperature curves are at intervals of half-billion years after accretion 4.5 billion years ago. Initial temperatures are computed from the condensation temperature, radiative equilibrium during growth and adiabatic compression. The melting curve for iron is from Higgins and Kennedy[19]. The radioactivity of the nucleus and the mantle is eucritic and chondritic, respectively. The radioactivity of the iron shell decreases from eucritic to zero.

course, but the accretion time should
not be less than the cooling time of
the nebula, an upper limit for which
has been estimated to be $10^4$ yr (ref.
6). The central part of the body is
relatively cold in any case.

The iron core was presumably at
least partially molten $3 \times 10^9$ yr or
longer ago judging from the remanent
magnetism in ancient rocks. In our
model melting of iron commences at
about $0.4 \times 10^9$ yr, and it would begin
earlier and proceed more rapidly for
shorter accretion times. We have
ignored latent heats of melting and
convection; the very large temperature
rise in the deep interior is caused by
the high radioactivity in this region.
We have adopted the values U = 9.1 x
$10^{-8}$ g/g, Th/U = 4 and K/U = $2 \times 10^3$
for the nucleus. The temperature
minimum at the core mantle boundary is
a consequence of the zero concentration
of U, Th and K assumed there. So early
melting of the iron core is a consequence of the properties of the initial
condensates.

The nucleus and the protocore are
not in gravitational equilibrium
because iron is denser than the
nucleus. As melting of the iron
commences the nucleus, if perturbed,
will attempt to rise, either in
fragments or in toto. The rise of the
nucleus to the top of the molten iron
zone and, eventually, to the base of

the mantle will not be symmetric
because of the convective pattern in
the core which is controlled by the
rotation of the Earth. The nucleus
will presumably rise preferentially
along one of the poles of rotation.

When the nucleus leaves the centre
of the Earth it will be replaced by
nickel-iron which, before its descent,
is close to its melting point. Because
of the steep slope of the melting curve
relative to the adiabatic compression
curve it will refreeze; this explains
the seismic evidence regarding the
solidity and the composition of the
inner core. The release of gravitational energy resulting from the
overturn of the original nucleus-core
configuration should result in an
average temperature rise of not more
than several hundred K. The present
size of the inner core may therefore
represent the initial size of the
nucleus.

The solidity of the inner core was
inferred initially from indirect
arguments and was verified in 1968 by
free oscillation calculations[13,14]. It
has a very high Poisson ratio,
consistent with its high temperature
and pressure[6,15]. The density and
velocity jump from the outer to the
inner core implies that the inner and
outer cores are not of the same
composition; the inner core is
deficient in the low density, high

velocity alloying component. Its properties are consistent with solid iron-nickel. Our model leads us to suggest that part of the extra material in the molten outer core may be residue from the nucleus, that is, Ca-Al rich oxides and silicates either in solution or in suspension. If so, the high radioactivity of this material may provide part of the energy for driving the dynamo.

We estimate that the density of the nucleus at the core-mantle boundary will be less than the density of the "normal" mantle at this boundary. This, together with its high radio-active content, provides mechanisms by which it might melt its way to the Earth's surface and initiate continental growth, preferentially in one hemisphere, some $3.8 \times 10^9$ yr ago. The change in the U/Pb ratio at $4.1 \times 10^9$ yr (ref. 20) and the Precambrian anorthosite event[21] may also be related to the emergence of the nucleus. As noted earlier, anorthosite is one of the early condensates. Such an emergence of the protonucleus clearly depends on the early mechanical properties of the mantle.

The size of the nucleus is about 2 to 3% of the mass of the Earth. This is approximately the amount of material that condenses from a cloud of solar or chondritic composition before iron. For comparison the continental crust

accounts for 0.35% of the Earth and the lower mantle transition region for 1.5 to 3%. Residual protonuclear material, mixed with iron, may account for the properties of the transition region. We would expect the transition region to vary in thickness and, perhaps, to be entirely missing in some regions. This would lead to topography at the core-mantle boundary, a feature that is desired to explain certain features of the Earth's gravity and magnetic fields[16,17]. The presence of patches of Ca-Al-U rich silicate material at the base of the mantle may also be the source of deep mantle convection plumes and hot spots[18].

This research was supported by a National Aeronautics and Space Administration grant.

[1]Birch, F., Bull. Geol. Soc. Amer., 76, 133 (1965).

[2]Birch, F., J. Geophys. Res., 70, 6217 (1965).

[3]Hanks, T. C., and Anderson, D. L., Phys. Earth Planet. Interiors, 2, 19 (1969).

[4]Ringwood, A. E., Geochim. Cosmochim. Acta, 30, 41 (1966).

[5]Turekian, K. K., and Clark, jun., S. P., Earth Planet. Sci. Lett., 6, 346 (1969).

[6]Clark, jun., S. P., Turekian, K. K., and Grossman, L., in The Nature of the Solid Earth (McGraw-Hill, New

York, 1972).

[7] Anderson, D. L., Sammis, C. G., and
     Jordan, T. H., Science, 171, 248
     (1971).

[8] Larimer, J. W., and Anders, E.,
     Geochim. Cosmochim. Acta, 31,
     1239 (1967).

[9] Blander, M., and Katz, J. L., Geochim.
     Cosmochim. Acta, 31, 1025 (1967).

[10] Anderson, D. L., Comments on Earth
     Sciences: Geophysics (in the
     press).

[11] Anderson, D. L., and Kovach, R. L.,
     Earth Planet. Sci. Lett. (in the
     press).

[12] Marvin, U., Wood, J., and Dickey,
     jun., J., Earth Planet. Sci.
     Lett., 7, 346 (1970).

[13] Anderson, D. L., and Smith, M. L.,
     Trans. Amer. Geophys. Union, 49,
     282 (1968).

[14] Derr, J., Trans. Amer. Geophys. Union,
     49, 282 (1968).

[15] Anderson, D. L., Mineral. Soc. Amer.
     Spec. Paper No. 3, 85 (1970).

[16] Doell, R. R., and Cox, A., in The
     Nature of the Solid Earth (McGraw-
     Hill, New York, 1972).

[17] Hide, R., and Horai, R., Phys. Earth
     Planet. Interiors, 1, 305 (1968).

[18] Morgan, W. J., Nature, 230, 42 (1971).

[19] Higgins, G., and Kennedy, G., J.
     Geophys. Res., 76, 1870 (1971).

[20] Obersby, V. M., and Gast, P. W., J.
     Geophys. Res., 75, 2097 (1970).

[21] Hertz, N., Science, 164, 944 (1969).

# Moon-making in three dimensions

## W.H. McC. (1972)

Almost every theory of the formation of the planetary system requires the existence of a "solar nebula" in the broadest sense of some diffuse material in the vicinity of the Sun at some early stage in its evolution. Most theories regard the material as having originally about the same relative abundances of the elements as in the Sun. They generally regard it as occupying about the region of the present system in the sense that the material had to be in this region for anything relevant to happen to it. Otherwise theories differ enormously in the picture of the solar nebula, treating it as a regular distribution of material having some simple motion of circulation about the Sun, or as forming a pattern of large vortices, or as highly turbulent material, or as part of a more extensive cloud with which it continues to exchange material and momentum, or as material controlled by a magnetic field, and so on.

In his article on the origin of the Moon on page 263 of this issue of Nature, Don L. Anderson uses the essentially simplest picture of the initial state of the nebula, but it is well to remember that even as a starting point it is only one of the array of possibilities just mentioned. On the adopted picture, cosmogonists discuss the distribution of temperature and pressure and thence infer the extent to which elements and compounds (metals, carbon, silicates, ices, and so on) condense out of the gaseous phase to form small solid grains, or to which gaseous components may be driven out of the nebula. Thus they derive for the relevant epoch a chemical composition depending primarily upon distance from the Sun. Each planet is then supposed to "accumulate" from the material of the nebula in an appropriate zone, the different compositions of the planets being supposed to depend primarily upon the differences between the zones, although time-dependences of some of the process may also have an effect.

Coming to satellites, theories differ greatly depending upon whether they emphasize similarities or differences between the planetary system and a satellite system. Again, theories of the origin of the Moon differ according to whether they emphasize its similarity to the other principal satellites regarding absolute dimensions, or its difference from them regarding dimensions relative to those of the associated planet (the ratio of the mass of the Earth to that of the Moon being almost ten times greater than for any other such pair). Anyhow, at present interest is mostly focused upon the Earth-Moon system simply because empirical knowledge about it is incomparably greater than about any like system. Also the Apollo missions have almost given the illusion that as much is now known about the Moon as about the Earth.

One obvious problem presented by the pair is that of their exceedingly different composition, which places the "accumulation" theory in a serious dilemma. For if Earth and Moon accumulated in the same zone, we should expect them to have similar compositions. On the other hand, if the Moon accumulated somewhere else in the system and was then captured by the Earth, it would be an example of a body whose composition is not characteristic of the zone where we now find it; but this almost contradicts the case for explaining the composition of bodies by a zoning of materials.

Anderson finds a way out of this dilemma effectively by going into an extra dimension. He points out that the calculation of the composition of the solar nebula would give a dependence upon distance from its median plane as well as distance from the Sun. Whereas the Earth may be supposed to have accumulated close to the plane, if the Moon formed by accumulation it would necessarily have drawn its material from a much larger range of distance on either side of the plane. Anderson obviously finds that the pursuit of this idea and its testing taxes all the manifold resources of the relevant chemistry and mineralogy, but he evidently draws encouragement from the results, so far as they go.

Anderson's ideas are in a general way similar to those of A. E. Ringwood (J. Geophys. Res., 75, 6453; 1970), which have been developed by other authors as well. Ringwood takes account of the time-sequence of various processes, as does Anderson, but Anderson's idea of also taking account of the variation of composition away from the median plane is apparently new in this context.

The fission theory (in any of its forms) and the capture theory of lunar origin have met with such severe criticisms that it has become a challenge to the "accumulation" theory to show that it can do better. Anderson's work and the like certainly shows that it meets with no obvious contradiction. But it then becomes so complicated that it is hard to see how any crucial test could be proposed within the frame of the work itself. Hopes of some decision seem to lie in two directions--first, the mechanics of the process: it requires to be shown that the accumulation process would actually operate and that it would be expected to produce one Moon with the correct mass, spin momentum, distance from the Earth and so on; second, comparative studies: it requires to be shown whether the processes proposed would account for the Galilean satellites of Jupiter, Titan (Saturn) and Triton (Neptune), all of which have mass and size very similar to those of the Moon.

*

# The origin of the Moon

## D.L. Anderson (1972)

Recent studies of the dynamics and thermodynamics[1-5] of the early solar nebula have provided a basis for a theory of the origin of the Moon which is consistent with presently developed views of the origin of the Solar System. The immediate objective of this article is to extend the hypothesis of inhomogeneous planetary accretion[5]. I suggest that the anomalous properties of the Moon, such as its enrichment in Ca, Al, Ti and other refractories and its depletion in iron and volatiles, can be explained if the bulk of the Moon represents a high temperature condensate.

### CONDENSATION OF THE SOLAR SYSTEM

I start from the commonly held view that the solid objects in the Solar System formed by condensation in a cooling disk-shaped nebula. According to Hoyle and Wickramasinghe[1] the surface temperature of the Sun during the overluminous convective stage is maintained between 3,000 K and 3,500 K during most of the contraction process but eventually reaches ~4,000 K as the main sequence is approached. The planets probably represent a small fraction of the original planetary material which is supposed to be of solar composition. Within the disk the temperature falls off roughly as the square root of distance and different substances condense at distances and times which depend on their concentrations and vapour pressures. Thus the refractory silicates and iron condense in the inner part and the volatiles in the outer part of the Solar System. Hoyle and Wickramasinghe suppose that the accretion of the Earth was substantially complete during the ~$10^4$ yr of the high temperature phase and that the volatiles were a later addition, being brought in by debris during the period of ~$10^7$ yr required for the Sun to reach the main sequence.

Opacity, density, pressure and temperature in the nebula vary with distance above the median plane. This implies that at any stage during the condensation process, the concentration and composition of the condensed material will be a function of location. Once condensation has started, accretion into planets may keep pace with subsequent cooling and condensation[1,5], and, indeed, the composition of the bodies in the Solar System is consistent with their having been formed from material that condensed to a given point in the condensation sequence. Planets accreting in high pressure, or low temperature, regions of the nebula will be further advanced at any given time than those accreting in low pressure, high temperature regions.

## COMPOSITION OF MOON AND EARTH

How are the differences of composition between the Moon and the Earth to be explained on the nebular condensation theory? I now suggest that the special features of the Moon arise because of the strong dependence of condensation temperature on pressure in the nebula, and the rapid decrease in pressure away from the median plane of the disk. The inclination of the orbits of the Moon and pre-lunar material must have been greater in the past and decreased with time owing to gas drag and tidal interactions. It is even possible that at the time of formation the plane of the Moon's orbit was nearly perpendicular to the ecliptic[6,7], and within the framework of the nebular condensation theory, the composition of the Moon was determined by the processes of condensation occurring at a distance from the Sun similar to that of the Earth's orbit but at the low pressures obtaining at some distance from the plane of the ecliptic.

The results of analysis on lunar samples indicate that the Moon is enriched in such refractories as Ca, Al, Ti and the rare earth elements and depleted in elements more volatile than iron. Fig. 1 gives chondritic-normalized lunar abundances as a function of condensation temperature. The condensation temperatures[2,3] are for the element or the first condensing compound containing this element. The condensation temperatures of the REE, Ba, and Sr are unknown but the average enrichment of these elements, shown by the dashed line, is about the same as other refractories. In the lower part is shown the fraction of available material that has condensed as a function of temperature[3]. The initiation of condensation of various compounds is indicated by arrows. Note the rapid increase in the fraction of material which has condensed as the temperature drops to the condensation temperature of iron and the magnesium silicates.

The enrichment factor in Fig. 1 is

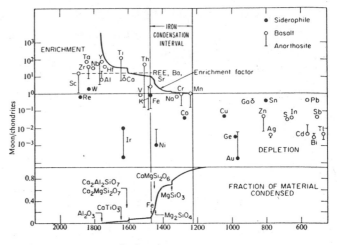

**Fig. 1** Chondritic normalized lunar abundances[14-17] as a function of condensation temperature[2-4] of the element or the first condensing compound containing this element at $10^{-3}$ atmospheres total pressure. Ratios greater than unity indicate enrichment and less than unity indicate depletion.

the reciprocal of the condensed fraction. It indicates the expected enrichment of the trace refractories in the condensed phase relative to chondritic abundances. The refractories are enriched in the Moon to about the level expected in the pre-iron condensates. As the temperature decreases below the condensation temperature of iron the enrichment factor rapidly approaches unity, that is chondritic abundances.

CONTENT OF MOON

I propose that the Moon is composed chiefly of compounds that condense before iron and that the volatile content of the Moon was brought in as a thin veneer after the solar nebula dissipated. The latter part of this proposal is not new; Ganapathy et al.[8] have suggested that the volatiles were

brought to the surface by chondritic material. A refractory outer layer was proposed[9,10] to explain some aspects of the geochemistry of the lunar surface material but the present proposal is novel in that it suggests that the whole Moon is enriched in

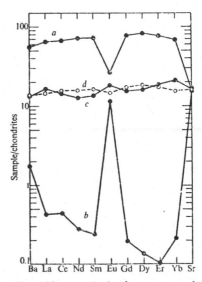

**Fig. 2** Chondritic normalized refractory trace elements for lunar basalt (a)[14], lunar anorthosite (b)[16,17], Ca-Al rich inclusions from the Allende meteorite (c)[13], and a mixture of 0.22 basalt + 0.78 anorthosite (d).

refractories such as Ca, Al, Ti, REE, Ba, Sr and U. The bulk composition may be similar to the Ca-Al rich inclusions in Type III carbonaceous chondrites[11] which are thought to represent the early condensate of the cooling nebula. There is trace element support for this conjecture. Fig. 2 shows the chondritic normalized abundances of Ba, Sr and the REE in a lunar basalt, a lunar anorthosite and the Ca-Al rich inclusions of the Allende meteorite. If lunar basalts and anorthosites were ultimately derived from extensive melting of a source region having the composition of the Allende inclusions, a mixture of the two should yield a level pattern and approximately the same abundances as the inclusions. These constraints are met with the mixture shown. The average enrichment of both the mixture and the Allende inclusions is 16x chondritic, about that predicted on the basis of Fig. 1. A Ca-Al rich interior has been dismissed on the grounds that the high pressure assemblage of such material would be greater than the mean density of the Moon. If, however, a large part of the Moon is above the equilibrium temperature of the high pressure phases, this argument no longer applies.

The accretion during condensation hypothesis seems capable of explaining the differences between the Earth and the Moon even if they formed at the same distance from the Sun. Condensation temperatures depend on pressure as shown in Fig. 3. The pressure in the nebula varies with distance from the Sun and with height above the median plane. The condensation intervals of the refractories, forsterite, and iron overlap at high pressures and diverge at low pressures. The line labelled 1 AU is the temperature in the vicinity of the Earth's orbit during the high luminosity phase of the Sun[1]. If the Earth accreted in a dense, high pressure part of the nebula, such as the median plane, it would have iron and the magnesium silicates available

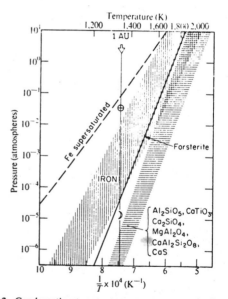

Fig. 3 Condensation temperature versus pressure in the solar nebula[2,3,18,19]. The temperature at 1 AU is appropriate for a solar surface temperature of 3,000 K (ref. 1). The Earth and the Moon are plotted at the pressure which will explain their bulk compositions. I assume that the Earth and the Moon accreted from material that condensed while cooling from high temperatures to the temperature shown.

for incorporation into its interior as well as the early condensing refractories. Away from the median plane the temperature remains almost constant (A. G. W. Cameron and M. R. Pine, unpublished) but the gas pressure decreases. Condensation, therefore, occurs at lower temperatures, and, in a cooling gas, at later times. As represented in Fig. 3, the Earth will be more than 50% assembled before the Moon starts and will have swept up most of the iron. After the Moon nucleates the Earth will still get most of the remaining iron and later condensates, since it spends all of its time in the median plane. Material condensing off the median plane will initially be in inclined orbits which get tilted toward the plane by gas drag and collisions.

If accretion was rapid and efficient the Moon may have been, and may still be, a chemically zoned body. In order to provide a Ca-Al rich source region with high abundances of refractory trace elements, it has been proposed[9] that at least the outer shell is enriched in the refractories, including Ca, Al, Sr, Ba, U and the REE. In the present model the interior is devoid of iron and the iron content increases with radius. The low-pressure, high temperature mineralogy of the deep interior would initially be perovskite, diopside and spinel grading into anorthite and ferro-magnesium silicates at shallower depths. The drop in electrical conductivity of the Moon at 250-350 km depth[12] is most reasonably interpreted as a sudden decrease in iron content at this depth, a possible relic of the initial layering or the result of iron enrichment in a residual melt. The mean density and the moment of inertia can be satisfied with the Ca-Al rich compositions discussed here. The low electrical conductivity of iron-free silicates leads to higher interior temperatures than inferred by Sonett et al.[12].

## OVERALL MODEL

The lunar igneous rocks and, by implication, their source regions[13], are enriched in refractories and depleted in elements more volatile than iron. The enrichment of refractories is consistent with that expected for the pre-iron condensates in a partially condensed cloud of solar composition. The volatile content can be explained by a small increment of carbonaceous material brought in after the main accretion[8]. I propose that the bulk of the Moon is composed of material that condensed prior to the condensation of iron. This can be accomplished if the Moon accreted from material that condensed at much higher temperatures or at much lower pressures than the other terrestrial planets and the chondritic meteorites. One likely source for this material is outside the plane of the ecliptic and this favours a lunar orbit of initially high inclination.

Because it accreted in a relatively high pressure part of the nebual, the median plane, the Earth starts to accrete earlier and grows faster than bodies which accrete in highly inclined orbits. Most of these latter bodies are brought into the median plane and swept up by the Earth. The composition of the Moon can be understood if it formed relatively late, after most of the condensable material near the median plane had been swept up by the Earth. This delay in the start of formation of the Moon may be a consequence of the strong pressure dependence of the condensation tempera- ture. The condensation temperatures of the early condensing refractories, iron and the magnesium silicates increase and converge as the pressure increases. This alone can qualitatively account for the differences in size and com- position of the Earth, Mercury, Venus and the Moon. Another effect sets in when a planet grows to sufficient size to retain an atmosphere. Such a planet is more efficient in retaining material that it encounters, and it also triggers condensation in its vicinity.

I propose that the high luminosity phase of the Sun prevented complete condensation throughout the disk at

1 AU.

E. Anders, H. Mizutani, S. Clark, and A. Cameron provided critical reviews of an earlier version of this manuscript. This research was supported by a National Aeronautics and Space Administration grant.

[1] Hoyle, F., and Wickramasinghe, N. C., Nature, 217, 415 (1968).

[2] Larimer, J. W., Geochim. Cosmochim. Acta, 31, 1215 (1967).

[3] Gossman, L., Geochim. Cosmochim. Acta (in the press).

[4] Clark, jun., S. P., Turekian, K. K., and Grossman, L., in The Nature of the Solid Earth (edit. by Robert- son, E. C.), 3 (McGraw-Hill, 1972).

[5] Turekian, K. K., and Clark, jun., S. D., Earth Planet Sci. Letters, 6, 346 (1969.

[6] Gerstenhorn, M., Z. Astrophys., 26, 245 (1955).

[7] Singer, S. F., Geophys. J. Roy. Astron. Soc., 15, 205 (1968).

[8] Ganapathy, R., Keays, R. R., Lanl, J. C., and Anders, E., Geochim. Cosmochim. Acta, Suppl. 1, 1117 (1970).

[9] Gast, P. W., and McConnell, jun., R. R., Third Lunar Science Conference Abstracts, 257 (1972).

[10] Wasserburg, G. J., Turner, G., Tera, F., Podosek, F. A., Papanastassiou, D. A., and Hineke, J. C., Third Lunar Science Conference Abstracts, 695 (1972).

[11] Clarke, jun., R. S., Jarosewich, E., Mason, B., Nelen, J., Gomez, M., and Hyde, J. R., Smithsonian Contrib. Earth Sci. No. 5 (1970).

[12] Sonett, C. P., Schubert, G., Smith, B. F., Schwartz, K., and Colburn, D. S., Proc. Second Lunar Science Cong., 2, 2415 (MIT, 1971).

[13] Gast, P. W., Hubbard, N. J., and Wiesmann, H., Geochim. Cosmochim. Acta, Suppl. 1, 1143 (1970).

[14] Mason, B., and Melson, W. G., The Lunar Rocks, 179 (Wiley-Interscience, 197).

[15] Paul, J. C., Morgan, J. W., Ganapathy, R., and Anders, E., Proc. Second Lunar Science Conf., 2, 1159 (MIT, 1971).

[16] Wakita, H., and Schmidtt, R. W., Science, 170, 969 (1970).

[17] Hubbard, N. J., Gast, P. W., Meyer, C., Nyquist, L., Shih, C., and Wiesmenn, H., Earth Planet. Sci. Letters, 13, 71 (1971).

[18] Lord, H. C., III, Icarus, 4, 279 (1965).

[19] Blander, M., and Katz, J. L., Geomchim. Cosmochim. Acta, 35, 61 (1971).

*

# Is the Moon hot or cold?

**D.L. Anderson and T.C. Hanks (1972)**

INTRODUCTION

Baldwin (1) has summarized some of
the evidence for the early and present
internal thermal states of the moon.
He argued that the interior of the
moon, below about 200 to 300 kilometers, is presently "hot," that is,
has interior temperatures close to or
exceeding the lunar solidus curve. The
evidence for a rapid differentiation of
the moon about $4.6 \times 10^9$ years ago, the
extensive igneous episode resulting in
mare formation $3.7 \times 10^9$ to $2.8 \times 10^9$
years ago, the depletion of the moon in
volatiles and its enrichment in calcium, aluminum, and the trace refractory
elements all argue for a hot origin and
high initial temperatures. A straightforward consequence of the lunar
thermal inertia is that if the interior
of the moon were ever hot it would remain hot to the present. The high
surface concentrations of uranium,
thorium, and potassium, and the Apollo

15 heat flow value of 33 ergs per
square centimeter per second (2),
indicate high present-day temperatures
in the lunar interior (3).

In apparent contradiction to these
conclusions, recent interpretations of
the lunar conductivity profile, the
nonhydrostatic shape of the moon, the
existence of mascons, the remarkable
aseismicity of the moon, and the
absence of present-day volcanism have
been interpreted as suggesting that the
lunar interior is presently cold and,
by implication, has always been cold.

Various attempts have been made to
reconcile the conflicting evidence by
postulating that the interior is cold
and the surface manifestations of
differentiation apply only to the outer
reaches of the moon. By implication,
the deep interior has always been cold
and always deficient in uranium,
thorium, and potassium. Remarkable
constraints must then be placed on the
accretional process on the moon with
respect to bulk and trace element
geochemical zonation (4).

We find that the basic observations do not demand a presently cold

moon and are, in fact, consistent with a hot moon. We find that an iron-deficient, highly resistive, hot lunar interior, capped by a cool, rigid lunar lithosphere with a thickness of several hundred kilometers, can explain the relevant observations and is a reasonable model of the moon today.

## STRENGTH OF THE MOON

The nonequilibrium shape of the moon and the existence of lunar mascons suggest that the lunar interior possesses a long-term resistance to relaxation of the implied elastic stress differences. This, in turn, suggests that either a finite elastic strength or a high viscosity is a basic property of the lunar interior, but neither is consistent with its possessing hot temperatures (5). Several aspects of this crucial interpretation have not been generally appreciated.

It has been pointed out several times that the apparent departure of the moon from hydrostatic equilibrium is not so remarkable as might be supposed (6). Because of the smaller gravity on the moon and its lower mean density, it should be able to support much larger excess mass loads than the earth. In fact, the inferred mass excesses and the implied stress differences are less on the moon than on the earth.

Kaula (7) has shown that the moon is gravitationally smoother than the earth, provided that gravity anomalies are considered equivalent if the corresponding surface mass anomalies produce equal stresses. This is the "equal stress implication" (8) according to which gravity anomalies are scaled with the gravitational field of the appropriate planetary object. In terms of the second degree harmonics of the gravitational fields, Lorell et al. (9) have shown that the moon is several times smoother than the earth, and that Mars is considerably rougher than either. If these results are indicative of the bulk strengths of the planetary interiors, as implied by Urey and MacDonald (5), the qualitative conclusion is that the earth is stronger than the moon. If this measure of strength is, in turn, indicative of interior temperatures, the conclusion is that the moon is hotter than the earth, not colder.

Similarly, the lunar mascons, representative of the higher degree harmonics of the lunar gravitational field, are hardly dramatic features by terrestrial standards. Kaula (7) notes that there are at least 20 terrestrial "mascons" larger than the largest lunar mascon at Mare Imbrium, and that the largest of these is an order of magnitude larger than that at Mare Imbrium.

Moreover, it is not essential that the lunar gravitational anomalies, of whatever degree, be supported by the deep lunar interior. While the non-equilibrium shape of the moon implies stress differences of 15 to 20 bars if it is supported throughout the lunar interior, it implies stress differences of only 28 to 50 bars if it is entirely supported by an exterior shell 400 km thick overlying a deeper interior devoid of strength (10).

The shapes of the positive gravity anomalies associated with the lunar mascons imply mass excesses at depths of 25 to 125 km (11), in terms of a model of layered slabs of high-density maria fill. Depending on the density contrast, it is estimated that the thickness of the maria pile is 15 to 30 km. This model implies that the exterior of the moon is capable of supporting shear stresses in the 50-bar range at a depth of about 80 km. Urey and MacDonald (5) obtained similar stress differences in about the same depth range, 70 to 120 km, to support the higher degree harmonics of the lunar gravitational field.

More recent data on the lunar gravitational field strongly suggest that the lunar mascons are near-surface features (12). Hulme (13) has shown that the positive gravity anomalies associated with circular maria can be maintained even if the moon cannot support stress differences below depths of the order of 100 km. Baldwin (14) estimated that the crater shapes imply stress differences of 30 to 50 bars in the upper 50 km of the moon.

One alternative to the interpretation of Urey and MacDonald (5), then, is the existence of an exterior shell several hundred kilometers thick, capable of supporting long-term stress differences of 50 to 100 bars. This estimated strength is substantially less than measured rock strengths, which are of the order of several kilobars or more (15). It is also less than the strength inferred for the earth's crust and upper mantle, which ranges from 100 bars to over a kilobar (16). Estimates of the minimum strength required to maintain the non-hydrostatic shape of the earth range from 20 to 97 bars (17,18).

The implied stress differences have probably not been maintained for billions of years by finite elastic strength; they have most likely been relaxing with time through slow anelastic deformation of the stressed exterior shell. To a first approximation, isostatic compensation can be modeled as a viscoelastic process. The time scale of the persistence of these features can then be translated into a minimum viscosity, variously estimated to be $10^{26}$ to $10^{27}$ poises (5,7) for the case of a uniform moon. If the moon is modeled as a highly viscous layer over an interior of low viscosity (19), its nonequilibrium shape and the persistence of mascons imply viscosities of $2 \times 10^{24}$ to $5 \times 10^{25}$ $cm^2$/sec for a shell 200 km thick. These estimates are greater than the viscosity of the earth's upper mantle, $10^{21}$ $cm^2$/sec, but are less than or comparable to theoretical and experimental estimates of the viscosity of the terrestrial lithosphere (20). If the viscosity of a lunar lithosphere 200 km thick is $10^{25}$ poises, the relaxation times for the nonhydrostatic shape and the mascons are $2.5 \times 10^9$ and $1 \times 10^9$ years, respectively.

On the other hand, any estimates of relaxation times based on a linear relation between stress and strain rate (Newtonian viscosity) can be misleading; the stress-strain rate relation is probably highly nonlinear. Experimentally obtained creep laws for rocks have the form $d\varepsilon/dt \propto \sigma^5$, where $d\varepsilon/dt$ is the strain rate and $\sigma$ is the stress (21). This leads to high strain and rapid stress release at high stress levels, and slow stress relaxation at low stress levels. If the stress differences were originally of the order of 1 kbar in the moon and have decayed to 50 bars in $4 \times 10^9$ years, one can calculate that the stress difference decayed to 500 bars in $4 \times 10^5$ years, to 100 bars in $2.5 \times 10^8$ years, and to

70 bars in $10^9$ years. Another $2 \times 10^9$ years only results in a further decrease in stress of 16 bars. Stresses of a few tens of bars can therefore be regarded as essentially permanent (22).

The various estimates of implied stress differences in the moon are, of course, subject to some uncertainty. However, the observations discussed above are consistent with the existence of a rigid, cool lithosphere overlying an interior with little or no strength.

## LUNAR ELECTRICAL CONDUCTIVITY PROFILE

Electrical conductivity profiles (23,24) have seemingly provided the most direct and compelling evidence for a cold lunar interior. Apart from the uniqueness of the electrical conductivity profile itself, the nonuniqueness associated with inferring temperature from conductivity does not seem to be generally appreciated (24).

Laboratory measurements of electrical conductivity are generally interpreted with the relation $\sigma^E = \sigma_0^E$ $\exp(-A/kT)$, where $\sigma^E$ is the electrical conductivity, $\sigma_0^E$ is a constant that depends strongly on the bulk and impurity compositions and oxidation states (25), $A$ is the activation energy, $k$ is the Boltzmann constant, and $T$ is the absolute temperature. For constant $\sigma_0^E$ the electrical conductivity is a strong function of temperature, low values of $\sigma^E$ implying low temperatures.

Sonett et al. (23) have presented electrical conductivity profiles for the moon; they interpreted these profiles in terms of laboratory measurements of $\sigma^E$ for lunar basalts and terrestrial samples of olivine and peridotite, and obtained temperatures less than $1000^{\circ}C$ throughout the moon. On the basis of the claim that olivine and olivine peridotite "represent the least conducting geological material,"

they suggested that $1000^{\circ}C$ is an upper bound on temperatures in the moon. More recent electrical conductivity measurements, however, have revealed still lower conductivities for olivine (Fig. 1).

Figure 1a is a summary of laboratory measurements of $\sigma^E$ for several different materials. The hatched region represents the range of values obtained for the Apollo 11 and Apollo 12 lunar basalts (26). The unlabeled lines are for olivine and peridotite, from England et al. (27). These are the conductivity measurements on low-

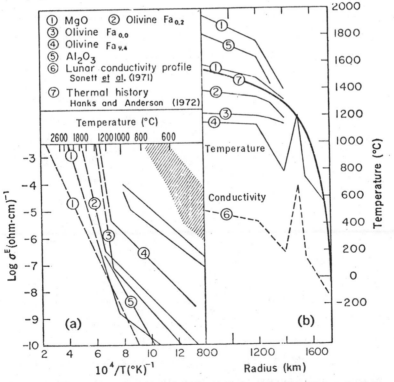

Fig. 1. (a) Electrical conductivities of MgO, $Al_2O_3$, forsterite ($Mg_2SiO_4$), and olivines of varying fayalite ($Fe_2SiO_4$) contents (40). (b) Electrical conductivity of the moon and inferred temperatures. The numbers on the curves refer to the compositions used in inferring the temperatures and correspond to the numbers on the electrical conductivity curves. Curve 7 is a thermal history model from Hanks and Anderson (3).

resistivity materials that have been used in determining the lunar temperatures. The curves labeled 1 to 5 in Fig. 1a are laboratory measurements for magnesium oxide, aluminum oxide, and three members of the olivine series with differing fayalite contents.

As shown in Fig. 1b, if the lunar conductivity profile is interpreted with any of these five materials, the corresponding lunar temperatures are considerably higher than those obtained by Sonett et al. (23) and Dyal and Parkins (24). Indeed, the temperature distributions so obtained are consistent with present-day temperatures obtained through thermal history calculations (3) and a presently hot lunar interior.

Of particular importance in the interpretation of the lunar conductivity profile are the amount and oxidation state of iron (28). The electrical conductivities of the end members of the olivine series differ by six orders of magnitude, forsterite being the more restive. The moon is clearly depleted in iron relative to terrestrial, solar, or chrondritic abundances. In terms of the olivine series, the likely contents of ferrous oxide in the lunar interior would still allow for a variation of three orders of magnitude in the resulting conductivity.

Equally important is the oxidation state of the iron. Material 4 of Fig. 1 is an olivine of 9.4 mole percent fayalite content, which is essentially free of the ferric ion. This material has a smaller electrical conductivity than other olivine samples possessing less total fayalite but more $Fe^{3}+$. The near absence of $Fe^{3}+$ in lunar materials suggests that they crystallized at extremely small partial pressures of oxygen. Thus, even if the bulk chemistry of the lunar interior indicated an FeO content equivalent to 10 mole percent fayalite, the lunar interior would still be highly resistive because of the lack of $Fe^{3}+$.

If so, the lunar conductivity profile is consistent with interior temperatures at or near the solidus curve (28). The inferred temperatures at depth are even higher if the FeO content of the lunar mantle is lower than that of the crust. The lunar conductivity profile suggests that this is the case. In the conductivity profile of Sonett et al. (23), the conductivity decreases by nearly three orders of magnitude between depths of 250 and 350 km. This is an indication of a compositional change, a phase change, or both, at this level. Sonett et al. (23) concluded that the material below the conductivity peak must be several orders of magnitude less conducting than dunite and that "such material is not known on earth."

Actually, the conductivities of MgO and $Mg_2SiO_4$ at about $1150^{\circ}C$ are close to those measured at a radius of 1400 km. The conductivity of more reasonable mineralogies such as calcium-aluminum garnet, diopside, spinel, and enstatite cannot be estimated at present. The precipitous decrease in electrical conductivity can most reasonably be interpreted as a decrease in FeO content from about 15 to 5 percent or from about 10 percent to essentially zero. This is consistent with the idea that the bulk of the moon is composed of the refractory materials that condensed before iron in the cooling solar nebula (3,29).

Since the average FeO content of the lunar crust is approximately 10 percent and since little $Fe^{3+}$ is in evidence from the lunar samples, it might be expected that material 4 gives a fair measure of the electrical properties of the outer part of the moon. Indeed, the temperatures inferred from the conductivity measurements on this material are remarkably consistent with those obtained from the thermal history calculation for the outer several hundred kilometers. The precipitous drop in lunar conductivity at a radius of 1500 km, however, gives a physically unacceptable local temperature minimum. This drop is probably indicative of a change in composition at a radius between 1400

and 1500 km. It is interesting that the condition for smooth temperature gradients can be met with any of the materials low in iron. A possible interpretation of this very rapid decrease in the lunar conductivity profile is that the FeO content of the lunar lower mantle is considerably less than that of the upper mantle or the crust, and that this decrease occurs in the depth interval from 200 to 300 km. If this is the case, the lunar interior must be highly resistive and hot to explain the observed conductivity profile. In the outer several hundred kilometers, any relative depletion of $Fe^{3+}$ implies higher temperatures than those obtained in the same region by Sonett et al. (23).

RELATIVE ABSENCE OF PRESENT-DAY VOLCANIC AND SEISMIC ACTIVITY

The low level of lunar seismicity (30) has also been used as an argument for a cold interior (31), although the details of the argument have not been presented. Terrestrial experience indicates the contrary. Laboratory results show that at low temperatures rocks fail by brittle fracture. At high temperatures they yield by steady-state sliding or aseismic creep. Thomsen (32) has noted that estimates

of the lunar pressure-temperature
regime would place the outer several
hundred kilometers of the moon well
within the experimental stable-sliding
regime.  The absence of lunar seismicity can be attributed to lack of suitable stresses or to the nonbrittle
behavior of the interior.  The latter
implies high temperatures.

The inverse correlation between
seismic activity and proximity to the
melting point is well known (33).
Most terrestrial earthquakes occur in
the cold, upper part of the crust.  The
absence of earthquakes at depths
greater than 16 to 20 km in California,
the absence of terrestrial earthquakes
below 700 km, and the minimum in earthquake activity in the vicinity of the
low-velocity zone can all be explained
in terms of high-temperature phenomena.

On the other hand, the low level
of lunar seismicity must mean that the
stress differences necessary to cause
fracture over dimensions of kilometers
or more are not available in the moon.
The existence of a rigid lithosphere
that is capable of supporting stress
differences of 50 to 100 bars does not
conflict with the observation that many
terrestrial earthquakes commonly have
stress drops between several tenths of
a bar and several hundred bars.  The
stress drop of an earthquake in no way
constrains the preexisting stress
except that the stress drop is a mini-

mum estimate of the preexisting shear
stresses.

It has also been suggested that
the absence of lunar igneous activity
after about $3.0 \times 10^9$ years ago implies
that the interior has been cold, at at
least not partially molten, since that
time (34).  Igneous activity certainly
implies partial melting at depth in a
planet, if impact melting is ignored,
but the absence of igneous activity does
does not imply the absence of partial
melting.  We would not expect partial
melting at depth to result in extrusive
igneous activity if the melt content is
small (less than 5 percent) or if
stresses, either internal or external,
are insufficient to break the overlying
layer, the lithosphere.  As noted above
above, the implied stress differences
in the lunar lithosphere are far below
the breaking strength of unfractured
rock; from another point of view, the
low level of lunar seismic activity
suggests that suitable stress differences, including the effective
dimension, are simply not available to
rupture a lithosphere of several
hundred kilometers.

Igneous activity on the earth is
generally restricted to areas where the
lithosphere is thin or where plates are
colliding.  It is absent, for example,
on old shields which, on the basis of
geophysical data, have thick lithospheres.  Igneous activity is presum-

ably shut off when the lithospheres gets too thick or the melt content of the interior gets too small, or both. Thermal history calculations show that the lithosphere thickens with time. Extrusive activity might also be triggered by rupturing of the lithosphere by large impacts or tidal stresses. These would both have been more effective in the early history of the moon.

On the earth, both seismicity and volcanism are closely related to the boundaries between moving plates. Shallow seismicity and basaltic magmatic activity occur in regions of plate separation; deep seismicity and andesitic volcanism occur when plates converge.

With respect to the moon the absence of seismicity and volcanism need only reflect the absence of relative plate motions, or plate tectonics as we know it on the earth. The absence of these phenomena in continental shield areas does not imply a cold terrestrial interior, but more likely reflects the great thickness of the lithosphere, 100 to 200 km. Plate mobility and penetration of the lithosphere by magma are probably consequences of the large thermal and mechanical inertia of the interior relative to the plate. The earth has a lithosphere that is negligible compared to the mass of the planet. The reverse

is obviously true for the moon if the lithosphere is several hundred kilometers thick.

## THERMAL HISTORY CONSIDERATIONS

The most decisive evidence favoring a hot lunar interior--the evidence for a rapid differentiation of the moon, the extensive igneous episode resulting in mare formation, and the high surface concentrations of uranium, thorium, and potassium--is somewhat ambiguous: these phenomena need only have involved the outer 300 to 400 km of the moon, the deeper interior playing only a minor part in the development of surface phenomena. The distribution of the radioactive heat sources, however, greatly constrains the number of possibilities.

If the deep lunar interior has uranium concentrations greater than approximately 30 parts per billion it must be at least partially molten at the present time, even if it started cold. The surface concentrations of uranium are 10 to 100 times larger than this. If the single lunar heat flow value is at all representative of the average value, similar concentrations of uranium must persist to depths of 50 to 200 km. The average uranium concentration of the moon which is consistent with the surface concentration and the observed heat flow is 0.09 part per million (3).

The surface material of the moon and, by inference, the source region of the lunar basalts is enriched in the refractory elements relative to carbonaceous chondrites. If the lunar basalts and anorthosites are mixed in the proportions required to remove the europium anomaly (35), the average uranium content is 0.12 ppm, which is 12 times the average chondritic value. The moon is apparently depleted in materials more volatile than iron. If we assume that the moon is composed of only those materials more refractory than iron, we estimate that it should be enriched in such refractories as uranium, barium, strontium, and the rare earth elements by a factor of about 10 relative to carbonaceous chondrites, which are assumed to have their full complement of condensable material. This leads to an estimate of 0.1 ppm for the uranium abundance in the moon. Corresponding enrichments in thorium can be expected. The same reasoning leads to a depletion in potassium which, however, contributes only a minor amount of radioactive heating. Eucrites provide a close match to the lunar surface chemistry; they contain 0.1 ppm of uranium. These estimates are remarkably consistent with each other and suggest that uranium and thorium are strongly enriched in the moon relative to their chondritic or cosmic abundances. There is certainly no reason to believe that they might be depleted.

## OTHER CONSIDERATIONS

Whether the deep interior of the moon is hot or cold has an important bearing on the overall composition of the moon and its origin. The chemistry of the lunar surface materials requires that their source region be enriched in calcium, aluminum, and the refractory trace elements such as uranium, thorium, barium, strontium, and the rare earth elements, and depleted in volatiles (36). The amount of this material, the inferred degree of partial melting, and the evidence for the great depth of the source region suggest that a substantial fraction of the moon is enriched in these refractory elements. The lunar seismic experiment has yielded data that is also consistent with the enrichment in calcium and aluminum of at least the outer 100 km (37). On the other hand, it has been suggested (38) that the whole moon cannot be enriched in calcium and aluminum because of the density associated with the garnet-rich assemblage that is stable at modest pressures. A ferromagnesian or chondritic interior has therefore been proposed (4), and it has been suggested that the exterior of the moon is composed of the initial high-temperature condensates. However, if temperatures are high the whole moon can be composed of high-temperature phases rich in calcium and aluminum, and there is no requirement for an ad hoc initial chemical layering of the above type. High temperatures move the stability field of plagioclase deep into the interior of the moon and this serves to decrease the mean density. A small core composed of the high-pressure phase assemblage can be tolerated without violating the mean density or moment of inertia. This removal of a constraint on the internal composition of the moon has been discussed by Anderson (39).

## REFERENCES AND NOTES

1. R. B. Baldwin, Science 170, 1297 (1970).

2. M. G. Langseth, S. P. Clark, J. L. Chute, S. J. Keihm, A. E. Wechsler, Moon 4, 390 (1972).

3. T. Hanks and D. L. Anderson, Phys. Earth Planet. Interiors 5, 409 (1972).

4. P. W. Gast, Moon 5, 121 (1972).

5. H. Urey and G. J. F. MacDonald, in Physics and Astronomy of the Moon, Z, Kopal, Ed. (Academic Press, New York, ed. 2, 1971), p. 213.

6. W. M. Kaula, J. Geophys. Res. 74, 4807 (1969); D. L. Anderson and R. L. Kovach, in Proceedings of the Caltech-JPL Lunar and Planetary Conference, 13-18 September 1965 (Jet Propulsion Laboratory, Pasadena, Calif., 1966), p. 84.

7. W. M. Kaula, Phys. Earth Planet. Interiors 2, 123 (1969).

8. ----, in Trajectories of Artificial Celestial Bodies, J. Kovalevsky, Ed. (Springer-Verlag, Berlin, 1966), p. 247.

9. J. Lorel, G. Born, E. Christensen, J. Jordon, P. Laint, W. Martin, W. Sjogren, I. Shapiro, R. Reasenberg, G. Slater, Science 175, 317 (1972).

10. M. Caputo, J. Geophys. Res. 70, 3993 (1965).

11. J. E. Conel and G. B. Holstrom, Science 161, 680 (1968).

12. W. L. Sjogren, P. M. Muller, W. R. Wollenhaupt, Moon 4, 411 (1972).

13. G. Hulme, Nature 238, 448 (1972).

14. R. B. Baldwin, Icarus 8, 401 (1968).

15. A useful tabulation of laboratory measurements of rock strengths appears in the article by J. Handin in "Handbook of Physical Constants" (S. P. Clark, Jr., Ed. Geol. Soc. Amer. Mem. 97 (1966), pp. 223-289). A few representative values are given here.

| Material | T ($^{o}$C) | Confining pressure (kbar) | Ultimate strength (kbar) |
|---|---|---|---|
| Anorthosite | 150 | 1.0 | 5.9 |
| | 500 | 5.1 | 9.4 |

| Basalt | 300 | 5.0 | 13.8 |
|--------|-----|-----|------|
|        | 800 | 5.1 | 2.6  |
| Granite | 150 | 1.0 | 3.3 |
|        | 500 | 5.1 | 8.3  |

16.  F. Birch has estimated the strength of the earth's mantle to be of the order of 100 bars by using data from geodesy, gravity, and geology. Under some high mountains it may reach several kilobars. Caputo (17) estimated the minimum strength required to maintain the global departures from hydrostatic equilibrium to be 20 to 70 bars. Kaula (18) has also analyzed the gravity field of the earth and derived maximum stress differences of 97 and 300 bars for the lower mantle and the crust, respectively. See F. Birch, in State of Stress in the Earth's Crust, W. R. Judd, Ed. (Elsevier, New York, 1964), pp. 55-80; H. Jeffreys, The Earth (Cambridge Univ. Press, London, ed. 4, 1959), p. 420.

17.  M. Caputo, J. Geophys. Res. 70, 955 (1965).

18.  W. M. Kaula, ibid. 68, 4967 (1963).

19.  Y. Shimazu, Icarus 11, 455 (1966).

20.  J. Weertman, Rev. Geophys. Space Phys. 8, 145 (1970); R. I. Walcott, J. Geophys. Res. 75, 3941 (1970).

21.  C. B. Raleigh and S. H. Kirby, in Mineral. Soc. Amer. Spec. Pap. 3, B. A. Morgan, Ed. (1970), pp. 113-124.

22.  This estimate for stress differences now being supported in the lunar crust is close to the yield strength calculated by Baldwin (14). From an analysis of variations in lunar crater dimensions as a function of diameter, he calculated that lunar rocks would behave elastically below stresses of 30 to 50 bars and distort viscously at higher stresses. His analysis covered crater diameters from fractions of a kilometer to 100 km. Therefore, the inferred strength of the crust is relatively constant for loads varying by four orders of magnitude in effective wavelength. This suggests that loads have decayed to this value and that the present implied stress differences are relatively permanent features of the moon. The estimates of viscosity in this article and elsewhere, where zero strength is assumed for long-term processes, are therefore upper bounds. If the moon's crust has a finite permanent strength, then the viscosity of the crust can be arbitrarily low.

23.  C. P. Sonett, D. S. Colburn, P. Dyal, C. W. Parkin, B. F. Smith,

G. Schubert, K. Schwartz, Nature 230, 359 (1971).

24. P. Dyal and C. W. Parkins, J. Geophys. Res. 76, 5947 (1971).

25. The dependence of electrical conductivity on sample quality, sample size, oxidation state, and impurity level are well known. A conductivity of $10^{-4}$ (ohm-cm)$^{-1}$, the value found by Sonett et al. (23) at a lunar radius of 1500 km, yields temperatures from 920$^\circ$ to 1790$^\circ$K, even if we restrict attention to olivine single crystals (26). The electrical conductivity of lunar basalt is two or three orders of magnitude greater than that of typical terrestrial olivine (24), although it is unlikely that the intrinsic bulk conductivity was being measured. The range of temperature inferred at a radius of 1500 km, under the unlikely assumption that the material at this depth is basalt, is 700$^\circ$ to 900$^\circ$K. These low temperatures are typical of those inferred in (23) and (24).

26. F. C. Schwerer, G. P. Huffman, R. M. Fuher, T. Nagata, Moon 4, 187 (1972).

27. A. W. England, G. Simmons, D. Strangway, J. Geophys. Res. 73, 3219 (1968).

28. A. Duba, H. C. Heard, R. N. Schock, Earth Planet. Sci.

Lett., in press; A. Duba, J. Geophys. Res. 77, 2483 (1972).

29. D. L. Anderson, Nature 239, 263 (1972).

30. G. Latham, M. Ewing, J. Dorman, D. Lammlein, F. Press, N. Toksoz, G. Sutton, F. Duennebier, Y. Nakamura, Moon 4, 373 (1972).

31. Lunar Science Analysis Planning Team, Science 176, 975 (1972).

32. L. Thomsen, Nature, in press.

33. D. L. Anderson, Sci. Amer. 207, 52 (July 1962).

34. D. A. Papanastassiou and G. J. Wasserburg, Earth Planet. Sci. Lett. 11, 37 (1971).

35. Europium is enriched in the moon relative to carbonaceous chondrites differently than the other rare earth elements.

36. N. J. Hubbard, P. W. Gast, C. Meyer, L. E. Nyquist, C. Shik, H. Wiesmann, Earth Planet. Sci. Lett. 13, 71 (1971); N. J. Hubbard and P. W. Gast, Geochim. Cosmochim. Acta 2 (Suppl. 2), 999 (1971).

37. D. L. Anderson and R. L. Kovach, Phys. Earth Planet. Interiors, in press.

38. G. W. Wetherill, Science 160, 1256 (1968); A. E. Ringwood and E. Essene, in Proceedings of the Apollo 11 Lunar Science Conference, A. A. Levinson, Ed. (Pergamon, New York, 1970), p. 769.

39. D. L. Anderson, J. Geophys. Res., in press.

40. T. J. Shankland, in The Application of Modern Physics to the Earth and Planetary Interiors, S. K. Runcorn, Ed. (Wiley-Interscience, New York, 1969), pp. 175-211; R. M. Hamilton, J. Geophys. Res. 70, 5679 (1965); S. P. Mitoff, J. Chem. Phys. 31, 1261 (1959). See also (24-28).

41. Supported by NASA grant NGL 05-002-069. Contribution No. 2188, Division of Geological and Planetary Sciences, California Institute of Technology.

# Dynamically plausible hypotheses of lunar origin

## W.M. Kaula and A.W. Harris (1973)

The necessity that at least half the primordial Moon be enriched in refractory aluminium and calcium silicates[1,2], as well as this outer half being depleted in volatiles and the entire Moon being depleted in iron, has led to a renewed emphasis on the capture hypothesis of lunar origin[3-7]. This emphasis in part springs from the need to have a high temperature (or low pressure) environment in which condensation, and hence presumably accretion, of substances less volatile than iron predominates[8,9]. The evident reasoning is that the Moon was made in a different place than it is now; the philosophy seems to be that celestial mechanical improbabilities are more tolerable than are thermochemical improbabilities, in view of the fewer bodies involved.

All those who hypothesise capture acknowledge that it is an improbable event. So most[5-7] suggest that the Moon was made somewhere near the Earth, which seems contrary to the thermochemically inspired wish to make the Moon in a different part of the Solar System. All but one[4-7], however, explicitly require the entire Moon to be captured at once, which is the most severely implausible form of the hypothesis. The reason for this implausibility is the extreme weakness of the only known energy sink for pure capture, tidal friction.

It is therefore worthwhile to emphasise more strongly than previously[10] why tidal friction is an implausibly weak energy sink. A way to demonstrate this weakness is to assume that most, but not all, of the Moon was captured; to hypothesise that the energy sink for this capture was collision with matter already in orbit around the Earth; and then to ask how much matter is needed to make such collisions a more effective energy sink than tidal friction. For a given dissipation factor  $1/Q$ and perigee distance $r_p$ , the energy dissipation per pass by tidal friction in the initial highly eccentric orbit is easily estimated, since, as shown by Gerstenkorn[11], for plausible rheologies this dissipation is mainly in the radial tides in the Moon.  So we can write the dissipation in terms of the work done[12,13] on a homogeneous Moon in a single approach from a distance to perigee:

$$\Delta E = \ldots = \frac{3h(1 + k)R^5 GM^2}{10 \ Qr_p^6} \quad (1)$$

(part of calculation omitted - eds.)

where $h$, $k$ are the displacement and potential Love numbers, about 0.033 and 0.020 for the present Moon[14], $R$ is the Moon's radius, $G$ is the gravitational constant, and $M$ is the Earth's mass. Assuming a $Q$ of 10 and a minimum perigee distance $r_p$ of 1.5 Earth radii, equation (1) gives $0.5 \times 10^{33}$ erg dissipation per pass.

For collision of a Moon of mass $m$ and velocity $v$ with another body of mass $m'$ and velocity $v'$

$$\Delta E = 1/2 \left| mv^2 + m'v'^2 - (m + m')v_0^2 \right| \quad (2)$$

$$mv - m'v' = (m + m')v_0 \quad (3)$$

For $v = 10$ km s$^{-1}$ and $v' = 7$ km s$^{-1}$, the use of the momentum conservation equation (3) to eliminate $v_0$ in equation (2) and $0.5 \times 10^{33}$ erg for

$\Delta E$ yields $1.4 \times 10^{21}$ g for $m$:  that is, the Moon needs to collide with only $2 \times 10^{-5}$ of its own mass for collision to be a more efficient energy dissipator than tidal friction.

Of course, in a single pass the Moon would have collided with only a minor portion of the matter in orbit about the Earth; a more realistic estimate requires a hypothesis as to the distribution of this matter.  For a uniform disk of thickness $b$ much less than radius $c$, an orbit of inclination $I$ appreciably greater than $b/c$ will collide with a volume $2b\pi R^2/\sin I$ per pass, or $2R^2/(c^2 \sin I)$ of the total. For a $c$ of, say, three Earth radii, this ratio is still more than 0.015. So for this plausible model, a total disk mass of only 1/600 the Moon's mass would have been a more effective energy sink than tidal friction.

It is sometimes argued that solar perturbations may keep the Moon close to the Earth long enough for tidal friction to be effective[6,7]; however, the effectiveness of collision dissipation will be equally increased by the same mechanism, so that its comparative advantage is maintained. Another mechanism proposed by Alfven and Arrhenius[7] is the locking of the Moon in synchronisation with a longitudinal variation in the Earth's gravitational field.  This synchronisation cannot be 1 : 1, since for the primordial day of 5 h the orbit would be within the Roche limit. Higher order synchronisations depend on the orbital inclination $I$ or eccentricity $e$. The torque exerted by the fixed variation in the gravity field must then be greater than the tidal torque, or [13,15]:

$$a(GM/r)(R/r)^l \, J_{lm} e^{-iI^j} \geq (k/Q)(Gm/r)(R/r)^2(R/r)^3 \quad (4)$$

where $R$ is now the Earth's radius, $J_{lm}$ is the dimensionless spherical harmonic potential coefficient, $a$ is a factor of order unity, and the exponents $i$, $j$ depend on the order of the commensurability. For example, for a 2 : 1 commensurability $i + j \geq 2$ for $l = 2$ and $i + j \geq 1$ for $l = 3$. At a 2 : 1 commensurability for the 5-h day, $r/R$ is 3.7. Thus

$$J_{22} \geq (k/Q)(m/M)(R/r)^3(1/ae^2) \approx 0.7 \times 10^{-4}/(Qe^2)$$

or

$$J_{31} \geq (k/Q)(m/M)(R/r)(1/ae) \approx 0.3 \times 10^{-3}/(Qe)$$

There is a marginal possibility the $J_{22}$ commensurability would be stable until radial tides pumped down the eccentricity. This says nothing about the probability of capture into the resonance, which depends on the change in tidal torque with passage through commensurability[16].

Returning to the main theme, a mechanism is needed for the Earth to acquire the 1/600 Moon to catch a whole Moon. In any model of planet formation from a gas and dust nebula, at the stage where protoplanets become sufficiently massive for gravitational capture to be significant, there still will be an abundance of smaller bodies: a hierarchy of planetesimals of number density varying inversely with mass over a wide range of sizes. The attraction of the protoplanet for these smaller bodies will cause an increase in the number density of small bodies in the vicinity of the protoplanet, and thus an increased probability of mutual collisions between small bodies. The calculation of this probability is somewhat intricate; it has been done best by Ruskol[19]. In any case, given such a collision in the vicinity of a protoplanet, the chance is quite good that the bodies will lose enough energy to be captured by the protoplanet, but retain enough angular momentum to go into orbit around the protoplanet, rather than falling in.

Once any matter was in orbit around a protoplanet, it acted as an effective trap for further infalling matter, thus increasing the mass of the circumplanetary ring. Most of this matter was in the form of planetesimals of radius less than 300 km, the consequence of gravitational instability of condensed matter in the nebula, followed by a period of collisions, as derived by Safronov[17] and later by Goldreich and Ward[18]. So the probability is greatest that satellites are created out of many smaller bodies in orbit around the planet: this is essentially the model of Ruskol[19-21]. But the anomalies in planetary rotation, Venus and Uranus, the larger mean eccentricities and inclinations of the small planet orbits (0.175 and $7.2^{\circ}$ for Mercury[22], 0.242 and $15.9^{\circ}$ for Pluto[23]), as well as the high Moon: Earth mass ratio, all indicate that the population of planetesimals included some sizable bodies, perhaps even larger than the Moon. Related to this is the apparent overstability of the Solar System: the generous spacing between the planets which is the essence of Bode's law. From models which depend on relatively close interaction for dynamical evolution[17,18], it is difficult to avoid the conclusion that there should be several more planets than observed. The clearing out of the intermediate bodies must have depended on longer term perturbations associated with the perihelion and node revolutions. Bodies midway between successive terrestrial planet extreme aphelia and perihelia with eccentricity oscillations comparable to those in the present system[22] would not survive. Bodies between major planets would need eccentricity oscillations three or four times those of major planets to collide with the planets.

The principal message is that collision with pre-existing satellite matter is the most effective means of capturing a moon, and that the "normal" class of models of evolution from dust through planetesimals to planets would have led to circumstances where matter was orbiting about protoplanets. There are limits, of course, to how much energy can plausibly be expended by collision. Furthermore, the farther away from the Earth a body originates, the narrower the range of conditions will bring it within a capturable energy relative to the Earth. Thus, for example, a near-minimum energy trajectory from Mercury to the Earth, required in Cameron's hypothesis[4], will approach the Earth with a relative velocity of about 7.5 km $s^{-1}$ ($2 \times 10^{37}$ erg to dissipate); the distance of this approach will vary 10 Earth radii for a change of only 0.006 km $s^{-1}$ in initial velocity at Mercury.

Given that the Moon's matter arrived in orbit about the Earth in many chunks, the realistic alternatives are: (1) one chunk was much larger than the rest, perhaps considerably more than half the total mass; and (2) all the chunks were small compared with the total mass of the Moon. Perhaps the best evidences of the latter are that the outer half of the Moon is severely depleted in volatiles and enriched in refractory silicates, and that at least the outer fourth of the Moon was heated sufficiently to differentiate a crust very soon after its formation[1,2]. Given that there were volatiles entrapped when the materials of certain bodies (such as the Earth and the parents of carbonaceous chondrites) were last put together into, say, 10 m or larger pieces, then the only way to remove it thoroughly from other matter in the same part of the Solar System is to break the latter down into small pieces, requiring many collisions. Furthermore, the only certain way to achieve the primordial heating evidence in the Moon's outer parts is by rapid infall in the terminal stages[24,25]. Both these circumstances--frequent collisions and rapid terminal infall--could not be obtained on the relatively long time scale of formation in heliocentric orbit[17,18], but could be on the short time scale of formation in geocentric orbit. As discussed by Ruskol[21], there were probably repeated cycles of coalescence and fragmentation of the circumterrestrial matter, because of repeated disruption by further matter falling in from outside the Earth-Moon system. In this manner, the formation of the Moon may have been delayed as much as $10^8$ yr. But once the protolunar matter was left undisturbed, the formation would have proceeded rapidly. To give the

initial inclination between the lunar orbit and the Earth equator, one or a few fair-sized planetesimals (less than 1/10 lunar mass) must have hit the Moon or the Earth from outside the Earth-Moon system after the Moon was formed formed[26].

A circumterrestrial regime of repeated collisions would certainly have been conducive to loss of volatiles and enrichment in refractory materials[27,28]. The depletion of the protolunar matter in iron relative to the Earth may have been due, however, to an earlier differentiation in the nebula, favouring enrichment of iron in earlier accreting bodies for either thermochemical[29] or mechanical[30] reasons.

Dynamical plausibility suggests that satellites have different compositions than their associated planets both because of later formation and because of the frequent collisions and rapid coalescence in a circumplanetary swarm. It also suggests that collisions are the most effective means of energy dissipation for orbital changes; tidal friction is too slow, and gravitational effects are too temporary. The retrograde satellites of Jupiter are sometimes mentioned as examples of gravitational capture[7]; however, they are too far inside the marginal Jacobi constant[31], and collisions seem the most reasonable explanation even for their capture[32].

Our notion of plausibility entails minimal change from the present Solar System, of course. If the solar nebula had a mass on the order of a solar mass, as hypothesized by Cameron[33], most of the dynamical considerations discussed here would not constrain events before the dispersal of the nebula by the T Tauri phase of the Sun. One difficult thing to understand about Cameron's model is that if the solid matter in the central plane of the nebula was dense enough to allow growth of major planets in a few thousand years, then in this same time most of the matter should have collected by gravitational instability into bodies of radius a few kilometres[17,18], too large to be carried by gas drag or to be swept away by any T Tauri wind. Furthermore, the number of these bodies and the high gas density would have led to many Moon sized bodies, and perhaps some terrestrial planet sized, too large to be broken up by collision. In other words, it is difficult to make the time scale of gravitational instability of solid matter longer than the time scale of gravitational capture of gas, as would be necessary if Cameron's massive nebula has a solar proportion of iron, silicates and so on.

A possible necessary condition strongly affecting the solar nebula and planet formation therefrom may be the short term existence of other stars only a few tens of astronomical units away. The cloud collapse calculations of Larson[34,35] indicate that if angular momentum is retained, multiple star systems result. Such systems are often unstable, usually ejecting one member at a time until a binary, or, less often, a stable triplet is attained[36]. If isolated stars such as the Sun are normally ejected from multiple systems, then no solution of planet formation, angular momentum distribution, and so on, would be complete without considering the influences of other stars in the system.

[1]Gast, P. W., The Moon, 5, 121 (1972).

[2]Gast, P. W., Fourth Lunar Sci. Conf. Abstracts, 275 (1973).

[3]Anderson, D. L., Nature, 239, 263 (1972).

[4]Cameron, A. G. W., Nature, 240, 299 (1972).

[5]Urey, H. C., The Moon, 4, 383 (1972).

[6]Singer, S. F., The Moon, 5, 206 (1972).

[7]Alfven, H., and Arrhenius, G., The Moon, 5, 210 (1972).

[8]Lewis, J. S., Earth planet. Sci. Lett., 15, 286 (1972).

[9] Grossman, L., Geochim. cosmochim. Acta, 36, 597 (1972).

[10] Kaula, W. M., Rev. Geophys. Space Phys., 9, 217 (1971).

[11] Gerstenkorn, H., Icarus, 11, 189 (1969).

[12] Munk, W. H., and MacDonald, G. J. F., The Rotation of the Earth, 207 (Cambridge, 1960).

[13] Kaula, W. M., Rev. Geophys. Space Phys. 2, 661 (1964).

[14] Harrison, J. C., J. geophys. Res., 68, 4269 (1963).

[15] Kaula, W. M., Theory of Satellite Geodesy, 37 (Blaisdell, Waltham, Mass., 1966).

[16] Goldreich, P., and Peale, S. J., A Rev. Astr. Astrophys., 6, 287 (1968).

[17] Safronov, V. S., Evolution of the Protoplanetary Cloud and Formation of the Earth and the Planets (Israel Program for Scientific Translations, Jerusalem, 1972).

[18] Goldreich, P., and Ward, W. R., Astrophys. J. (in the press).

[19] Ruskol, E. L., Soviet Astron., AJ, 4, 657 (1961).

[20] Ruskol, E. L., Soviet Astron., AJ, 7, 221 (1963).

[21] Ruskol, E. L., Soviet Astron., AJ, 15, 646 (1972).

[22] Brouwer, D., and Clemence, G. M., in Planets and Satellites (edit. by Kuiper, G. P., and Middlehurst, B. M.), 31 (Chicago, 1961).

[23] Williams, J. G., and Benson, G. S., Astr. J., 76, 167 (1971).

[24] Ruskol, E. L., The Moon, 6, 190 (1973).

[25] Mizutani, H., Matsui, T., and Takeuchi, H., The Moon, 4, 476 (1972).

[26] Safronov, V. S., and Zvjagina, E. V., Icarus, 10, 109 (1969).

[27] Ruskol, E. L., in The Moon (edit. by Urey, H. C., and Runcorn, S. K.), 426 (Internat. Astr. Soc., 1972).

[28] Anderson, D. L., Fourth Lunar Sci. Conf. Abstracts, 40 (1973).

[29] Turekian, K. K., and Clark, S. D., Earth planet. Sci. Lett., 6, 346 (1969).

[30] Orowan, E., Nature, 222, 867 (1969).

[31] Henon, M., Astr. Astrophys., 9, 24 (1970).

[32] Colombo, G., and Franklin, F. A., Icarus, 15, 186 (1971).

[33] Cameron, A. G. W., Icarus, 18, 407 (1973).

[34] Larson, R. B., Mon. Not. R. astr. Soc., 156, 437 (1972).

[35] Larson, R. B., in Symposium on the Origin of the Solar System (edit. by Reeves, H.), 142 (Cent. Nat. Res. Sci., Paris, 1972).

[36] Harrington, G. S., Bull. Am. astr. Soc. (in the press).

*

# Update on Mars:
# clues about the early solar system

## W.D. Metz (1974)

Mars has been studied for centuries with optical telescopes, and more recently with radar and with instrumented spacecraft on "flyby" trajectories. But the wealth of data obtained from the Mariner 9 spacecraft during nearly 700 orbits around the red planet has vastly improved knowledge of that body and has provided a third reference point--in addition to the moon and the earth--for understanding planetary physics. In the 14 months since the Mariner spacecraft first reached Mars, it has become clear as never before that the sun's fourth planet is a geologically active body with volcanic mountains and calderas larger than any on the earth. Interactions of wind, dust, and surface materials are now considered to account for the changes in the planet's appearance

which have puzzled observers for so long. And there are strong indications that water, probably trapped in the polar caps, at one time flowed freely over part of the planet's surface, a possibility that has raised anew hopes that some form of life may exist on Mars.

Between 13 November 1971, when the Mariner spacecraft became the first man-made object to orbit another planet, and 27 October 1972, when depletion of the supply of gas used to keep the spacecraft positioned with its radio transmitter directed toward the earth caused it to be shut down, Mariner 9 took more than 7000 photographs in mapping the entire surface of the planet with its television cameras. Thermal and chemical maps of the surface, as well as studies of surface pressure, atmospheric composition, and the martian gravity field, were made with ultraviolet and infrared spectrometers, an infrared radiometer, and the telemetry signals

from the spacecraft to Earth.
Officials at the National Aeronautics
and Space Administration count the
mission a tremendous success, as do
the scientists who are now digging
into the vast amount of new information and offering the first
tentative accounts of the planet's
history.

Both atmospheric and surface
features of the planet were studied.
The planet-wide dust storm that
obscured the surface in the early
months of the mission proved a boon
to those studying the martian
atmosphere. Carl Sagan of Cornell
University mapped streaks of dust
extending from the leeward side of
many craters and found that they
formed a distinct global pattern.
The streaks appear to have been laid
down during the period of strongest
winds and hence to give an indication
of the surface winds during the dust
storm. The circulation pattern,
according to Conway Leovy of the
University of Washington, agrees with
theoretical models of meteorology at
low martian latitudes and seems to be
peculiar to its southern summer
solstice--during which energy input
from the sun is concentrated in a
belt near $20^{\circ}$ to $30^{\circ}$ south of the
equator. It appears likely that the
coincidence of the summer solstice
and the closest approach of Mars

to the sun (perihelion) gave rise to
the conditions that resulted in the
dust storm. An initiating factor
in the dust storm, according to
Leovy, may have been the increased
absorption of the sun's radiation
due to dust particles raised by the
strong winds that blow along the
edges of the south polar cap during
its yearly retreat.

## Water in the Polar Caps

The polar caps had been thought
to be carbon dioxide, which is the
main component of the martian atmosphere. Measurements with the
infrared spectrometer, however,
showed the presence of water vapor
in the atmosphere. At times, the
atmosphere near the north pole was
saturated with water vapor, although
the total amount, because of the low
temperatures, was small. Variations
in the amount detected were correlated with the retreat of the northern
polar cap, suggesting that it is the
source of the water vapor. Other
evidence is consistent with the idea
that the residual polar caps are
largely water.

Ozone was detected in the martian
atmosphere with the ultraviolet
spectrometer. The observed variations
in ozone may add to the understanding
of the stability of the protective

ozone layer in the earth's stratosphere. The Mariner results show that ozone concentrations decreased in the summer when more water vapor was present in the atmosphere, presumably because of photochemical reactions involving water and ozone.

Several cloud systems were observed in the martian atmosphere. Among the most spectacular, according to Leovy, were those observed during summer on the slopes of the large volcanoes. The clouds were composed of water-ice crystals and had a cellular structure indicating convective activity, possibly the result of being heated from the surface, although the possibility that the clouds arise from vapors emitted by the volcanos is also being studied.

Wind speeds during the dust storm are believed to have averaged at least 30 meters per second in the equatorial region, and some investigators believe that velocities twice that high may have prevailed at the beginning of the storm. These high winds seem consistent with a martian surface marked by widespread erosion. Indeed, according to Hal Mazursky of the U.S. Geological Survey laboratory in Flagstaff, Arizona, eolian erosion is a dominant feature of Mars and is apparently so intense in some areas as to have completely eroded away preexisting volcanos. The edge of the largest existing volcano on Mars, Nix Olympica, is apparently being rapidly eaten away, exposing a 1- to 2-kilometer cliff around its base.

Nix Olympica and the three volcanos along the Tharsis ridge area are huge by any scale. One of the latter rises some 26 kilometers above the surrounding floor. As indicated by the relative scarcity of craters, these volcanic structures are also geologically recent. One estimate by William Hartmann of Science Applications Incorporated in Tucson, Arizona, puts the age of Nix Olympica at 100 million years and that of the Tharsis area at 300 million. But volcanic activity on Mars is not just a recent phenomenon. The Mariner photographs show many older volcanic structures, ranging from small calderas and volcanic vents to volcanic flows reminiscent of the lunar maria, and including at least one large primitive volcano. Many of these older structures are heavily cratered and severely eroded, and they appear to date back as far as 3 billion years, if crater counts are any indication.

East of the Tharsis volcanos is found a high plateau, and still further to the east, a large rift system that stretches nearly 5000 kilometers along the equator. The plateau is broken by faults that appear to indicate vertical movement of the crust. The faults coalesce to form the rift valley, a feature that presumably resulted from tensional forces in the crust whose cause is still unknown. Whatever its origin, the rift valley is the most dramatic evidence of tectonic activity on Mars. On a smaller scale, however, there appears to be evidence of faulting and other tectonic activity over much of the planet.

Perhaps the most startling finding of the Mariner effort has been the discovery of channels in the martian surface which appear to have been cut by running water. Three types of channels have been observed, all of them quite distinct from the lava channels seen on the moon and also on Mars. The largest of the putative stream beds emerge from the foot of landslide areas north of the rift valley and may have originated in the melting of permafrost uncovered in the slide debris. Smaller sinuous channels that originate nowhere in particular and run downhill, coalescing with other channels and becoming broader, have also been found. A third type is composed of a complex, interwoven pattern of channels that, according to Mazursky, is characteristic of intermittent stream flow. All except the first type of channel originate in flat terrain and seem to imply rainfall runoff as the source of the water.

The channels appear to be geologically very recent but not all of the same age, so that their formation could not have been due to a one-time event, and they occur largely in the warmer equatorial regions of Mars. That they exist at all is quite remarkable, because under present conditions water could not exist on the martian surface in liquid form--the liquid phase is unstable at the prevailing temperatures and would immediately freeze or evaporate. The concensus of those who have examined the evidence, however, is that other explanations for the channels are even less likely; liquid carbon dioxide for example would require nearly 1000 times the existing atmospheric pressure on Mars. Hence the evidence is suggestive that the past environment of Mars must at some time have differed considerably from present conditions.

Another peculiar feature of the martian terrain is the laminated structures exposed by the retreating polar caps. What appears to be a series of overlapping plates, each composed of dozens of continuous layers of alternately dark and light material, presents a banded appearance. A convincing hypothesis to many investigators is that the alternating layers represent deposits of dust and ice or frozen carbon dioxide, although what formed the intricate pattern of the plates is a subject of considerable debate. The number and uniformity of the layers, however, suggests a regular pattern of dusty and dust-free epochs in the planet's recent history.

There is a great deal of dust on Mars. Shifting dust seems to have been confirmed as the agent responsible for changes in the dark spots visible on the martian surface as well as for other variable surface features. The source of the dust, which seems to be composed of grains comparable in size to sand, is thought to be the continuing erosion of rocks. Erosion may be most intense in the equatorial regions. From there, dust may be carried poleward, deposited, and gradually redistributed over the planet.

The largest portion of the planet's surface is old, heavily cratered rock that some investigators believe to be the ancient crust. The incidence of craters that can be discerned is considerably less, however, than on the martian moon Phobos. Exactly how far back martian geologic history can be traced, if intense erosive processes have always been present, is uncertain.

Bruce Murray of the California Institute of Technology has proposed that the planet was without any significant atmosphere until recently. He believes that the planet is just beginning to be geologically active and that it created its present atmosphere volcanically during the formation of the Tharsis and Nix Olympica structures. Murray is correspondingly pessimistic that life could have evolved under the harsh, moonlike conditions that in his view prevailed over most of martian history.

Murray offers no explanation for the formation of the channels, but he does propose that the regular pattern of the polar laminated terrain is associated with periodic alteration in the martian climate. He finds that the planet's orbit changes shape from nearly circular to more elliptical with a period of about every 2 million years, perturbations caused by the influence of other planets, primarily Jupiter and Saturn. The resultant variation in the sun's radiant energy reaching the poles could under some circumstances, Murray speculates, lead to changes in the growth and sublimation of the polar caps. If dust is laid down alternately with layers of frost, thin laminae of the type observed could be formed.

On the question of the planet's ancient history, other investigators are less inclined to agree with Murray and cite evidence of volcanism and erosive processes stretching back over much of the planet's lifetime. Hartmann, for example, believes that in the past the planet must have had more rather than less atmosphere than it now has. Hartmann agrees, however, that tectonic activity of the type well known on the earth may just be getting started. Mars thus contrasts strongly with the moon, whose evolution shows no evidence of horizontal movement by crustal plates, and the earth, where convection currents in the mantle have been vigorous enough to completely remodel the surface over much of its history. Hartmann believes that the martian crust in the Tharsis region has been pushed upward in the last few hundred million years, possibly by a mantle convection current.

Others have suggested more explicit analogies with tectonic patterns on earth. Mazursky, for example, believes that the high plateau adjacent to the Tharsis ridge represents light, continental-type rock that overlies heavier materials found in nearby low-lying areas. The heavier material, Mazursky suggests, may be comparable to the basaltic rock of the earth's ocean basin floors. In this admittedly speculative view, the line of volcanos along the Tharsis ridge may represent the incipient stages of a tectonic process in which a plate of the heavier basaltic material is being thrust under the lighter continental plate. Others, such as John McCauley of the U.S. Geological Survey laboratory in Flagstaff, Arizona, do not see any evidence in the martian photographs of the compressional movements that such a process would entail. He points out that the density of Mars is less than that of the earth, that the tectonic activity on Mars happened late in its history, and that its apparent lack of vigor may indicate that no further crustal movements are in prospect.

In regard to the more recent evolution of the planet, the most puzzling problem appears to be the climatic changes that could have led to the formation of the channels by liquid water. Sagan has proposed that the climatic instability responsible for this is not inconsistent with that suggested by Murray for the formation of the polar laminae. If the polar caps were to entirely evaporate, Sagan calculates, the atmospheric pressure on Mars would be about 1 bar, the same as on the earth. Only about one-tenth this much atmosphere, assuming it was 1 percent water vapor, would be sufficient to permit liquid water as a stable phase at the daytime temperatures characteristic of the martian equatorial region. Hence Sagan proposes that rainfall and rivers may be recurring phenomena on Mars during each successive "interglacial" period.

Sagan believes that an advective instability in the martian atmosphere is the most probable cause of the drastic climatic changes he envisions. Changes in the absorption of sunlight at the poles--for example, by a layer of dust deposited in a major storm-- would cause the pole to heat up and the total atmospheric pressure to increase slightly. With a more dense atmosphere, the normal circulation patterns that carry heat poleward from the equator would operate more efficiently, heating the poles still further. Continuation of the process would eventually lead to conversion of the polar caps to atmosphere and rivers. The process is also reversible, Sagan believes, because liquid water could serve as a trap for dust, cleaning the atmosphere and allowing it to recondense at the poles.

If indeed the environment on Mars alternates between wet and dry, at least in the tropics, then speculations about life on the planet become more interesting. Life forms that could go into extended repose while awaiting the availability of liquid water, for example, would seem well within the realm of possibility, if such forms exist. Experiments that expose martian soil to liquid water and test for biological activity might therefore be of particular interest. As it happens, two such experiments are planned for the Viking spacecraft that will attempt to land on Mars in 1976.

*

# Plate Tectonics and Hot Spots
## Section 2.

The extinct crater of Diamond Head with Honolulu in
the background, Hawaii. Photograph by Jere H. Lipps.

The idea that parts of the earth's
crust have changed positions relative
to one another through geologic time is
an old one. Various people, like
Francis Bacon, Francois Placet and
Alexander von Humboldt speculated in
the 1600's and 1700's about the origin
of the Atlantic Ocean, and the similar
shapes of continental coastlines on
each side of it. A Frenchman, Antonio
Snider-Pelligrini, suggested in 1858
that the continents on both sides of

93

the Atlantic were once joined together and had drifted apart, but no-one paid any attention to the idea until the German, Alfred Wegener, put together a massive amount of evidence in favor of it early in this century. Wegener noted similarities on both sides of the Atlantic in shapes of coasts, continuity of mountain ranges and other geologic structures, distribution of sedimentary rock types and rock ages, occurrences of distinctive fossils, and similarities in the pattern of ancient climates, particularly a late Paleozoic glaciation. The full controversy over continental drift was thus launched. One side, mainly geologists familiar with rocks from South America, Africa and India, and biologists concerned with the distribution patterns of modern organisms, favored drift, but they were opposed by most geologists in the Northern Hemisphere and by geophysicists who could not accept the mechanisms proposed by Wegener for propelling the solid continents through the earth's solid crust. The latter group assumed that Wegener's correlations were merely coincidental inasmuch as there was no understandable way of moving continents. The controversy rested there, with most geologists and geophysicists doubting that the continents had ever drifted. Their reasoning could not be faulted scientifically, for the evidence available was not com-

pelling enough.

In the 1950's measurements were made of former positions of the magnetic poles of the earth, and it was found that their positions had changed relative to the continents. Now this could be explained in one of two ways: either the magnetic pole had changed its position for some reason, or it had remained comparatively constant and the continents had moved. It turned out that positions for the north magnetic pole determined from North American rocks were different from those obtained from European rocks, and that difference was consistently greater in older rocks. The different "polar wandering path" for each continent indicated that it was the continents that had moved in different directions, since one magnetic pole cannot move in two directions at once. Yet even this evidence did not convince many geologists of the reality of continental drift, and they still believed that there must be a simpler explanation of the evidence.

All this changed very rapidly in the 1960's. Harry Hess and Robert Dietz speculated about the addition of new crust at midocean ridges as the seafloor spread laterally. Magnetic measurements showed bands of different magnetization parallel to midocean ridges. In 1963 Fred Vine and D.H. Matthews suggested that these were caused by solidification of new magma at the midocean ridge as the seafloor grew during changes in the earth's magnetic field; bands of different magnetization would then be spread laterally as the midocean ridge rifted. In 1964 Vine published a landmark paper in which he presented evidence that sea floor spreading was going on in various parts of the oceans. A flood of papers followed as marine geologists, structural geologists, seismologists, geophysicists, sedimentary petrologists and paleontologists re-evaluated old and new data in support of the new ideas. By 1968 hundreds of papers in many scientific journals allowed the ideas to be brought together into the theory of "plate tectonics", which has by now revolutionized the thinking of all earth scientists. In our readings, we reproduce a summary of the concept in two papers by HAMMOND (1971), who gives an account of the idea and some of its implications.

The fact that ocean crust was being created at the midocean ridges and destroyed at the oceanic trenches was clear by the end of the 1960's, but the mechanisms which might be involved were still the subject of debate. Hammond summarizes some of the possible mechanisms in the first reading.

One of the proposed mechanisms, the so-called "Hot Spot" model, has set off yet another controversy. MORGAN (1968) presented the idea in a volume dedicated to Harry Hess. He suggests that "plumes" of hot mantle material from deep in the earth, generated or maintained by radioactivity, rise and fan out at certain points at the bottom of the lithospheric plates that form earth's outer layer. The plumes cause "hot spots" to form on the earth's surface, usually marked by volcanos such as those in Hawaii. Morgan seems convinced that the plumes originate deep in the mantle and have remained stationary for long periods, while the lithospheric plates have drifted over them to produce long chains of volcanic mountains or islands. Morgan also believes that the plumes may drive the movement of the plates.

There is great disagreement about this. In a summary report in Science, METZ expressed the divergent opinions of various earth scientists. BURKE, KIDD & WILSON, and MOLNAR and ATWATER, all present evidence which suggests that the hot spots move relative to one another, casting doubt on some of Morgan's idea, though not all of it. As P.J.S. notes in a news report in Nature, conflicting data presented by geophysicists might result in resolving the controversy. By considering that the earth's mantle might move, as well as the lithospheric plates, some anomalies disappear. This idea is expressed by HARGRAVES and DUNCAN, who suggest that the mantle moves or rolls beneath the lithosphere but over a different axis of rotation, carrying the plumes with it.  D.D. points out that their paper and that of Molnar and Atwater indicate that the earth's crust and mantle are each moving in quite different and independent ways, not correlated with the earth's axis of rotation.

The controversy stands there as we write this introduction, having evolved through several crises and yet being far from resolved.  It is clear, however, that earth science has moved from a view of the earth as a rather static solid ball to a view which considers the earth as an exceedingly dy-namic, mobile body that has undergone a spectacular course of evolution. While the last two decades have been perhaps the most exciting ever in earth science as a result of the emergence of this dynamic viewpoint, more excitement will be forthcoming as we begin to understand the dynamic processes within the depths of the earth that control the surface events we now understand.

FURTHER READINGS

WILSON, J.T. (ed.), 1972.  Continents Adrift.  Readings from Scientific American.  W.H.Freeman, San Francisco.

# Plate tectonics:
# the geophysics of the Earth's surface

## A. Hammond (1971)

A major conceptual revolution has been taking place in the earth sciences during the last 5 years. The central idea of the new theory--that the earth's surface consists of a small number of rigid plates in motion relative to one another--has gained increasing acceptance among geophysicists and geologists as a result of marine magnetic and seismic studies. Plate tectonics, as the theory is called, includes earlier notions of continental drift and seafloor spreading as consequences of the relative plate motions which involve velocities as high as 10 centimeters per year. Plate motions and interactions are thought to be responsible for the present positions of the continents, for the formation of many of the world's mountain ranges, and for essentially all major earthquakes. But the forces that cause the plates to move are not yet understood; a lack of

information about the earth's mantle, where the driving mechanisms are believed to originate, is a key difficulty.

Plate tectonic theory has simplified and unified widely separated fields in the earth sciences, from volcanology and seismology to sedimentary geology. In the process, many older ideas--that all marine sediments found on continents must have been deposited in shallow water, or that mountains were formed in place by lifting and folding of preexisting continental material--have been substantially modified. The way in which most geologists and geophysicists now view the evolution of the earth's surface differs so greatly from earlier ideas, that, as one scientist pointed out, existing textbooks in these subjects will have to be rewritten.

That continents drift had been proposed early in this century, but the idea was not widely accepted. The concept of large crustal movements was kept alive, however, especially by geologists in the Southern Hemisphere

Reprinted from Science 173, 40-41 by permission of the publisher. Copyright 1971 by the American Association for the Advancement of Science.

who were impressed by similarities in
the rock formations and fossil remains
of Africa and South America. In the
1950's, paleomagnetic work on the
location of the ancient magnetic poles
reawakened interest in the possibility
of continental drift. Exploration of
the ocean floor suggested that it is
relatively young, compared with the
continents, and discovered the fracture
zones associated with the mid-ocean
ridges. In the early 1960's, sea-floor
spreading was proposed as a mechanism
for continental drift. Supporting
evidence was obtained from the pattern
of magnetic anomalies that form symmetric stripes on either side of the
ridges and from the seismic records of
earthquakes whose epicenters were
located along the ridge fracture zones.
By 1967, there was strong evidence that
oceanic crust was being created at the
mid-ocean ridges. Data recorded from
the deep earthquakes associated with
the oceanic trenches indicated that the
corollary process of crustal destruction was occurring as slabs of crust
were shoved down into the mantle.
Shortly thereafter, the theory that the
earth's surface and its motions could
be understood in terms of rigid plates
was explicitly formulated.

## Plate Theory

The theory of plate tectonics has
now been extensively developed. The
plates--as few as six plates in some
models and as many as 20 plates in
others are postulated to constitute all
of the earth's surface--are believed to
be 50 to 100 kilometers thick and to
slide on the warmer, less rigid material of the upper mantle. The boundaries of the plates do not generally
follow continental boundaries, so that
plates often include both oceans and
continents. Tectonic activity, according to the theory, is concentrated
along plate boundaries, which are of
three types: spreading centers, where
new crust is created as two plates move
away from each other, such as at a mid-
ocean ridge; subduction zones or
trenches where one plate is thrust
under another as they move toward each
other; and faults where two plates
slide past each other. Plate models
have been developed and used by Jason
Morgan of Princeton University, X. Le
Pichon of the Centre Oceanologique de
Bretagne in France, and others, in
their reconstructions of past crustal
motions (1).

A survey of worldwide seismic data
by Brian Isacks, Jack Oliver and Lynn
Sykes of Lamont-Doherty Geological
Observatory in New York provides
additional evidence in support of the
plate concept (2). The study shows
that most earthquakes are confined to
narrow belts (the plate boundaries)
surrounding large areas where earthquakes are infrequent; the directions

of motion of the plates, as determined from the seismic data, are in good agreement with plate tectonic models and with magnetic evidence from the sea floor. The agreement between theory and two independent types of evidence is to many earth scientists convincing proof of the plate tectonic concept, at least in its broad outline.

Geophysicists are now focusing a great deal of research on the driving motions of the plates and the details of plate movements. In the latter process, refined models of the earth's surface are constructed using data from the worldwide mapping of magnetic anomalies on the sea floor and from the detailed analysis of earthquakes in the trenches. Walter Pitman and Manik Talwani of Lamont, for example, have used magnetic data to reconstruct the evolution of the Atlantic ocean, which apparently opened in several stages. The central Atlantic, for example, seems to have opened before the south Atlantic, and the motion of the American plate relative to Europe has been in a different direction from that relative to Africa. The relative motions of Europe and Africa have consequently been very complex. The Lamont geophysicists have inferred this motion by working out the motion of each continent individually with

respect to the mid-Atlantic spreading center.

Some progress has also been made in the understanding of how plates evolve. Plates are continuously being created at the ocean ridges and destroyed in the trenches, and, as plates interact, new trenches or ridges may open up. In several locations three plates come together in what is known as a triple junction; Dan McKenzie of Cambridge University and Morgan have shown that the orientation of plate boundaries at such a point determines whether or not such junctions retain their geometry as the plates move. The migration of triple junctions, they find, is associated with many changes in plate geometry which otherwise would appear to have been caused by a change in the direction of relative motion between two plates. In the north Pacific, for example, triple junctions appear to have played a major role in the complicated geological history of the west coast of North America and the surrounding sea floor.

Although the geometry of the plates and the kinematics of their motions are now reasonably well known, the mechanisms that drive their motion, and in particular the nature of the driving forces, are not understood. Several conflicting models have been proposed to explain plate motion; most models are based on the assumption that some form of thermal convection within the mantle is responsible. But whether the convection is restricted to the upper few hundred kilometers of the mantle, or whether it also involves the lower mantle, is actively debated.

The uncertainty concerning the driving mechanism reflects present ignorance about the composition and properties of the mantle itself, for which very little quantitative evidence is available. Early estimates of the viscosity of the lower mantle (below 700 km), for example, had indicated that it was too high to allow convection. But the evidence for such a viscous mantle seems inconclusive, and attempts have been made to calculate the rheological properties of the mantle directly with the use of models derived from solid state physics.

Here again, however, there are conflicting theoretical results. Some models of mantle composition assume that its response to stress involves creep by means of mass diffusion of atoms, such as proposed by R. Gordon of Yale. Such models predict a high viscosity for the lower mantle. In contrast, J. Weertman of Northwestern University believes that creep proceeds according to a nonlinear law by means of a dislocation motion at high stresses; his calculations predict a substantially lower viscosity that would permit deep convection (3). The application of either model to the geophysical situation is complicated because the composition of the mantle is not well known. Laboratory experiments that simulate mantle conditions may provide information that will help distinguish between the two theories.

Because of gaps in the data on the mantle, it is hard to do more than show that a given model of the driving mechanism is consistent with observations; predictions are difficult to check. It is known, for example, that the heat flux from the earth is unusually high at the mid-ocean ridges. McKenzie estimates that the anomalous heat flux can be explained by the upwelling of molten mantle to fill the gap made as two plates pull apart. Hence, he believes, the heat flux data provide no information about the earth's interior that could be used to distinguish between competing models. Seismic studies have also provided evidence of inhomogeneities in the mantle, but very few undisputed facts have emerged.

## Driving Mechanisms

Whatever the driving mechanism, an enormous amount of energy--at least $10^{26}$ ergs per year, according to Leon Knopoff of the University of California at Los Angeles--is needed to move the huge plates. Many geophysicists believe that the source of energy is the heat released by the radioactive decay of uranium, thorium, and potassium that occur in the mantle in trace amounts. Others think that energy released in changes of phase associated with formation of the core, tidal forces from the moon, or gravitational forces are also involved.

Thermal convection can occur in a fluid heated from below, and over long periods of time the mantle can apparently behave like a fluid. One type of theoretical model for plate motion is based on convection cells similar to those studied by Lord Rayleigh. According to model calculations by Don Turcotte and Ken Torrance of Cornell University, the drag of the moving fluid on the bottom of the surface plates could cause their motion. The numerical calculations include the effects of

temperature-dependent viscosities on the flow pattern. The heat flux at the earth's surface that is predicted by these simplified models agrees approximately with the observed flux.

Seismic data and laboratory experiments on the properties of matter at high temperature and pressure indicate that the mantle material undergoes at least one phase change at a mean depth of 400 km, resulting in a density change of about 7 percent. Some geophysicists have assumed that this phase change would bar any flow across it and hence that only shallow convection could occur. Recent calculations by Turcotte and by Gerald Shubert of UCLA, however, indicate that the phase change may have a destabilizing effect and thereby increase the convection, which according to their model would involve material in the mantle down at least to 700 km (4).

A second type of mechanism to explain the movement of the plates depends on the assumption that the weight of the relatively cold surface plate descending (in a trench) into the warmer and less dense material of the mantle would help to pull along behind the portion of the plate still on the earth's surface. Calculations by McKenzie have indicated that the mechanism is feasible and could play an important role in determining plate motion, although it might be only an auxiliary driving force. Bryan Isacks and Peter Molnar of Lamont have analyzed earthquakes in the trenches, and their results give some support to the idea that the descending plate is pulling its surface portion behind it.

A third type of mechanism is based on the assumption that the convection is occurring deep within the lower mantle. Morgan, for example, has suggested that plates are driven by a small number of hot spots that represent convection plumes rising from the lower mantle. The rising material in this model spreads out in the upper mantle to provide the stresses on plate bottoms; the return flow is accomplished by a gradual settling throughout the mantle. The resulting flow pattern for deep convection, Morgan believes, is thus more analogous to that of a cumulus cloud than to the roll or cell-like pattern visualized for shallow convection. The hot spots in Morgan's model are assumed to be fixed with respect to the mantle and are located near present-day sites of volcanism, such as Hawaii and Iceland. Apparent differences in the types of basalts found in the mid-ocean ridges and oceanic islands are explained by this model as the result of island chain formation by the motion of a plate across a fixed hot spot. The ages of the Hawaiian Islands, for example, increase toward the northwest, and the present active volcano is at the southeast end of the chain. Morgan calculates that the orientation of this and other island chains is consistent with past motions of the Pacific plate.

Despite present uncertainties about the driving forces, the plate concept has had a remarkable unifying effect on the earth sciences, and has stimulated renewed activity in continental as well as marine geology. A second article will describe recent geological studies of mountain building and formation of continents.

## REFERENCES

1. W. J. Morgan, J. Geophys. Res. 73, 1959 (1968).
2. B. Isacks, J. Oliver, L. Sykes, ibid., p. 5855.
3. J. Weertman, Rev. Geophys. 8, 145 (1970).
4. G. Schubert and D. L. Turcotte, J. Geophys. Res. 76, 1424 (1971).

# Plate tectonics (II): mountain building and continental geology

## A. Hammond (1971)

The development of plate theory and the growing recognition that the earth's surface has been markedly rearranged throughout its history by plate movements has prompted geologists to reexamine many of their ideas about geological processes. As a result, a new theory of mountain building has been developed, relating mountain building to the destruction of oceanic crust in a trench and the collision of continental masses. Geologists are using the new theory to explain the structures of mountain belts all over the world--from the Alps and the Appalachians to the Urals--and to relate their structures to the movements of the crustal plates. According to some geologists, there is evidence of plate movements and interactions that took place as far back as 2.5 billion years

ago. Eventually, geologists hope to understand how platelike blocks of crust first formed on the cooling earth and thus how continents began.

The way in which geologists use plate concepts in their research depends partly on the period of the earth's history under study. For the most recent period, backward from the present to the breakup of Gondwanaland and the opening of the Atlantic about 200 million years ago, data from magnetic studies of the ocean floor can be used to infer past movements of the plates; plate motions can then be used to interpret geological formations on the continents.

Prior to this, however, a different approach is necessary, because the older oceanic crust, having been destroyed in trenches or incorporated into mountain belts, is no longer available for study. In studying the period between 200 million years ago and the time when crustal plates were first formed, perhaps 2.5 billion years ago, geologists are using evidence from continental rocks to locate regions of past tectonic activity corresponding to former plate boundaries. By identifying rock assemblages that are thought to be old oceanic crust, geologists hope to determine plate motions at least qualitatively as far back as the end of the Precambrian era (600 million years ago) and to gain new insight into events throughout much of the Precambrian. In the period before 2.5 billion years ago--extending back to the age of the oldest known rocks (almost 3.5 billion years)--plates similar to those of today may or may not have existed; but some geologists have speculated that tectonic mechanisms such as trenches and spreading centers may have played a role in the history of the thinner, less rigid crustal pieces that are thought to have been present at that time.

Mountain Building

Mountain belts are commonly characterized by their relatively thick layers of sedimentary rocks from successive geological periods which are exposed as parallel "stripes" along the belt. The complexity of the pattern had puzzled geologists for years. Before plate tectonics, it was usually assumed that such mountain belts were formed in place as a result of sinking and sedimentation, followed by compression, folding, and uplift as a result of inexplicable forces.

In contrast, the plate tectonic theory of mountain building, as developed by John Dewey and John Bird of the State University of New York at Albany, and others, assumes that much of the material in mountain belts was formed elsewhere and eventually was incorporated into the belt as a result of plate motion (1). According to their theory, mountains are formed at the margins of continents, as a result of the consumption of a plate in a trench or the collision of the continent with other pieces of continental crust. When a trench opens to consume oceanic crust near a continental margin, thermal processes are believed to cause material to rise from the region of the mantle above the descending plate and to form a mountain belt such as the Andes. Continental rocks, however, are apparently too light to descend into a trench, so that, when two continents or a continent and an island arc approach each other, the resulting collision creates mountains by mechanical processes. The Himalayas, for example, are thought to be the result of a collision between the Indian and Asian plates.

Attempts to interpret mountain belts in terms of plate tectonics involve reexamination of the existing literature as well as new field studies of critical areas. Geologists are trying to integrate the geological record of whole regions with plate motions. For example, the Alps, which are one of the most complicated structural belts in the world, are now thought to be a result of repeated collisions between the African and European plates during the opening of the Atlantic Ocean. Therefore geologists are now restudying the Alps; by correlating particular rock assemblages with the margins of the crustal plates as they existed at the time the rocks were formed, geologists are trying to work out a detailed history of how the Alps were formed. A knowledge of the plate motions, inferred from marine magnetic data, and of the geological sequences of rock types in the mountain belt will make it possible for geologists to piece together the system of trenches, spreading centers, and collisions that formed the Alps.

The Caribbean region has also been the site of mountain building for the last 120 million years. The history of the region, like that of the Alps, depends on the inter-

actions of two plates--those of North and South America--whose relative motions are still incompletely known.

Ranges such as the Alps and the Andes are relatively young and are associated with current plate boundaries. But mountain building by plate motions prior to the last 200 million years is suggested by studies of older mountain ranges such as the Appalachians and the Urals. Rock sequences in these mountain belts appear to have a strong resemblance to those in more recent mountain ranges. Analysis of the Appalachian range and of the corresponding Caledonian range in Great Britian--by Dewey, Bird, and William Kidd of Cambridge University-- indicates that these mountains were formed in part from oceanic crust and in part from sediments that were originally deposited in a "proto-Atlantic" ocean. This predecessor to the present-day Atlantic Ocean seems to have opened in the late Precambrian, then closed to form the mountains in the late Paleozoic era (about 300 million years ago). According to this interpretation, some of the rocks in the Appalachians represent pieces of old oceanic crust that was entrapped in collisions and trench processes very similar to those occurring now.

Similarly, the Ural Mountains in central Russia are thought to be associated with the convergence and collision of the Russian and Siberian plates some 300 million years ago, and the disappearance of the intervening ocean into a series of trenches. According to Warren Hamilton, of the U.S. Geological Survey in Denver, the convergence process was a gradual one; each subcontinent apparently grew oceanward as island arcs collided with them, and the trenches on the continental margins approached each other. The regions between the trenches and the continents, where volcanic activity often occurs, also migrated oceanward. Eventually, according to Hamilton's analysis, the continental pieces collided, thereby incorporating oceanic crust and the debris from trenches and island arcs into the present mountain chain.

The internal evidence within a mountain belt can show such a sequence of events, according to William Dickinson of Stanford University, when interpreted in terms of the theory of trench processes on continental margins. Although many of the details are not yet understood, magmatic rock is believed to form above and slightly inland from a trench in such a way that its

potassium concentration will vary systematically away from the trench. This potassium gradient can be found in the volcanic rocks of most mountain belts. Further inland in a mountain belt, there is typically a lowland belt of sediments that have been deformed from the side nearest the volcanic belt. On the oceanic side of the volcanic belt, a third belt of oceanic material that has been altered by subduction into a trench is often found. These parallel belts are now believed to be good indicators of past plate margins, and, when dated and analyzed for rock types, to allow at least a qualitative reconstruction of the processes by which the mountains were formed.

## Formation of Continents

The relative motions of the crustal plates have dramatically altered the map of the world in past eras. Asia, for example, may be an agglomerate of as many as 15 plates that have been joined together within the last 500 million years. The process is continuing today; the Australian plate is thought to be in the process of colliding with that of Southeast Asia.

The geologic history of the western United States is also

beginning to be unraveled.  Tanya
Atwater of the University of
California at San Diego has studied
the relative motions of the American
and Pacific plates and examined their
implications for the geology of the
regions.  Magnetic data obtained
from the seafloor off the California
coast shows that about 30 million
years ago the westward movement of
the continental plate overrode a
trench, the continuation of which
still exists to the south off the
South American coast.  Atwater
examined two models of plate motions
since the disappearance of the
trench, one of which is based on the
assumption that the Pacific and
American plates were fixed with
respect to one another until they
broke along the San Andreas fault
system 5 million years ago.  A
second model, which she thinks is
more probably correct, assumes a
constant relative motion between
the plates and predicts a greater
amount of deformation of the region
than does the first model.  The
present relative motion along the
fault, according to her analysis,
is at the rate of 6 centimeters per
year.

The geologic history of the
western United States, and in
particular of the Rocky Mountains,
goes back beyond 200 million years.

One of the earliest events appears
to have been a rifting process that
tore off a triangular piece of the
continent from the southwestern
corner in the late Precambrian.
The missing piece, according to
Hamilton, eventually collided with
Asia and is now part of northeastern
Siberia.  After the rifting event,
material that became the western
mountain belts is thought to have
accumulated along the continental
margin as island arcs moved in and
collided with the west coast.

Even the older Precambrian rocks
that constitute the bulk of the
central continental masses are all
very much deformed, and, according
to Dickinson, are probably related
to earlier episodes of continental
rifting, repositioning, and collision.
Of particular interest is the
period before about 2.5 billion
years ago, when the first large
areas of continental crust were
apparently forming and when plate
concepts may not be valid to
interpret the geology.  Although
ideas about the formation of
continents are still very speculative,
the success of plate theories in
explaining more recent geology has
convinced some geologists that
continents evolved in a continuous
process; previously, many had
thought that continents were formed

in a single episode very early in
the earth's history.  A related
problem concerns the geochemical
differentiation of the earth--how
the lighter elements were extracted
from the mantle and aggregated in
the continental crust.  The crust
in this early period was apparently
much thinner than it is at present,
and volcanism was much more common.
But there may still have been "soft"
crustal slabs (which later hardened
into plates) that moved relative to
each other and interacted in trench
systems, and this process may have
been important in the gradual
buildup of continental crust and
the extraction of elements from the
mantle.

Plate theory may also have an
impact on some fields of biology.
As the history of the earth's
surface, especially episodes of
rifting and collision of continents,
becomes better known, it may help
paleoecologists clarify the history
of life that is recorded in the
worldwide distribution of fossils
and allow them new insights into
the process of evolution.  A single
community of organisms that was
divided and separated by the splitting
of a continent should follow
somewhat different evolutionary
sequences.  Correspondingly, groups

continental collision would face
sudden competition for the available
ecological niches.  In this way
geological history could become a
laboratory for the study of evolution,
and E. M. Moores and J. W. Valentine
have proposed that the fossil
record could be used to study such
phenomena as species diversification
and the general problem of the
geophysical influence on evolution
(2).

To many geologists, the
excitement of the last few years lies
in being able to explain in detail,
with the plate tectonics mechanism,
a mountain belt that they had been
studying without much progress for
years.  Specialization in the
regional geology of only one part
of the world is being replaced by
an emphasis on the global applicability of the plate tectonic theory and
by the recognition of the similarities between widely separated regions.
The problems of applying the plate
model to particular circumstances
are by no means all solved, but
plate tectonics has provided
geologists with a unifying concept
for interpreting what they see; so
far it appears to have been remarkably successful.

REFERENCES

1.  J. F. Dewey and J. M. Bird, J.
    Geophys. Res. 75, 2625 (1970).
2.  J. W. Valentine and E. M. Moores,
    Nature 228, 659 (1970).

*

# Plate motions
# and deep mantle convection

**W. Jason Morgan (1972)**

BASIC MODEL

    Let us suppose there is convection deep in the mantle. The arguments presented here do not depend on the depth of such convection--any depth from just beneath the asthenosphere to the core-mantle boundary would suffice --but, for present purposes, let us say such convection extends to a 2,000 km depth. It is common knowledge that such deep convection is improbable due to the efficiency of heat transport by radiation at this depth, but let us explore the possibility of such convection and then come back to the heat flow "proofs" of impossibility. Suppose there are several (approximately 20) plumes of deep mantle rising upward and the rest of the mantle is slowly sinking downward in a pattern analogous to a thunderhead or a coffee percolator. To add concreteness, suppose there are several "pipes" in

the rigid middle mantle and that very hot lower mantle is coming upward in these pipes and being added to the asthenosphere. The more rigid middle mantle, including the "walls" of the pipes, is slowly moving downward to fill the void created below in the more fluid lower mantle, and this rigid middle mantle is being added to at its top as the asthenosphere cools and welds itself to the mesosphere. The 400 or 600 km discontinuity may mark this boundary between mesosphere and asthenosphere.

    Such a model has the following features. There are about 20 pipes to the deep mantle bringing up heat and relatively primordial material to the asthenosphere (Fig. 1). Within the asthenosphere, there will be horizontal flow radially away from each of these pipes. These points of upwelling will have unique petrologic and kinematic properties, but there will be no corresponding unique downwelling points, as the return flow is assumed to be uniformly distributed throughout

Reprinted from Mem.Geol.Soc.Am. 132, 7-22 by permission of the author and publisher. Copyright 1972 The Geological Society of America.

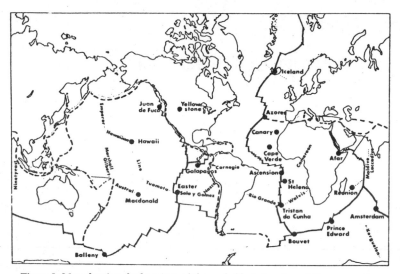

Figure 1. Map showing the locations of the probable hot spots and the names of some features cited in the text.

the remainder of the mantle. The pattern of localized upwelling without localized downwelling was suggested by the gravity map (Fig. 8, to be discussed later). How will such a flow pattern interact with the crustal plates above? A plate will respond to the net sum of all stresses acting on it (the shear stress acting on its bottom due to currents in the asthenosphere plus the stresses on its sides due to its motion relative to adjacent plates). It appears that the plate-to-plate interactions are very important in determining the net forces on a plate, that is, the existing rises, faults, and trenches have a self-perpetuating tendency. This claim is based on two observations: (1) rise crests do not commonly die out and jump to new locations (Labrador and Rockall

are the only places for which the evidence strongly suggests extinct rise crests), and (2) points of deep upwelling do not always coincide with ridge crests (for example, the Galapagos and Reunion upwellings are near triple junctions in the Pacific and Indian Oceans; asthenosphere motion radially away from these points would help drive the plates away from the triple junction, but there is considerable displacement between these pipes to the deep mantle and the lines of weakness in the lithosphere which enable the surface plates to move apart). Also note the toughness of the plates as exemplified by the fact that the upwelling beneath Hawaii has not torn apart the Pacific plate.

This model is compatible with the observation that oceanic island basalts are different from oceanic ridge basalts (Gast, 1968). Island type basalts, as on Iceland or Hawaii, would have access to relatively primordial material from deep in the mantle. In contrast, the ridge crests tap only the asthenosphere--the asthenosphere passively rising up to fill the void created as the plates are pulled apart by the stresses acting on them. The oceanic ridge basalts are known to be relatively low in potassium and in some trace elements. We may account for this by claiming that the asthenosphere source has been reworked and cleaned out of lighter elements in previous sea-floor spreading episodes, or that the lighter elements have had sufficient time to migrate upward (to the bottom of the lithosphere) and are not present to rise to the ridge crest where the plates are pulled apart. The oceanic island basalts are rich in potassium and have a rare earth distribution implying more fractionation, in accord with their deep primordial source. If we relate the observed island basalt fractionation to the composition of the parent rock, we should have a new picture of the composition of the deep mantle. Such an estimate will undoubtedly be higher in potassium than those estimates based on ridge basalts. The implied increased estimate of radiogenic heat production is desirable in

this scheme, in that the deep convection requires more heat production at depth than radiative transport alone can cope with.

As the Pacific plate moves over the upwelling beneath Hawaii, the continuous outpouring of basalt from this point produces a linear basaltic ridge on the sea floor--the Hawaiian Islands. Likewise, the excessive flow from Tristan da Cunha has produced the Walvis and Rio Grande Ridges in the South Atlantic. Here we require Africa to drift to the northeast (parallel to the Walvis Ridge) and South America to drift roughly northwest (parallel to the Rio Grande Ridge). Note that the transform faults between Africa and South America trend east-west; the transform faults show the <u>relative</u> motion of the African and South American plates, the Walvis and Rio Grande Ridges show the <u>absolute</u> motion of Africa and South America (plus the effects of the migration of the hot spot, to be discussed later).

We assume that all such aseismic ridges are produced by plate motion over hot spots fixed in the mantle. Thus the aseismic ridges indicate the trajectories of the plates over fixed points and we may reconstruct continental positions with both latitude and longitude control, an important addition to paleomagnetic reconstructions. This interpretation of the

aseismic ridges and island chains is
identical to that presented by Wilson
(1963, 1965) except that here we
attribute a more fundamental nature to
the hot spots--we associate the hot
spots with major convection deep in the
mantle, providing the motive force for
sea-floor spreading.

We shall now examine three aspects
of the worldwide pattern of island
chains and aseismic ridges consistent
with the concept of plate motions over
fixed hot spots.

ISLAND CHAINS IN THE PACIFIC

There are only two presently active
volcanos in the interior of the Pacific
plate, Hawaii and Macdonald Seamount
(Johnson, 1970).  It has long been

noted that the active Hawaiian volcano
is at the southeast extreme of the
Hawaiian chain and that there is a
linear progression of the age of these
islands as they become farther from
Hawaii (Fig. 2).  Johnson has noted
that Macdonald Seamount (29.0° S.,
140.3° W.) is likewise situated at the
southeastern extreme of the Austral
Islands chain.  Could both of these
island chains have been generated by a
single motion of the Pacific plate over
these two hot spots?  The Hawaiian
chain is terminated by the Emperor
Seamounts; is there an analogous
feature for the Austral chain?  The
answer to these questions is shown in
Figure 3.  We have assumed three fixed
hot spots located at 19° N., 155° W.
(Hawaii), at 29° S., 140° W.

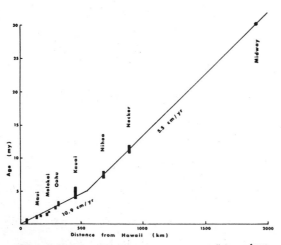

Figure 2. The ages of Hawaiian volcanos versus distance from
the present active volcano at Hawaii. The point at Midway is based
on Miocene fossils; the other ages are K-Ar results reported by
Funkhouser and others (1968).

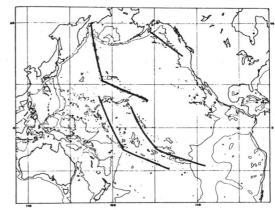

Figure 3. Hot spot trajectories constructed by rotating the Pacific
plate 34 degrees about a pole at 67° N., 73° W., then 45 degrees
about a pole at 23° N., 110° W.

(Macdonald), and at 27$^O$ S., 114$^O$ W. (where the East Pacific Rise intersects the Tuamotu and Sala y Gomez Ridges). The solid lines show the points which would pass over these hot spots if the Pacific plate were rotated 34 degrees about a pole at 67$^O$ N., 73$^O$W. and then 45 degrees about a pole at 23$^O$ N., 110$^O$ W. (The fourth solid line in Figure 3, extending from the Juan de Fuca Ridge to Kodiak Island, will be discussed later.) We thus note the similarity of the Hawaiian-Emperor, Tuamotu-Line, and Austral-Gilbert-Marshall chains with the lines generated from present-day active hot spots. In particular, the Marshall-Gilbert Islands do not coincide with the proposed locus of the Pacific over the fixed hot spot. We shall use this to measure the constancy of the fixed spots, but first let us estimate ages along these chains.

We have the rate of recent motion of the Pacific plate past the fixed points from the K-Ar ages of the Hawaiian Islands shown in Figure 2, and we see that the rate may be variable with about 10 cm/yr motion for the past 5 m.y. and about 5 cm/yr before that time. (Note that all three of these island chains are nearly 90 degrees from the pole at 67$^O$ N., 73$^O$ W. and so all have essentially the same velocities.) Another point on this curve is the age of Midway Island,

which has been dated pre-Miocene from drill holes through the coral cap. We estimate the age of the Hawaiian-Emperor "elbow" two ways. (1) From a linear extrapolation of the Hawaii to Midway distance and age difference, we estimate the elbow to have an age of 43 m.y. (2) The Nazca Ridge--Sala y Gomez Ridge intersection presumably represents the equivalent feature in the eastern Pacific. The hot spot which which made the Sala y Gomez-Nazca and the Tuamotu-Line Ridges is directly on the crest of the spreading rise, so the magnetic anomaly pattern adjacent to these features will directly give the age during which each feature was made. From Morgan and others (1969) we see that anomaly 13 (38 m.y.) is near the Nazca-Sala y Gomez conjunction. A third line of evidence which could have bearing on the age of this change in trend is the study by Menard and Atwater (1968) of the changes in the fracture zones pattern in the northeast Pacific. They do not find a major change at about anomaly 13 (although they note a change near the coast of California at the time of anomaly 11); the major change in this pattern occurred at anomalies 21 to 24 (55 m.y.). We shall assume that the bend in the Hawaiian-Emperor chain was made 40 m.y. ago, while keeping in mind that a 55 m.y. age for this feature cannot be ruled out. What is the age of the northern end of the Emperor or Line features? Again using the assumption that the Taumotu-Line chain was generated at a ridge crest, we infer from the nearness of anomaly 32 that 100 m.y. is a good estimate for the age of the northernmost features.

These age assignments will now be compared to ages determined by drilling or dredging on atolls and guyots in the Pacific. The Mid-Pacific Mountains and Magellan Seamounts will be featured in this discussion, so we first present our interpretation of these in terms of hot spots. The Mid-Pacific Mountains (or Marcus-Necker Ridge), the Magellan Seamounts, and the Caroline Islands are here regarded as east-west island chains formed from 100 to 150 m.y. ago by a rotation of the Pacific plate about a pole near the present North Pole. This motion was not displayed in Figure 3 because all of the island chains are close together and do not have a geometry to accurately determine the pole, and also because of the complications introduced by the "wandering" hot spot, which will be discussed later. The Mid-Pacific Mountains and the Magellan Seamounts are regarded as continuations of the Tuamotu-Line and the Austral-Marshall-Gilbert chains and the Caroline Islands as the continuation of another hot spot chain not present today. Hamilton (1956) reports the following ages (here

converted from his age classification name to millions of years) for dredge and core samples obtained from five guyots in the Mid-Pacific Mountains: Hess Guyot ($18^{\circ}$ N., $174^{\circ}$ W.), 120 m.y.; Cape Johnson Guyot ($17^{\circ}$ N., $177^{\circ}$ W.), 120 m.y.; Guyot 20171 ($21^{\circ}$ N., $171^{\circ}$ W.), 80 m.y. On two of the guyots sampled, Horizon ($19^{\circ}$ N., $169^{\circ}$ W.) and Guyot 19171 ($19^{\circ}$ N., $171^{\circ}$ W.), no age older than about 55 m.y. was obtained. A younger age does not contradict these guyots being formed at about 100 m.y., as only surface samples were obtained and much older sediments may not be exposed. Hamilton and Rex (1959) summarize the fossil ages found in the Marshall Islands. Bikini ($12^{\circ}$ N., $165^{\circ}$ E.) and Eniwetok ($12^{\circ}$ N., $162^{\circ}$ E.), just west of the Marshall chain, have been drilled and dated. The age at at the bottom of the Bikini hole (about halfway to the basalt basement) is 35 m.y., and the oldest age sample dredged from the adjacent Sylvania Guyot is 55 m.y. More important, two drill holes on Eniwetok penetrated the coral cap and reached the basalt basement. The fossils at the bottom of these holes are about 55 m.y. old. This implies that Eniwetok had a different history than that suggested by its position in the supposed hot spot trajectory. We also note that two of the Japanese Seamounts discussed in the following section on paleomag-

netics (near 28° N., 148°E.) have been
dredged and dated by the K-Ar method.
Their ages, 80 m.y., also contradict
their position on the western end of
the Mid-Pacific Mountains. We thus
have conflicting evidence in the
western Pacific and some additional
factors must be found if we are to
reconcile this with the simple hotspot
pattern that we observe farther east.

The solid lines generated by
rotating a rigid Pacific plate over the
hot spots do not exactly follow the
island chains. We may use this systematic departure to estimate the rate of
migration of the hot spots relative to
one another. The trajectories follow
the Hawaiian-Emperor and Tuamotu-Line
chains fairly exactly, so we may use
the departures of the Austral-Marshall-
Gilbert chain to measure mobility. The
measured distance from Macdonald Seamount to the turning point is 10 percent longer than the predicted
distance; the distance from the turning
point to the northernmost of the
Marshall Islands is 25 percent less
than the corresponding prediction. The
rate of plate motion over the "fixed"
hot spots is about 7 cm/yr (30 degrees
in 40 m.y. and 40 degrees in the
following 60 m.y.). Thus this hot spot
moving at about 1 cm/yr relative to the
others is a good measure of its
mobility in the deep mantle. If we had
chosen trajectories based on a more

compromise set of rotations which did
not agree so well with the Hawaiian-
Emperor or Tuamotu-Line chains, then
each of the hot spots migrating at
about .5 cm/yr in this reference frame
would match the observations.

Why are the Hawaiian Islands
islands? In the simple model presented
above the continuous eruption of deep
material should make a smooth continuous ridge--what geological complexities
must we introduce to get isolated
episodic volcanos? We adapt Menard's
(1969) model of growing volcanos to this
this problem. We suppose that the
light fractionation from the deep plume
continuously flows up but is trapped by
the asthenosphere-lithosphere "interface" (not a sharp boundary but a
gradual transition in rigidity). This
trapped island-type basalt accumulates
in pockets, analogous to oil trapped by
certain formations, and its unstable
situation causes vents to the surface
to form, which tap the reservoir and
cause volcanos at the surface. This
complexity has the possibility of
answering a number of questions. (1)
The plume may plaster the asthenosphere-lithosphere boundary over an
area 100 mi square, but a single vent
to the surface can tap this reservoir
and concentrate this into a single
volcano (as opposed to a continuous
ridge). (2) The motion of the lithospheric plate eventually displaces

the vent from the area above the deep plume sufficiently far so that a new vent forms. The old vent then taps only the remains in the reservoir in its immediate vicinity and soon dies out. We thus might expect to find a simple relation among the spacing of volcanos, the rate of plate motion, and the magnitude of the hot spot (as measured by the volume of the volcanic chain). (3) The activity of each island ends with alkali-rich eruptions. This different chemistry may result from the remains in the old vent after it has migrated and has been cut off from the hot spot. Does the volume of alkali-rich basalt agree with this? Is the chemistry compatible with this less than 100 km origin? Does the start of the alkali eruptions on an old volcano coincide with the start of a new volcano next in line? (4) This model allows a volcano to continue to grow for millions of years even after it has left its source area, in agreement with the model described by Menard (1969). The bulk of Menard's data supporting this growing seamount model comes from the Juan de Fuca Ridge region. We claim this region has two minor hot spots creating the line of seamounts and guyots between Cobb Seamount and Kodiak Island and the Explorer Sea-mount-Pratt-Welker guyot string farther north. (We thus limit the applicability of the growing seamount model to regions of hot spots.) The question as to whether all off-ridge seamounts are produced by minor hot spots raises interesting possibilities, but we shall sidestep this generality.

It is said, based on dredge samples, that seamounts such as Cobb Seamount are capped with an alkali-rich basalt. We claim that such a seamount is not primarily made of ocean ridge-type basalt capped in its last stages of growth with a more alkaline skin, but that it is made of the island-type basalt throughout. Dredging cannot answer this question; only deep drill-ing can distinguish between an alkali-rich coating or island-type throughout basalt.

Having minor hot spots at the Juan de Fuca Ridge offers an explanation as to why this ridge exists in the first place. The North American and Pacific plates could quite logically have their present motions without there even being being an oblique Juan de Fuca Ridge; but placing one of the world's driving mechanisms here assures the continuing existence of a spreading ridge at this location--a ridge that may change its orientation but must pivot about this hot spot.

## PALEOMAGNETISM

Francheteau and others (1970) have presented a polar wandering diagram for the Pacific plate based primarily on studies of the magnetic field around seamounts. Figure 4a is a reproduction of their Figure 17 with the following changes. First, we assign definite, though of course possibly inaccurate, ages to each pole position based on our understanding of their discussion of the possible ages of each seamount. The number in the name of each pole position shows our estimate of its age in millions of years. Second, we have greatly enlarged the error circle of the Midway data point. Francheteau and others used the Midway determination of Vine (1968), and Vine (personal commun.) states their estimate of the error is too small. The measured drill core samples were all from the same flow and the scatter reported represents differences in the single flow, not the scatter that might result from polar migration about the average

dipole if many flows had been sampled. In addition to the paleomagnetic data points, Figure 4a also shows a predicted polar wander curve for the Pacific plate made from the rotations needed to make the hot spot trajectories shown in Figure 2 (0.85°/m.y. for 40 m.y. about a pole at 67° N., 73° W.; 0.75°/m.y. for 60 m.y. about a pole at 23° N., 110° W.). Each dot on the polar wander curve shows the predicted position for successive 10 m.y. ages. The observed paleomagnetic pole positions should coincide with the dot for its age if (1) the motion of the Pacific plate is as described above, and (2) the magnetic pole does not migrate relative to the fixed hot spots. Figure 4b shows the paleomagnetic data corrected for the predicted motion of the Pacific plate. In principle we have taken the inclination and declination of the

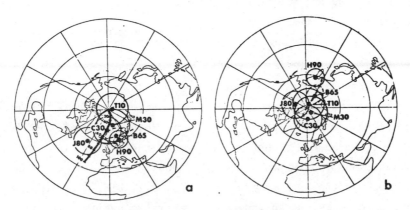

Figure 4. (a) The Pacific paleomagnetic pole positions of Francheteau and others (1970) and the polar wander curve predicted by the motion shown in Figure 3. (b) The paleopoles are "corrected" for the presumed motion of the Pacific plate; the paleopoles of all ages should now coincide at the north pole.

original measurement, rotated the
Pacific plate back to its orientation
at the time the feature was magnetized,
and computed the position of the paleo-
pole at that time.  Ideally all data
points would form a tight cluster about
the "north" pole.  We see that, except
for the Hawaiian Seamounts of presumed
90 m.y. age, there is excellent agreement between the predicted polar wandering and the observed paleomagnetic
positions.

The African plate offers another
test of a paleoreconstruction based on
plate motion over hot spots versus
paleomagnetic data.  The Walvis Ridge
is the most conspicuous aseismic ridge
in this region, but there are also submarine ridges trending northeast away
from present-day active volcanos at
Reunion Island, Bouvet Island, Ascension Island, and the Cape Verde and
Canary Islands.  A similar trend exists
for St. Helena Island and the Cameroon
trend, but here we have the peculiar
situation of an active volcano at both
ends of the trend.  Perhaps the trapped
basalt at the asthenosphere-lithosphere
boundary has taken nearly 100 m.y. to
find a vent to the surface at Mt.
Cameroon--if so, we have a mechanism to
account for the anomalously young ages
(Eocene) of much of the activity in the
Marshall and Gilbert island chains as
discussed above.  A rotation of 27 degrees about a pole at $25^o$ N., $55^o$ W. was

found to best fit this data, and the
trajectories of the hot spots on the
African plate based on this are shown
in Figure 5.  The time at which a hot
spot was beneath points on these trajectories was computed by assuming linear
interpolation between 110 m.y. and the
present--the uniformity of this motion
is based on the JOIDES results in the
South Atlantic.

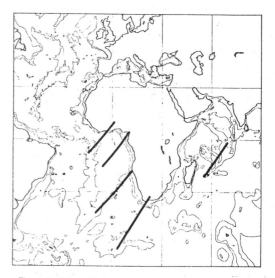

Figure 5. Hot spot trajectories constructed by rotating the African plate 27 degrees about a pole at 25° N., 55° W.

Figure 6a shows the paleomagnetic
pole determination of Africa as
tabulated by McElhinny and others
(1968).  The pole positions B14, B15,
B16, B17, B18, and B19 in McElhinny and
others' classification were used.
Numbers representing the age of the
site are used to identify each pole.

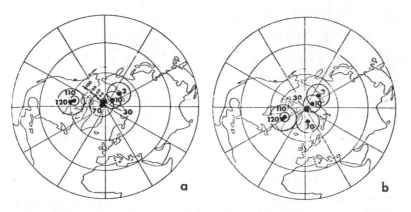

Figure 6. (a) African paleomagnetic pole positions of McElhinny and others (1968) and the polar wander curve predicted by the motion shown in Figure 5. (b) The paleopoles are "corrected" for the presumed motion of the African plate; the poles of all ages should now coincide at the North Pole. The discrepancy of the 110 and 120 m.y. poles may be due to a shift of the entire mantle shell relative to the core.

The solid line shows the polar wandering curve predicted for the motion of Africa shown in Figure 5.  Figure 6b shows the paleomagnetic data corrected for the presumed motion of Africa, analogous to how Figure 4b was obtained from 4a.  The clustering at the "north" pole is not as good as the Pacific data. data.  We may claim this is due in part to the slower motion of Africa, hence the migration of the hot spots would be more noticeable than in the Pacific. The African data does not lend support to the hypothesis presented here, but it could be reconciled with this model if there was an episode of rapid polar wandering between 90 and 110 m.y. ago. Such polar wandering would be a rapid shift of the whole mantle (in which all the hot spots would move in unison) to a new axis of rotation, as envisaged by

Goldreich and Toomre (1969).

No other plate contains a variety of aseismic ridges so that a direct test of the motions inferred from the ridge pattern and from paleomagnetism may be made.  However, we may use the deduced motion of Africa and the Pacific together with the known relative motions of the plates to infer the motion over the mantle for the other plates.  A tentative inference of the motion of North America, based on a counterclockwise rotation of Africa over the mantle about $30^{\circ}$ N., $60^{\circ}$ W. and the clockwise rotation of North America relative to Africa about $60^{\circ}$ N., $30^{\circ}$ W., shows that North America has rotated 30 degrees clockwise about the present north pole since mid-Cretaceous time.  North American Tertiary paleomagnetics cluster near

the present North Pole but the Cretaceous paleopoles are distinctly different, in the Bering Sea, in agreement
with the pattern shown here for Africa.

PRESENT MOTION OF THE PLATES

Table 1 lists the components of an
angular velocity vector for each
crustal plate.  The relative motions of
adjacent plates have been determined
from fracture zone strikes and spreading rates for most ridge systems and
Table 1 attempts a synthesis of this
data into a worldwide self-consistent
model.  Table 2 shows relative motions

computed from the vectors in Table 1
for most of the spreading pairs.  The
reader may compare these relative
motion poles and rates with his own
favorite data to judge accuracy of this
synthesis.  (A paper discussing the
data used to arrive at Table 1 is in
preparation.)

The relative motions are found by
subtracting vectors in Table 1, so any
constant vector may be added to all the
vectors in Table 1 without affecting
the relative motions.  We have added
that constant vector so that, in
addition to satisfying the relative
motion data, it also satisfies the
hot spot data.  The heavy vectors in
Figure 7 show the motion of the crustal
plates over each hot spot; in each case
the vector is closely parallel to an
aseismic ridge or island chain.  Thus
the hot spots form a reference frame
fixed in the mantle, and Table 1 and
Figure 7 show the absolute motion of

Table 1.   Absolute Motions of the Crustal Plates in Degrees/m.y.

| Plate Name | $W_x$ | $W_y$ | $W_z$ |
|---|---|---|---|
| AM American | .023 | -.022 | -.140 |
| PA Pacific | -.173 | .334 | -.702 |
| AN Antarctic | -.117 | -.033 | .268 |
| IN Indian | .459 | .315 | .350 |
| AF African | .149 | -.112 | .147 |
| EU Eurasian | -.050 | .052 | .039 |
| CH Chinese | -.145 | .176 | .223 |
| NZ Nazca | -.118 | -.314 | .616 |
| CO Cocos | -.693 | -.921 | .624 |
| CR Caribbean | -.180 | .430 | .090 |
| JF Juan de Fuca | .907 | 1.234 | -1.512 |
| PH Philippine | 1.295 | -.694 | -.859 |
| SM Somalian | .113 | -.143 | .181 |
| AR Arabian | .412 | -.025 | .352 |
| PR Persian | -.367 | .023 | -.141 |

Table 2.   Relative Plate Motions Deduced from Table 1

| | Latitude (°N.) | Longitude (°E.) | Spreading Rate (cm/yr) |
|---|---|---|---|
| EU-AM | 60 | 135 | 1.2 |
| AF-AM | 62 | -36 | 1.8 |
| AM-PA | 54 | -61 | 3.9 |
| PA-AN | -69 | 99 | 5.8 |
| AF-AN | -24 | -16 | 1.7 |
| IN-AN | 7 | 31 | 3.7 |
| AR-AF | 36 | 18 | 1.9 |
| AR-SM | 28 | 22 | 2.0 |
| IN-SM | 16 | 53 | 3.3 |
| SM-AN | -19 | -26 | 1.5 |
| CO-PA | 44 | -113 | 10.5 |
| CO-NZ | 1 | -133 | 4.6 |
| NZ-PA | 64 | -85 | 8.2 |
| NZ-AN | 51 | -90 | 2.6 |

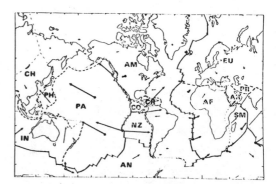

Figure 7. The present motion of the plates over the hot spots.
This motion is computed from Table 1 and agrees with the relative
motion data, as well as with the trends of the aseismic ridges-
island chains. The length of each arrow is proportional to the plate
speed.

each plate over the mantle. The good agreement between the arrows in Figure 7 and the trends of the island chains-aseismic ridges leads to two conclusions: (1) there has not been a major reorganization of plate motion in the past 40 m.y. or so, and (2) the hot spots have remained relatively fixed in the mantle. The slight disagreement is most pronounced in the Atlantic region where the slowly spreading plates are most vulnerable to "noise"-- the magnitude of this divergence suggests that each hot spot wanders at less than ~½ centimeter per year.

Kaula's (1970) recent gravity map of the earth is shown in Figure 8. This is an isostatic anomaly map computed for spherical harmonics of order 6 through 16, so the features of 1,000 to 10,000 kilometers length are displayed. Note that there are gravity highs over Iceland, Hawaii, and most of the other hot spots (Galápagos is a conspicuous exception). Such gravity highs are symptomatic of rising currents in the mantle; the less dense material in the rising current produces a negative gravity anomaly, but the satellite passes closer to the elevated surface pushed up by this current and the net gravity field is positive. From formulas in Morgan (1965) we can estimate the size of the rising current. Take 10 mgal excess and 1,000 km diameter as typical for the hot spots; such a geoid high could be produced by a mass deficiency of $10^{20}$ gm centered at about 300 km depth, or roughly a cylindrical plug 100 km in diameter extending from the surface to 600 km depth with a density deficiency of 1 percent.

Note the paradox: both rising

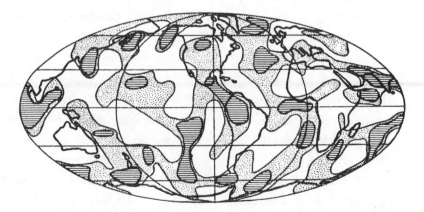

Fig. 8. Isostatic gravity map of the earth, redrafted from Kaula (1970), to emphasize positive anomalies. The shaded areas show regions of positive anomalies; the heavier shaded areas show regions where the anomalies are greater than +20 mgal. Note the correlations of the gravity highs with Iceland, Hawaii, and most of the other hot spots.

currents and oceanic trenches are associated with positive gravity anomalies. This behavior has been explained by Kaula and others as due to flow in a nonuniform viscous material. At the rises, the light ascending current buoys up the surface for a net positive gravity effect (and descending currents of the same pattern would pull down the surface for a net gravity minimum). However, a deep lithospheric plate may push down onto a hard bottom surface. If the lower surface supports part of the weight of the sinking plate, the top surface will not be depressed by the flow pattern and the satellite will sense only the excess mass of the plunging lithosphere for a positive gravity effect.

The gravity measurements appear to offer the best method to assess the strength of the different plumes. More measurements are needed as the geoid maps change dramatically each year (compare Kaula's 1970 statements with those of earlier years). In choosing the possible plumes shown in Figures 1 and 7; the gravity measurements were augmented by what is known locally as the "Hess Gravity Theorem," namely that one does not need a gravimeter to measure gravity, one needs only to look at the topography. This is a corollary of the statements made above; the flow patterns associated with positive gravity anomalies raise the surface, thus high topography means positive gravity and vice versa. We thus look for those abnormally shallow places in the oceans, such as the areas near the Galapagos, the Juan de Fuca Ridge, and Prince Edward Island. The National Geographic Society globe has contours at particularly apt intervals and spiderlike fingers radiate away from many of these topographic highs. Whether the unusually high Tibetan Plateau or southern Africa should be considered symptomatic of a subcontinental hot spot is an open question; the more uniform oceans are more amenable to this type of analysis. The best case for a present day subcontinental hot spot could be made for the Snake River flood basalts in analogy to the Deccan Traps of the early Reunion hot spot.

The data presented in this paper, the parallelism of the Pacific island chains, the agreement of this motion of the Pacific with the paleomagnetic results, and the agreement of the present relative motions of the plates with the trends of the island chains-aseismic ridges all substantiate that plate motion over mantle hot spots is a valid and useful concept but this data contributes little to the hypothesis that these hot spots provide the motive force for continental drift. The case for this association rests on three facts: (1) Most of the hot spots

spots are near a ridge, and a hot spot is near each of the triple ridge junctions; (2) the gravity and regional high topography suggests that more than just surface volcanism is involved at each hot spot; and (3) neither rises nor trenches appear capable of driving the plates, implying that asthenospheric currents acting on the plate bottoms must exist.

The symmetric magnetic patterns and the mid-ocean position of the rises suggest that the ridges are passive. The first deduction of plate tectonics was that if two plates are pulled apart, they split along some line of weakness and in response, asthenosphere rises to fill the void. With further pulling of the plates, the laws of heat conduction and the temperature dependence of strength dictate that future cracks appear right down the center of the previous "dike" injection. If the two plates are displaced equally in opposite direction or if only one plate is moved and the other held fixed, perfect symmetry of the magnetic pattern will be generated. The axis of the ridge must be free to migrate (as shown by the near closure of rises around Africa and Antarctica). If the "dikes" on the ridge axis are required to push the plates apart, it is not clear how the symmetric character of the rises is to be maintained.

The best argument against the sinking lithospheric plates providing the main motive force is that small trench-bounded plates such as the Cocos do not move faster than the large Pacific plate. Also, the slow compressive systems, as in Iran, would not appear to have the ability to pull other plates, such as the Arabian plate, away from other units. The pull of the sinking plate is needed to explain the gravity minimum and topographic deep locally associated with the trench system (see Morgan, 1965), but we do not wish to invoke this pull as the main tectonic stress.

We are left with sublithospheric currents in the mantle. The question now is whether these currents are great rolls--mirrors of the rise and trench systems--or whether they are localized upwellings, that is, hot spots. Also, how deep do such currents extend? The circumstantial evidence seems to favor the hot spot mode, but there are several tests which could answer this question. (1) The most dramatic proof would be to seismically detect the shadow cast by a deep plume (the large time delay of teleseismic events in Iceland may be a plume effect). (2) Assumptions as to the magnitude of each plume and of the stresses at rise, fault, and trench plate-to-plate boundaries could be made and, the directions of the resulting plate motions could be deduced from these simplified dynamics. (3) A re-evaluation of the heat flow problem may show that convection deep in the mantle is necessary to remove heat from the lower mantle. The near equality of the oceanic and continental heat flux may be explained in terms of hotter than normal asthenosphere flowing away from each hot spot. (4) Finally, a continuing study of the Cenozoic and Cretaceous sea-floor spreading may show that major reorganizations of the spreading pattern coincide with the disappearance or emergence of new hot spots.

## ACKNOWLEDGMENTS

Many of the points presented here arose during discussions with K. S. Deffeyes and F. J. Vine, and I thank them for their many suggestions. I also thank the organizers of the Birch Symposium at Harvard in 1970, as the near simultaneous presentation of papers by P. W. Gast, W. M. Kaula, and W. H. Menard directly lead to this model. This work was partially supported by the National Science Foundation and the Office of Naval Research.

## REFERENCES CITED

Francheteau, J., Harrison, C. G. A., Sclater, J. G., and Richards, M. L., 1970, Magnetization of Pacific seamounts, a preliminary polar curve for the northeastern Pacific: Jour. Geophys. Reseach, v. 75, p. 2035-2061.

Funkhouser, J. G., Barnes, I. L., and Naughton, J. J., 1968, Determina- of ages of Hawaiian volcanoes by K-Ar method: Pacific Sci., v. 22, p. 369-372.

Gast, P. W., 1968, Trace element fractionation and the origin of tholeiitic and alkaline magma types: Geochim. et Cosmochim. Acta, v. 32, p. 1057-1086.

Goldreich, P., and Toomre, A., 1969, Some remarks on polar wandering: Jour. Geophys. Research, v. 74, p. 2555-2569.

Hamilton, E. L., 1956, Sunken islands of the Mid-Pacific Mountains: Geol. Soc. America Mem. 64, 97 p.

Hamilton, E. L., and Rex, R. W., 1959, Lower Eocene phosphatized globigerina ooze from Sylvania Guyot: U.S. Geol. Survey Prof. Paper 260-W, p. 785-797.

Johnson, R. H., 1970, Active submarine volcanism in the Austral Islands: Science, v. 167, p. 977-979.

Kaula, W. M., 1970, Earth's gravity field: Relation to global tectonics: Science, v. 169, p. 982-985.

McElhinny, M. W., Briden, J. C., Jones, D. L., and Brock, A., 1968,

Geological and geophysical
implications of paleomagnetic
results from Africa:  Rev.
Geophysics, v. 6, p. 201-238.

Menard, H. W., 1969, Growth of drifting
volcanoes:  Jour. Geophys.
Research, v. 74, p. 4827-4837.

Menard, H. W., and Atwater, T. M.,
1968, Changes in direction of
sea-floor spreading:  Nature, v.
219, p. 463-467.

Morgan, W. J., 1965, Gravity anomalies
and convection currects:  Jour.
Geophys. Research, v. 70, p.
6175-6204.

Morgan, W. J., Vogt, P. R., and Falls,
D. F., 1969, Magnetic anomalies
and sea-floor spreading on the
Chile Rise:  Nature, v. 222, p.
137-142.

Vine, F. J., 1968, Paleomagnetic evidence for the northward movement
of the North Pacific basin during
the past 100 m.y. (abs.):  Am.
Geophys. Union Trans., v. 49,
156 p.

Wilson, J. T., 1963, Continental drift:
Sci. American, v. 208, p. 86-100.

---- 1965, Evidence from ocean islands
suggesting movement in the earth:
Royal Soc. London, Philos. Trans.,
v. 258 (Symposium on Continental
Drift), p. 145-165.

*

# Plate tectonics:
# do the hot spots really stand still?

**W.D. Metz (1974)**

With plate tectonics firmly established as a theory that explains to a
large extent why the earth looks as it
does, what do you do with the features
that the theory doesn't explain? For
instance, major oceanic ridges have
been identified where plates are
spreading apart, in the Atlantic,
Pacific and Indian oceans, causing many
earthquakes underneath the ridges. But
large ocean ridges are also found that
have no earthquake activity underneath.
How are they formed? What produced the
long chain of volcanoes, mostly underwater, in the northern Pacific from the
Hawaiian Islands almost all the way to
Siberia? This island chain is far from
the boundaries of the Pacific plate,
and does not parallel any of the geologic features of the ocean floor in the
region.

The tendency of most geophysicists
is to group such inexplicable features
together and say they were formed by
"hot spots" under the moving plates.
The Hawaiian chain, which includes
approximately 80 volcanoes, was the
first to be explained by a fixed hot
spot periodically erupting through a
plate. Now perhaps 200 features are
called hot spots by one researcher or
another, but they are not well understood. What causes the hot spots,
whether they are really fixed in the
deeper parts of the earth, and how applicable the term is to all the features
left over from global tectonics are
open questions.

The term hot spot sounds innocuous
enough, like a soft patch in an asphalt
road, but in fact it is the name for a
source of prodigious energy. The hot
spot in the earth's mantle under Hawaii
is capable of producing volcanoes equal
in height to Mount Everest, as measured
from the ocean floor. The volcanoes
that formed the Hawaiian archipelago
and its northward underwater continuation, the Emperor seamounts, are broad,

smooth structures called shield volcanoes which reach a diameter of 120 km at the base.  The chain appears to have been formed along a number of loci, often with more than one volcano active at a time (Fig. 1).  The oldest volcanoes are at the northern end of the chain and the youngest are on the island of Hawaii, where Mauna Loa and Kilauea are still active.

Radioactive dating of samples by G. Brent Dalrymple and his colleagues at the U.S. Geological Survey in Menlo Park, California, shows that the volcanoes are progressively older to the northwest, but not in a linear fashion. Until recently data were available for dating the chain only as far out as Midway, which is 2400 km from Kilauea and about 18 million years old.  Using data from this section of the chain, Dalrymple and his associates estimated that the average velocity of the plate over the hot spot was 13 cm/year, too fast for the hot spot to be considered fixed, according to most estimates.

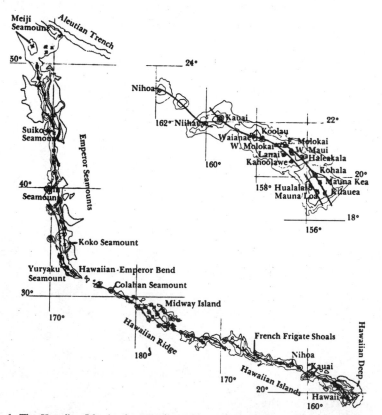

Fig. 1. The Hawaiian Islands, the Hawaiian Ridge, and the Emperor seamounts form a continuous chain that grows older to the northwest. The serpentine lines are loci on which volcanoes were formed, often two or more at a time.

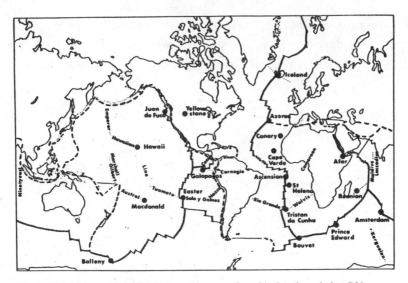

Fig. 2. Major hot spots are labeled on this map of world plate boundaries. Ridges are denoted by solid lines and trenches by dashed lines.

New data from volcanoes farther out in the Hawaiian-Emperor chain show that the age progression, averaged over a longer time, was considerably slower. Samples from the Koku and Yuryaku seamounts now indicate that the age of the Hawaiian bend is 41 to 43 million years. Those dates, obtained by Dalrymple, David Clague of the Scripps Institution of Oceanography in La Jolla, California, and R. Moberly of the University of Hawaii, Honolulu, indicate that the average rate of migration of volcanism has been about 8 cm/year, a number that jibes well with the motion of the Pacific plate at the present time (the last 10 million years). While they do not prove that the Hawaiian chain was caused by a fixed hot spot, the new dates seem to

many geophysicists to indicate that the data are more favorable than ever before.

The Hawaiian-Emperor chain is not the only string of volcanic islands in the Pacific that could have been produced by a hot spot. In 1971, W. Jason Morgan at Princeton University, introducing the plume hypothesis (see **below**), suggested that the Austral-Cook-Marshall Islands and the Tuamotu-Line Islands were also produced by hot spots (Fig. 2). These two island chains have a bend with the same general orientation as the Hawaii-Emperor chain. Morgan suggested that all three island chains, plus another near Alaska, were produced as the Pacific plate moved over fixed hot spots. The bends occurred, he postulated, because the plate changed its direction of motion. Fewer volcanoes have been dated in the Austral-Cook-Marshall and Tuamotu-Line Islands than in the Hawaiian chain, but as more data are gathered, the evidence for age progression seems to improve.

# Relative and latitudinal motion of Atlantic hot spots

**K. Burke, W.S.F. Kidd, and J.T. Wilson (1973)**

The hypothesis that hot spots[1] and their underlying plumes[2] all remain fixed with respect to one another, and perhaps also to the Earth's spin axis and axial dipolar magnetic field, has recently received attention[3-5]. Atwater and Molnar[6] have shown, by rotating plates back to their 13 and 38 m.y. position, that some hot spots (notably Iceland and St Paul/Amsterdam) have probably moved with respect to one another during this time. Using a different approach we present here evidence that during the past 120 m.y. some hot spots have moved significantly with respect to each other and to the magnetic field.

We use the term "hot spot" to describe succinctly a class of localized volcanism and associated uplift characteristically found within plates (Hawaii and Tibesti, for example), but also found on divergent plate boundaries (Iceland, for example). The term is used to describe the surface feature, with no intended implications about processes below the surface, in the same way as the term "island arc" is used. In the ocean a hot spot trace is a volcanic ridge or line of seamounts which leads away from a hot spot and which is considered to have been progressively generated as the plate moved relatively over or away from the underlying source of the hot spot. Where a hot spot occurs on a spreading ridge axis, a trace is generated on each plate. The former presence of a hot spot on continental crust is expressed by relatively localized alkaline volcanic piles or subvolcanic alkaline intrusives, sometimes accompanied by more extensive flood basalts. The former position of a hot spot at any time in the past is marked by the point of that age on its trace.

Reprinted from Nature 245, 133-137 by permission of the authors and publishers. Copyright 1973 Macmillan Journals Ltd.

Table 1   Position and Palaeolatitude Data for Some Atlantic Hot Spots

| Hot spot | Trace | Position of hot spot 120 m.y. ago | Present coordinates of hot spot | Present coordinates of hot spot position 120 m.y. ago | Palaeolatitude of position of hot spot 120 m.y. ago | 180 m.y. ago |
|---|---|---|---|---|---|---|
| Azores | Newfoundland Ridge (or "Fracture Zone"); (eastern trace in tectonized area) | South-east corner of Grand Banks | 38°N 27°W | 41.5°N 48°W | 26°N | 29-30°N |
| Colorado Seamount | New England Seamounts; Corner Rise Seamounts; ill defined to present Mid-Atlantic Ridge | South-eastern New England Seamounts on Bermuda discontinuity | 34°N 37.5°W | 36.5°N 59°W | 25°N | 25-27°N |
| Colorado Seamount | Interpolated from continental margin to Great Meteor Seamount; Cruiser Seamount-Mid-Atlantic Ridge | Just west of Azores--interpolated | 34°N 37.5°W | 29°N 19°W | | |
| Tristan da Cunha | Rio Grande Rise | Florianapolis | 37°S 12.5°W | 27°S 48°W | 29°S | |
| Gough Island | Walvis Ridge | Cape Fria | 40°S 10°W | 18.5°S 11.5°E | 30°S | |
| Discovery Seamounts | Poorly defined line of seamounts | Luderitz | 47°S 6.5°W | 27°S 14°E | 39°S | |
| Discovery Seamounts | None known except one seamount just off continental margin | East of Montevideo | 47°S 6.5°W | 35.5°S 53°W | 38°S | |
| Bouvet Island | Meteor Seamount chain | South of Cape Agulhas | 54.5°S 3.5°E | 35°S 18°E | 48°S | |
| Bouvet Island | North Falkland Plateau; ill defined to east | South-west Argentine Basin --interpolated | 54.5°S 3.5°E | 45°S 59°W | 48.5°S | |

## MISFIT

Figure 1 is a reconstruction of the Atlantic ocean and its surrounding continents about 120 m.y. ago[7], and five selected Atlantic hot spots which possess comparatively well defined traces are plotted on it at the positions that they occupied at that time. If hot spots stay fixed with respect to each other, it should be possible to superimpose the same hot spots in their present relative positions on their positions 120 m.y. ago. It is not possible to do this, and the misfit is conveniently represented by keeping one hot spot fixed, and plotting the others in their present positions relative to it. We have chosen to keep the Colorado Seamount hot spot fixed for the purposes of illustration, as it has a trace on the North American plate, and the reference frame on Fig. 1 is with respect to present North America. Although the Azores hot spot does plot at the relative position it occupied 120 m.y. ago, the three others shown from the South Atlantic fall about $20^{\circ}$ beyond their positions at that time. This indicates that relative motion between them and the two northern hot spots has occurred at an average rate of 1.8 cm yr$^{-1}$ during the past 120 m.y., in a direction approximately perpendicular to the general direction of relative plate motion.

Data used to plot the hot spot positions on Fig. 1, and for subsequent parts of this paper, are listed in Table 1. The positions of the two northern hot spots at 120 m.y. ago were picked from maps of seafloor age in the central Atlantic[7,8]. A certain amount of interpolation has been necessary for the Colorado Seamount hot spot traces. The early part is well defined only on

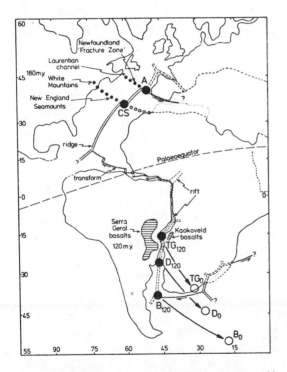

**Fig. 1** Sketch of configuration of Atlantic continents, with schematic plate boundaries, for about 120 m.y. ago (from ref. 7); reference frame fixed to North America. ●, Positions of five hot spots at this time as given by their traces; ●, earlier parts of traces of the two northern hot spots; ○, an interpolated older trace (see text); ○, present positions of the southern hot spots relative to Colorado Seamount hot spot kept fixed at its position relative to North America 120 m.y. ago. A, Azores; CS, Colorado Seamount; TG, Tristan da Cunha/Gough; D, Discovery Seamounts; B4, Bouvet. Large arrows indicate total motion of the three southern hot spots between 120 m.y. ago and the present relative to Colorado Seamount hot spot. Palaeoequator is around average pole at 69° N 180° W with respect to North America.

the western side, and the later part only on the eastern side of the Mid-Atlantic ridge, but the age ranges of the partial traces overlap, so the interpolation is valid. Some authors (for example ref. 9) prefer to generate the New England (Kelvin) and Corner Rise seamounts from the Azores hot spot. We do not favour this, because partial closure of the central Atlantic by appropriate rotations[8] connects the Corner Rise seamounts with a point of the same age on the trace of seamounts running east from Colorado Seamount.

The relative motion between the five hot spots is set out in Table 2, showing in quantitative form what we show diagrammatically in Fig. 1. The large motion between the two groups of hot spots is clear, but there does not appear to have been significant motion between the hot spots in each group.

## RELATIVE MOTION

The average rate of relative motion between the two groups has been about 1.8 cm $yr^{-1}$, but this rate has not been constant. Points representing particular ages on the traces from Colorado Seamount and Tristan/Gough can be picked from maps of seafloor age in the central[7,8] and southern[10,11] Atlantic. This identification of the age points along the two traces assumes that the two hot spots have stayed approximately on the axis of the spreading ridge, and is justified because hot spot traces extend away on both sides of the spreading ridge. Data are given in Table 3 and the change in great circle distance with time is shown in Fig. 2. Although the shape of the curve showing the relative motion is crudely exponential, we do

---

Table 2    Great Circle Distances between Hot Spots

| Hot spots | 180 m.y. | 120 m.y. | Present | Difference between 120 m.y. and present |
|---|---|---|---|---|
| Azores-Colorado Seamount | 10 | 10 | 9 | -1 |
| Colorado Seamount-Tristan da Cunha | -- | 57 | 75.5     77 | +20+2 |
| Colorado Seamount-Gough Island | -- | | 79 | |
| Gough Island-Discovery Seamounts | -- | 9 | (7 to) 10.5 | -2 to + 1.5 |
| Gough Island-Bouvet | -- | 19 | 16.5     18.5 | -2.5     -0.5 |
| Tristan da Cunha-Bouvet | -- | | 20.5 | +1.5 |
| Discovery Seamounts-Bouvet | -- | 9.5 | 10 | +0.5 |

All measurements in degrees (1 degree ≈ 110 km).

not attach significance to this, since it only concerns the motion between two of many hot spot groups. We think, however, that the variations in the rate of motion are significant, and this rate has varied between about 5 cm yr$^{-1}$ (100 to 80 m.y.) and about 0.5 cm yr$^{-1}$ (25 to 0 m.y.). It is intriguing that the time of maximum rate of relative motion coincides with a time of world wide accelerated spreading rates[7], but we have not been able to confirm this correlation for any other hot spots.

We take the Mascarene/Chagos-Laccadive Ridge and most of the Ninety-East Ridge to have been produced during the past 65 m.y. by the two hot spots now at Mauritius/Réunion and St Paul/Amsterdam Islands respectively. Tentative age data at two places along each of these traces[12,13] suggest that at present these two hot spots have the same separation as they had 65 m.y. ago (about 25°), but that their separation

was slightly greater (about 29°) 25 m.y. ago. No significant motion between the Mauritius/Réunion and the Tristan/Gough hot spots can be detected for the past 12 m.y. (ref. 14); the motion between 12 m.y. and 65 m.y. ago must amount to about 6° towards each other, on the basis of established plate motions[15].

Ideally, palaeomagnetic data can be used to establish how the relative motion between the hot spots has been distributed with respect to the magnetic field and by inference to the spin axis. We have estimated average

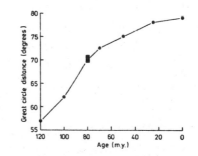

Fig. 2  Great circle motion between Colorado Seamount and Tristan/Gough hot spots during the past 120 m.y.

Table 3    Increments of Motion between Tristan/Gough and Colorado Seamount Hot Spots

| Age (m.y.) | Present coordinates of hot spot position at ages given Walvis Ridge-Gough | | Colorado Seamount eastwards | | Great circle distance (degrees) | Rate of motion cm yr$^{-1}$ |
|---|---|---|---|---|---|---|
| 120 | 18.5°S | 11.5°E | 29°N | 19°W | 57 | |
| | | | | | | 2.8 |
| 100 | 25°S | 6°E | 29-29.5°N | 23-24°W | 62 | |
| | | | | | | 4.1 to 5.0 |
| 80 | 32°S | 2°E | 30-31.5°N | 29°W | 69.5-71 | |
| | | | | | | 1.6 to 3.3 |
| 70 | 34°S | 3°W | 32.5°N | 30°W | 72.5 | |
| | | | | | | 1.4 |
| 50 | 36°S | 5.5°W | 33°N | 32°W | 75 | |
| | | | | | | 1.3 |
| 25 | 40°S | 10°W | 34°N | 35°W | 78 | |
| | | | | | | 0.44 |
| 0 | 40°S | 10°W | 34°N | 37.5°W | 79 | |

palaeomagnetic poles for various plates at particular times from data contained in a recent compilation[16]. We used only non-redundant poles in reliability category A, and only if the age of any pole is well defined to within 10 m.y. of the age required (exceptionally 20 m.y. for the 180 m.y. and 120 m.y. ages), and we excluded poles from areas that may have suffered subsequent tectonic displacement. As the poles selected have restricted age ranges, any possible polar wandering as distinct from plate motion can be neglected for each age (but not between). Known plate rotations[8,15,17,18] can then be applied to the groups of poles of each age, and a combined pole defined, which can then be similarly rotated to give a consistent result for each plate.

Palaeolatitudes derived from the average pole for 180 m.y., and perhaps the other average poles, may have errors larger than the differences we are attempting to detect as the selected data scatter over significant areas, up to 15 to 20° across. But as the average poles for 180 and 120 m.y. have been selected and then deliberately rotated so as to be compatible with total subsequent plate motion for those plates involved, palaeolatitudes derived from them may be more reliable than those obtained from separate average poles for each plate, which do not coincide when subsequent plate motion is removed. The average palaeomagnetic poles given in Table 4 should be regarded as present estimates, and only for the restricted age ranges given. No motion of the African plate with respect to the dipole field is required by the palaeomagnetic data for the past 25 m.y. (ref. 19).

Seventeen poles for the Eurasian plate 60 m.y. ago are almost all closely grouped, but the average pole they define is not compatible with an average pole for North America[5] when the appropriate North Atlantic motion[8]

Table 4    Average Palaeomagnetic Poles

| Plate | 180 | 120 | Age (m.y.) 65-60 | 50 | 25 |
|---|---|---|---|---|---|
| Africa | 67.5°N 104°W | 55°S 74°E | -- | -- | 90°N |
| North America | 72.5°N 94°E | 69°N 180°W | (77°N 114°E ?) 85°N 163°W (ref. 5) | -- | -- |
| South America | -- | 83.5°S 113°W | -- | -- | -- |
| India | -- | -- | 32°S 100°E | -- | -- |
| Indo-Australia | -- | -- | -- | 70°S 54°E | 76°S 89°E |
| Eurasia | -- | -- | 77°N 150°E? | -- | ? |

is removed (Table 4, in brackets).
Three poles in the Eurasian group plot
$20^{o}$ away from the main group, and their
position is more compatible with the
North American pole.  In view of this
uncertainty, we have not used the
60 m.y. poles to define palaeolatitudes
for the Atlantic hot spots, as the
error is potentially as much as $20^{o}$.
Phillips and Forsyth[20] have previously
discussed this discrepancy.

Palaeolatitudes derived from
average poles are probably of variable
reliability and only relatively small
changes in one or more of the average
pole positions may make a great deal
of difference to the inferred motion of
particular hot spots with respect to
the magnetic field and the spin axis.
The palaeolatitudes we derive show that
all the hot spots have moved differing

amounts with respect to the magnetic
field for at least some parts of the
last 120 m.y., but the precise amounts
for each hot spot we regard as less
certain.

Palaeolatitudes of the Atlantic
hot spot positions 120 m.y. ago, and
for the two northern hot spots at 180
m.y. ago, are given in Table 1.  The
discrepancies between the palaeolatitudes at 120 m.y. and the present
latitudes of the hot spots are set out
in Table 5, and shown graphically in
Fig. 3.  First, the results show that
the total motion between the two groups
of hot spots with respect to latitude
is nearly as great as their motion with
respect to each other during the past
120 m.y., indicating that aggregate
relative longitudinal motion has been
subordinate, at least for these

| Hot spot (trace east or west) | Latitude/palaeo-latitude change from 120 m.y. to present | Differences between hot spots (degrees) |
|---|---|---|
| Azores (W) | $12^{o}$N | +3 to +3.5 |
| Colarado Seamount (W) | $9^{o}$N | |
| | | |
| Colorado Seamount (E) | $8.5^{o}$N | +16.5 to +19 |
| Gough (E) | $10^{o}$S | 0 to -2 |
| Discovery Seamount (E) | $8^{o}$S | -1.5 to -3 |
| Bouvet (E) | $6.5^{o}$S | |
| | | |
| Tristan da Cunha (W) | $8^{o}$S | -- |
| Discovery Seamount (W) | $9^{o}$S | -- |
| Bouvet (W) | $6^{o}$S | -- |

Table 5    Latitudinal Motion of Hot Spots

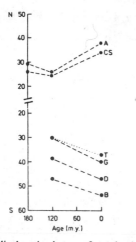

**Fig. 3** Latitudinal motion between five Atlantic hot spots during the past 180 m.y. Letters identifying hot spots are as in Fig. 1.

particular hot spots (see also Fig. 1). Second, each group of hot spots has moved an approximately equal distance away from the equator toward the poles of their respective hemispheres. Duncan and others[3] showed that Eurasian palaeomagnetic data for 60 m.y. ago can be interpreted as indicating a large northward motion of the Iceland hot spot (about 20°) since this time. In view of the uncertainty in the palaeomagnetic data, we would suggest that although relative northward motion of this hot spot is probable, the amount of motion may be more modest. We estimate from known plate motions[8], allowing for uncertainty across the Labrador Sea, that the Iceland hot spot has not moved more than 5° away from the Azores and Colorado Seamount hot spots during the past 60 m.y. On the basis of the North American average

palaeomagnetic pole[5], this indicates no more than about 8° of northward motion of the Iceland hot spot during this time.

MORE HOT SPOTS

We have attempted to extend this study to the two hot spots of Mauritius/Réunion and St Paul/Amsterdam Islands. The palaeomagnetic data perhaps suggest that Mauritius/Réunion has moved northwards about 10° in the past 65 m.y. and that all or almost all of this relative motion occurred before 25 m.y. ago. On the same basis St Paul/Amsterdam has moved northwards about 16° since 65 m.y. ago. If a very tentative estimate is made of its position 50 m.y. ago, extrapolating from a point suggested to be 45 m.y. old[13] and using the Australian palaeomagnetic pole, it is possible that the overall relative northward motion of this hot spot may have reversed for a while between 50 and 25 m.y. ago.

The only other hot spot for which data are available is Hawaii; the palaeomagnetic data cannot resolve any motion of this hot spot in the past 70 m.y. (ref. 4), but the equatorial sedimentation data[4] perhaps suggest it has moved slightly southwards[21], perhaps up to 5° in the past 40 m.y. By implication, the other hot spots on the Pacific plate which seem to be approximately fixed with respect to Hawaii[2,4] may be moving in a similar manner.

McElhinny[5] showed that the apparent motion of the Iceland hot spot with respect to the magnetic field[3] is more likely to be due to relative motion between hot spots than to a rotation of the mantle containing fixed plumes with respect to the lithosphere as a whole[3,22]. Our conclusion that some groups of hot spots have moved significantly with respect to each other confirms this suggestion. The rate of motion between two small groups of hot spots is comparable with rates of plate motion, and has varied by an order of magnitude over the past 120 m.y.; the maximum rate of separation for these two groups appears to coincide with a temporary global increase in spreading rates[7]. The two or three hot spots within each of these two groups do not, however, seem to have moved significantly with respect to their partners for the maximum lengths of time that they can be observed (180 and 120 m.y.). While moving apart and remaining essentially fixed internally, the two groups of Atlantic hot spots seem to have rotated somewhat relative to one another (Fig. 1). Differential motion of hot spots with respect to the magnetic field and spin axis occurs and for the

few hot spots we have studied it seems to have been predominantly in a pole-ward direction, though not all seem to move toward the pole of the hemisphere in which they occur. The poleward motion may also be intermittent, and even reversible. Data from other hot spots are desirable, but it is un-fortunate that although there are many hot spots[23], there are few, if any, others that have clearly marked traces with well defined ages along them and that can be observed for comparable periods of time to those in the central and southern Atlantic.

SATISFACTORY HYPOTHESIS

We think that the idea that deep mantle plumes[2,24] underlie hot spots is at present the only satisfactory hypothesis available to account for hot spots, especially to explain their persistence as discrete localized anomalies, in some cases for 180 m.y. This is particularly the case for those hot spots which are on the axes of spreading ridges, and which have remained so for long periods; these are epitomized by the hot spots on the axis of the central and southern Mid-Atlantic Ridge. Although these hot spots have moved with respect to one another, and to the position of the spreading ridge at 120 m.y. ago (Fig. 1), they are still on, or near[14], the

axis. The jumping of a restricted
length of spreading ridge axis is
probably a related phenomenon, and in
several cases this is clearly
associated with hot spots. It may
occur repeatedly in a consistent
direction, and this leads to apparent
asymmetrical spreading[25]. Examples
of ridge axis jumping occur at the
Galapagos[26,27], St Paul/Amsterdam
Islands[15], south of Australia[25],
and in Iceland[28]. The inference we
make is that whatever is under the hot
spots controls the position of the
ridge, and not the reverse; the
presence of underlying deep mantle
plumes seems to us the only available
hypothesis that will account for this.

We suggest that within small
groups of hot spots, the hypothesis
that they are fixed with respect to one
another may be valid, especially for
short periods of a few tens of millions
of years, but which hot spots belong to
a particular internally fixed groups
may be hard to prove. The hypothesis
that all hot spots are fixed with
respect to one another and form an
independent global reference frame is,
however, clearly not valid.

The groups of hot spots that we
have shown to remain essentially fixed
internally have an approximate maximum
horizontal dimension of 2,000 km. This
is of the same order as the large
swells that occur in East Africa and

the Red Sea area, which are about
1,500 km across. We interpret the
pattern of alkaline volcanism on abrupt
topographic and structural uplifts each
about 200 km across within these swells
as showing the presence of four or five
hot spots in each swell. The Cameroon
zone also defines an ellipse about
1,500 km across and contains at least
six discrete hot spots on uplifts,
although a large overall topographic
swell is not present. Apart from these
examples, it is difficult to make a
similar grouping of many of the
remainder of the 120 hot spots we
recognize, although some tentative
suggestions can be made. Menard[29] has
recently demonstrated the occurrence of
low amplitude positive topographic
anomalies 1,000 to 2,000 km across, and
without obvious associated hot spots,
in the North-east Pacific. We point
out that similar large swells without
associated hot spot volcanism are
especially prominent in southern
Africa, although the amplitude of these
features is larger, and all but one are
in continental lithosphere.

It may be that the large scale
swells and internally fixed groups of
hot spots reflect conditions at a depth
comparable to their horizontal
dimensions, and that the smaller
uplifts and hot spots, of which there
might be only one in some large swells
(Hawaii?), reflect conditions at

shallower depths. We speculate that
the large swell may represent a broad,
very slowly upwelling column, and that
the smaller uplifts and hot spots
within them represent smaller, more
strongly heated and more rapidly rising
columns within the large slowly rising
column. The implications of the poleward motion of the few hot spots we
have examined are not clear to us, and
we think speculation on this is perhaps
unjustified until well distributed data
from other hot spots become available.
If it is accepted that deep mantle
plumes[2,24] underlie hot spots, then
information relating to processes at
the core-mantle boundary, and perhaps
in the core, might eventually be
extracted from the distribution and
relative motion of hot spots. Our
observations of the motion of some hot
spots during the past 120 m.y. indicate
that, if hot spots relate to mantle
convection, this convection is probably
highly complex and perhaps unlikely to
be successfully modelled in terms of
a small number of Rayleigh-Benard
cells.

We thank Dr. E. Irving for advice
and suggestions.

[1] Wilson, J. T., Can. J. Phys., 41,
863 (1963).

[2] Morgan, W. J., Bull. Am. Ass. Petrol.
Geol., 56, 203 (1972).

[3] Duncan, R. A., Petersen, N., and
Hargraves, R. B., Nature, 239,
82 (1972).

[4] Clague, D. A., and Jarrard, R. D.,
Geol. Soc. Am. Bull., 84, 1135
(1973).

[5] McElhinny, M. W., Nature, 241, 523
(1973).

[6] Molnar, P., and Atwater, T., EOS-
Trans. Am. geophys. Un., 54,
240 (1973).

[7] Larson, R. L., and Pitman, W. C.,
Geol. Soc. Am. Bull., 83, 3645
(1972).

[8] Pitman, W. C., and Talwani, M.,
Geol. Soc. Am. Bull., 83, 619
(1972).

[9] Coney, P. J., Nature, 233, 462 (1971).

[10] Ladd, J. W., Dickson, G. O., and
Pitman, W. C. The South
Atlantic (edit. by Nairn, A. E.
M., and Stehli, F. G.) (Plenum,
New York, in the press).

[11] Mascle, J., and Phillips, J. D.
Nature, 240, 80 (1972).

[12] Joides, Deep Sea Drilling Project,
Leg 22, Geotimes, 17(6), 15
(1972).

[13] Joides, Deep Sea Drilling Project,
Leg 26, Geotimes, 18(3), 16
(1973).

[14] Burke, K., and Wilson, J. T., Nature, 239, 387 (1972).

[15] McKenzie, D., and Sclater, J. G., Geophys. J. R. astr. Soc., 24, 437 (1971).

[16] Hicken, A., Irving, E., Law, L. K., and Hastie, J., Publ. Earth Phys. Branch, Energy Mines Resources, Ottawa, 45, 1 (1972).

[17] LePichon, X., and Fox, P. J., J. geophys. Res., 76, 6294 (1971).

[18] Bullard, E. C., Everett, J. E., and Smith, A. G., Phil. Trans. R. Soc., A258, 41 (1965).

[19] Piper, J. D. A., and Richardson, A., Geophys. Jl R. astr. Soc., 29, 147 (1972).

[20] Phillips, J. D., and Forsyth, D., Geol. Soc. Am. Bull., 83, 1579 (1972).

[21] Clague, D. A., and Jarrard, R. D., EOS-Trans. Am. geophys. Un., 54, 238 (1973).

[22] Deffeyes, K., EOS-Trans. Am. geophys. Un., 54, 238 (1973).

[23] Kidd, W. S. F., Burke, K., and Wilson, J. T., EOS-Trans. Am. geophys. Un., 54, 238 (1973).

[24] Deffeyes, K. S., Nature, 240, 539 (1972).

[25] Weissel, J. K., and Hayes, D. E., Nature, 231, 578 (1971).

[26] Holden, J. C., and Dietz, R. S., Nature, 235, 266 (1972).

[27] Hey, R. N., Johnson, G. L., and Lowrie, A., EOS-Trans. Am. geophys. Un., 54, 244 (1973).

[28] Ward, P. L., Geol. Soc. Am. Bull., 82, 2991 (1971).

[29] Menard, H. W., EOS-Trans. Am. geophys. Un., 54, 239 (1973).

# And now the rolling mantle

## P.J.S. (1973)

It may never be easy to come to terms with a major new hypothesis, idea or way of looking at things during its early years.  For one thing, there is the purely practical point that it is likely to be difficult to see precisely what is going on.  The central idea itself may be only vaguely formulated or imperfectly expressed; its characteristics and implications may be unclear or in dispute; the data required to confirm or refute speculation may be incomplete or even non-existent; and what data are available may be interpreted in different ways by different people to give conflicting conclusions.  Ultimately, the idea will evolve to the point where it is either rejected or accepted as the conventional wisdom; but in the meantime the situation is likely to be confusing to participants and onlookers alike.

Such is the case with the concept of thermal plumes in the mantle; and one can understand Tozer's exasperation (Nature, 244, 398; 1973) at the

apparent use of the phrase 'thermal plumes' to indicate little more than 'states of mind'.  Do plumes exist or not?  If they exist, how do they differ from conventional views of rising magma?  Are the supposed plumes fixed to the mantle or are they in motion?  And if they move, do they do so randomly (with or without reference to the processes embodied in the new global tectonics), in small or large groups, or all together?  Does material from rising plumes replenish the asthenosphere?  How might plumes be used to infer something about, say, the possibility of polar wandering?  The list of questions is endless; and none has yet been answered satisfactorily.  For the time being one can do little more than fall back on the old clichés that it is early days yet and time will tell.

The state of uncertainty is well illustrated by some recent attempts to come to terms with the question of whether the supposed mantle plumes (or hot spots) move.  Last year, Morgan (Mem. Geol. Soc. Am., 132, 7; 1972) concluded that relative motion between plumes in the Pacific could have been as much as 1 cm yr$^{-1}$, and Duncan et al. (Nature, 239, 82; 1972), assuming plumes to be fixed with respect to the Earth's rotational axis, showed that European plate motion predicted by plume traces is inconsistent with palaeomagnetic data for the past 50 million years, and invoked polar wandering to explain the discrepancy. Minster et al. (Eos, 54, 238; 1973) have used observations of relative velocities between plates at plate margins to construct a numerical model of instantaneous plate tectonics which is consistent with the view that plumes (with the exception of Iceland) are fixed in the mantle. Molnar and Atwater (Eos, 54, 240; 1973), on the other hand, have rotated plates back to their 13 m.y. and 38 m.y. positions to show that some hot spots (especially Iceland and St Paul/Amsterdam) probably move with respect to each other. Clague and Jarrard (Eos, 54, 238; 1973) construct a rotational model for the Pacific plate based on the Hawaiian hot spot and, on the assumption that Pacific hot spots are fixed with respect to one another, successfully predicted the ages of other Pacific seamount and island chains (with the exception of two ages in the Austral chain).

A few weeks ago, however, Burke et al. (Nature, 245, 133; 1973) showed that all hot spots cannot maintain a stable configuration, although the members of small groups of hot spots may well remain fixed to one another. It may also be noted in passing that the same authors (Kidd et al., Eos, 54, 238; 1973) have concluded that all plumes are not permanent anyway; during the past 100 m.y., new plumes have appeared and others have disappeared, and there are some which exist but fail to make any volcanic mark on the surface surface even when beneath stationary plates.

Into this confused situation, Hargraves and Duncan have now introduced (on page 361 of this issue of Nature) what may turn out to be a reconciling factor--the possibility of mantle motion independent of the lithosphere (mantle roll)--largely in response to a dilemma posed by their earlier work (Duncan et al.) and that by McElhinny (Nature, 241, 523; (1973). The problem was that soon after Duncan and his colleagues had invoked polar wandering to explain the discrepancy between European plume traces and palaeomagnetic data, McElhinny used purely palaeomagnetic results from all lithospheric plates to show that the lithosphere has not moved as a unit with respect to the rotational axis for the past 50 m.y. In other words, if the European discrepancy is not to be attributed to polar wandering in its guise as movement of the whole lithosphere, what must be involved is movement of just the mantle (rather than just the lithosphere, or the mantle and lithosphere together) with respect to the rotational axis, or movement of the plumes with respect to the mantle.

By extending their analysis to more plume traces on other plates, Hargraves and Duncan now show that the discrepancy between palaeomagnetic and plume trace data is a feature of several plates and thus, ruling out lithospheric rotation as the cause, that the plumes must move in relation to the Earth's spin axis. Moreover, the discrepancies are different for different plume traces. The fact that the discrepancies are not equal suggests, on the face of it, that plumes move with respect to each other in the mantle, in which case there is no need to invoke any movement of the mantle with respect to the rotational axis. But Hargraves and Duncan have chosen to look at the problem in a different way and ask the following question. Is there any systematic motion of the mantle as a whole with respect to the rotational axis which, carrying the plumes with it, can account for the various different trace-palaeomagnetic discrepancies without destroying the stable configuration of the plumes?

The answer to this question seems to be positive, at least in the sense of least squares, for Hargraves and Duncan have been able to identify a best-fitting axis for mantle rotation which reduces the apparent motion between plumes to a minimum. In other words, accepting that the lithosphere as a whole remains fixed with respect to the rotational axis, they propose that the mantle may nevertheless rotate or roll about an axis which differs from the Earth's spin axis. As it does so, it carries the plumes with it in a near-stable configuration; relative motion between plumes is minimal, but the possibility of some interplume motion cannot be completely excluded. Because the mantle rotates, the existence of plume trace-palaeomagnetic discrepancies is expected; and because it rotates about an axis independent of the Earth's spin axis, the magnitude of any discrepancy will depend on the distance between the relevant plume and the pole of the mantle roll axis.

It is this ability to view the apparently disparate behaviour of mantle plumes as the result of a single process which makes the Hargraves-Duncan model of particular interest. The behaviour of the eight plumes used to derive the model must clearly be consistent with it. The obvious question to ask now is whether the hundred or more other plumes can be fitted into the same pattern.

*

# Does the mantle roll?

R.B. Hargraves and R.A. Duncan (1973)

With wide acceptance of continental drift as it is now conceived (the independent motion of individual lithospheric plates), consideration of the possibility of true polar wandering seems to have languished. For many Earth scientists polar wandering is associated with palaeomagnetism and is now assumed to be totally unnecessary in order to explain palaeomagnetic data. Divergence of the palaeomagnetic pole from the geographic pole (which is the present situation) is presumed adequate to explain any anomalies.

True polar wandering, sensu stricto, refers to a change in the position of the Earth as a whole or as outer lithosphere-mantle shell with respect to its rotation axis, which is presumed to be fixed in space. Such a change could conceivably result from redistributions of mass within the Earth (either by upward convection or downward subduction) whereby its moment of inertia axes shift[1]. Should such a change occur, it would have palaeoclimatic impact and would represent a component of plate motion with respect to the geographic pole, which is common to all plates. Such true polar wandering has no a priori connexion with palaeomagnetism unless, as is conventionally assumed, well determined palaeomagnetic poles coincide with the geographic pole.

Whether or not polar wandering has occurred is clearly a question of considerable geophysical significance. Pending the recognition of a frame of reference other than the rotation axis, and identification of a sufficiently precise measure of its position other than by palaeomagnetic studies, the assumption of a coaxial dipole model for the Earth's field must be made if evidence of true polar wandering is sought.

TESTING

McElhinny[2], following McKenzie[3], has recently presented a method for testing for true polar wandering, using

Reprinted from Nature 245, 361-363 by permission of the senior author and publisher. Copyright 1973 Macmillan Journals Ltd.

palaeomagnetic data only but requiring that such data be available for every lithospheric plate. The movement vectors (the apparent polar wander curves) for each plate, weighted by the area of the plate, are summed. By this means the relative motions between plates are cancelled and the resultant vector signifies the magnitude and direction of true polar wandering (that is, a motion common to all plates). As McElhinny has shown, the resultant vector for the past 50 m.y. has a magnitude $2.1^{\circ}$, which he concluded was not significantly different from zero.

McElhinny's method indicates no net movement of the rotation axis with respect to the lithosphere in the past 50 m.y. Duncan et al.[4], however, have shown that, under the assumption that plumes are fixed with respect to the Earth's spin axis, European plate motion predicted by plume traces is not in accord with palaeomagnetic data for the past 50 m.y. They invoked polar wandering to account for the discrep-

ancy. If this discrepancy is due to true polar wandering, then evidence of it should be found in the comparison of plume traces and palaeomagnetic data for every plate. We have accordingly explored this possibility. Table 1 presents the result of comparisons of eight plume traces with 50 m.y. polar wander curves for appropriate plates. The 50 m.y. palaeomagnetic poles are essentially those of McElhinny[2] and the plume trace data are described in the appendix.

The 50 m.y. latitude of the plume, as determined by the great circle distance between the 50 m.y. site of volcanism along a given plume trace and the appropriate 50 m.y. palaeomagnetic pole, should be identical to the latitude of the present site of plume activity if the plume has remained fixed with respect to the Earth's spin axis (=the magnetic pole). A discrepancy implies that the plume and/or the magnetic pole has moved. Assuming the validity of McElhinny's result--that

Table 1   Palaeomagnetic and Plume Data

|  |  | Present site Latitude | Longitude | 50 m.y. site Latitude | Longitude | 50 m.y. magnetic pole Latitude | Longitude | $a_{95}$ | 50 m.y. latitude | Discrepancy |
|---|---|---|---|---|---|---|---|---|---|---|
| (1) | Iceland | 64.0°N | 342.0°E | 6.20°N | 353.0°E | 77.0°N | 161.0°E* | 2.0 | 47.2°N | 14.8°N |
| (2) | Yellowstone | 45.0°N | 250.0°E | 38.0°N | 238.0°E | 85.0°N | 197.0°E | 14.0 | 41.7°N | 3.3°N |
| (3) | Tibesti | 20.0°N | 19.0°E | 32.0°N | 13.0°E | 81.0°N | 168.0°E | 4.5 | 23.8°N | 3.8°S |
| (4) | Hawaii | 19.0°N | 204.0°E | 37.0°N | 171.0°E | 71.0°N | 354.0°E | 13.4 | 18.0°N | 1.0°N |
| (5) | Balleny | 67.0°S | 165.0°E | 42.0°S | 145.0°E | 70.0°N | 306.0°E | 5.3 | 60.4°S | 6.7°S |
| (6) | Réunion | 21.0°S | 55.0°E | 5.0°S | 70.0°E | 70.0°N | 306.0°E | 5.3 | 15.8°S | 5.2°S |
| (7) | Tristan (E) | 37.0°S | 347.0°E | 37.0°S | 363.0°E | 81.0°N | 168.0°E | 4.5 | 45.7°S | 8.7°N |
| (8) | Tristan (W) | 37.0°S | 347.0°E | 34.0°S | 333.0°E | 79.0°N | 122.0°E† | 14.0 | 43.2°S | 6.2°N |

*This is the pole obtained directly from the Faeroes lavas[23]; McElhinny[2] used 75°N, instead of 77°N, in his analysis

†McElhinny[2] does not give a separate 50 m.y. magnetic pole for South America. This Tertiary pole is from Creer[24].

there has been negligible polar wandering during the past 50 m.y.-- then discrepancies can be explained only by movement of the plumes with respect to the Earth's spin axis.

The discrepancies calculated in Table 1 are of variable magnitude and some, at least, could be due to error in the location of the plume sites and appropriate palaeomagnetic poles 50 m.y. ago. The palaeomagnetic data are indeed few and of variable quality; they are, however, the same data used by McElhinny and constitute the best that are presently available.

Precise location of even present day "plumes" (assuming they exist) may be difficult; location of 50 m.y. sites poses additional problems. Uncertainty about present day location derives from (a) the long duration of igneous activity at individual sites (for example, 18 m.y. volcanics in the Tristan da Cunha group[5]) and (b) the wide extent of contemporaneous igneous activity associated with individual plumes (for example, contemporaneous, although petrographically distinct, activity along 300 km of the Hawaiian chain[6]).

These uncertainties are compounded by inadequate exposures and age control in identifying 50 m.y. plume sites. Where interpolation or extrapolation was necessary, a uniform spreading rate has been assumed (see appendix).

Although we cannot make a quantitative estimate, we feel that for some individual plumes such errors and uncertainties may perhaps cause discrepancies of as much as $10^o$, and yet we know of no reason why such an error should not be random from one plume to another. Nevertheless, some discrepancies are judged to be too great to arise from such errors, and for the purposes of our analysis all have been treated at face value. Two possible extreme interpretations are: (1) that the plume trace/palaeomagnetic discrepancies are due to random movement of plumes in time; the population of plumes should not, then, maintain a stable configuration; (2) that plumes are fixed with respect to one another but move in unison as the mantle rolls about some axis independent of the Earth's spin axis. Discrepancies between the plume traces and the palaeomagnetic polar wander paths would be expected, their magnitude varying as the distance of given plumes from such a "mantle roll" axis. A stable configuration of plumes, however, is preserved.

Morgan[7] concluded that interplume motion in the Pacific, at least, could have amounted to as much as 1 cm $yr^{-1}$. Minster et al.[8] conclude that their instantaneous plate motion model is compatible with the hypothesis that, with the conspicuous exception of

Iceland, all other plumes are stable with respect to one another.

## DISCREPANCIES

To test whether the discrepancies in Table 1 might be due primarily to random interplume motion, or to "mantle roll", the following method was used. If there is a "mantle roll" axis about which the mantle has rotated during the past 50 m.y., it should be identified by a least-squares fit of the data from Table 1. That is, the latitudes of the present plume sites can be changed, by rotation about an axis, to those 50 m.y. latitudes inferred from the palaeomagnetic evidence. Using an iterative program, one axis from a grid of mantle roll axes is chosen and the best fitting angle of rotation is found by the method of nonlinear least squares[9]. The success of the fit is measured by the sum of squares of residuals, which should be minimized at the best fitting axis. The sum of squares of residuals (SSR) $= \Sigma\, \theta_i - f_i$ $(\theta_p, \varphi_p, \underline{a})^2$ where $\theta_i$ are the 50 m.y. latitudes and

$$f_i(\theta_p, \varphi_p, \underline{a}) = \hat{\theta}_i$$ are the latitudes of the rotated present plume sites. $\theta_p$ and $\varphi_p$ are the latitude and longitude of the chosen axis and $\underline{a}$ is the rotation angle. In this calculation

$\theta_i$ and $\hat{\theta}_i$ are expressed in rad. This process can be continued over a grid of trial axes covering a hemisphere and the resulting array contoured to determine the best fitting mantle roll axis and angle of rotation.

The result of such a test using the data in Table 1 is illustrated in Fig. 1. The specific best fit axis is at $25°$N, $30°$E, with an angle of rotation of $12.1°$, which decreases the SSR from 13.1 to 3.1. It is more realistic to recognize a sigmoidal "region" of good fitting axes between $20°$ and $40°$E. This meridional region of good fitting axes about the best fitting axis is, however, well defined and distinct from the poorer fitting axes over the rest of the hemisphere (Fig. 1). The data seem to be insufficient to impose a clear latitudinal constraint (in terms of residuals) on the location of the best fit axis; but the angular rotations called for increase prodigiously for

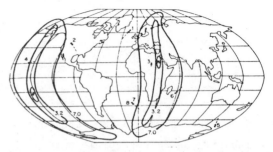

**Fig. 1** Contours on surface of "sum of squares of residuals" (SSR, see text) to show location of best fit rotation axis, and the pattern of increase in SSR away from it. Clockwise rotation of the mantle of $12.1°$ about an axis at 25°N, 30°E decreases the SSR from 13.1 to 3.1. Dashed lines with arrows indicate the motion of plumes (as numbered in Table 1) predicted by this mantle roll.

axes at the extreme latitudes.

Iceland accounts for the largest latitude discrepancy in Table 1; it is also the conspicuous exception to the stable plume model of Minster et al.[8]. To ascertain whether this discrepancy is the primary cause of the mantle roll calculated, we repeated the test omitting the Iceland data altogether. The longitudinally constrained ($20^\circ$–$40^\circ$E) sigmoidal region of best-fit mantle roll axes is again obtained, with a comparable improvement by a factor of 4 in the SSR (from 6.5 to 1.5). The specific best fit axis is now located at $45^\circ$N, $30^\circ$E, with a clockwise rotation of $12^\circ$.

In general, these results mean that $12^\circ$ clockwise rotation of the mantle about an axis emerging in the northern Sudan greatly improves the fit of 50 m.y. plume-site latitudes inferred from palaeomagnetic evidence to the latitudes of their present sites (see Table 2). A still better fit may be achieved by allowing interplume movement but the close fit under the requirement of whole-mantle roll suggests a high degree of common motion. By the same argument the result, we feel, is inconsistent with the discrepancies of Table 1 being due entirely to random error in palaeopole or plume location.

This leads to the curious picture of the lithosphere essentially fixed to the Earth's spin axis with the mantle beneath rolling about some independent axis. Roll of the mantle, uncoupled from the lithosphere, is compatible both with McElhinny's conclusion that polar wander with respect to the lithosphere has not occurred within the past 50 m.y., and with the variable

Table 2    Latitude Discrepancies Observed and Predicted by $12.1^\circ$ Clockwise Mantle Roll about an Axis at $25^\circ$N, $30^\circ$E

| Plume | Latitude discrepancy Observed | Predicted |
|-------|-------------------|-----------|
| Iceland | $14.8^\circ$N | $8.6^\circ$N |
| Yellowstone | $3.3^\circ$N | $7.9^\circ$N |
| Tibesti | $3.8^\circ$S | $2.0^\circ$N |
| Hawaii | $1.0^\circ$N | $0.3^\circ$S |
| Balleny | $6.7^\circ$S | $8.4^\circ$S |
| Réunion | $5.2^\circ$S | $5.3^\circ$S |
| Tristan (E) | $8.7^\circ$N | $6.6^\circ$N |
| Tristan (W) | $6.2^\circ$N | $6.6^\circ$N |

The N or S notation signifies that with respect to the Earth's rotation axis, the present plume site is (or predicted to be, with mantle roll) north or south of its position 50 m.y. ago.

discrepancies between plume traces and polar wander paths. Grommé and Vine[10] conclude that the Hawaiian plume has not changed its position significantly with respect to the magnetic pole since the formation of Midway Island (18 m.y. ago), whereas Duncan et al.[4] proposed that the Iceland and Eifel plumes have moved considerably. This conflict is possibly resolved by the fact that Hawaii is close to the proposed mantle roll axis (the antipode at $25^{\circ}$S, $150^{\circ}$W) and would not be expected to move a great distance (in fact the plume motion predicted is primarily westwards, rather than translatitudinal), whereas Iceland is more than $60^{\circ}$ away and should then show substantial movement. The possible causes or consequences of this hypothetical process of mantle roll will not be considered here. We feel, however, that the evidence for it is suggestive and encourages its consideration and the acquisition of data in sufficient quantity and quality for a more decisive test.

This research was in part supported by the National Science Foundation. We thank C. Powell, W. J. Morgan, T. H. Jordan, K. S. Deffeyes and M. W. McElhinny for helpful discussions.

APPENDIX

The following are the geochronological data for plume traces.

(1)    Iceland Plume: the Wyville-Thompson Ridge connects the present site of plume activity to the Faeroe Islands, ages 50 m.y.[11].

(2)    Yellowstone Plume: the trace to the Snake River plains, 25 m.y. ago, is conspicuous (R. L. Armstrong, W. P. Leeman and H. E. Malde, to be published) but its earlier manifestations (nearer the continental edge) are obscure. Extrapolation along the Yellowstone-Snake River trace at a velocity of 2 cm $yr^{-1}$ would locate its 50 m.y. position at San Francisco; this palaeoposition has been used.

(3)    Tibesti Plume: a conspicuous line of igneous centres curves generally north-west from Tibesti (very recent activity) toward Tripoli[12]. Although volcanism along this line has recurred up to the Pliocene (at least), the oldest activity near Tripoli has been dated at 52 m.y. (ref. 13). This has been taken as the palaeoplume site.

(4)    Hawaiian Plume: present activity at Hawaii is connected by a trend of volcanism to the Emperior Seamounts[6,7,14-16]. A 50 m.y. position has been interpolated from known ages and inferred plate

velocity[17].

(5) Balleny Plume: the presently active Balleny Islands sit at the south end of the South Tasman Rise, a conspicuous aseismic ridge connecting the Australian-Antarctic spreading ridge to Tasmania. Although the well known Tasmanian dolerites are Jurassic in age, seafloor anomalies indicate initiation of the ridge adjacent to Tasmania at about 50 m.y.[18]. The South Tasman Rise also parallels the Tasmantid Guyots[19] and eastern Australian volcanism.

(6) Réunion Plume: the Réunion-Laccadive-Maldive line[7] is interrupted by a recent spreading episode. The south end of the Chagos Plateau is proposed as the 50 m.y. site of volcanism based on magnetic anomaly evidence[20,21].

(7) Tristan (East) Plume: volcanism at Tristan da Cunha near the mid-Atlantic Ridge is linked by the Walvis Ridge to South-west Africa. Again, the intersection of anomaly 20 (~50 m.y.) with this aseismic ridge gives the 50 m.y. site of plume activity[21,22].

(8) Tristan (West) Plume: towards South America the plume appears as the Rio Grande Rise. Anomaly 20 marks the 50 m.y. site[21,22].

[1] Goldreich, P., and Toomre, A., J. geophys. Res., 74, 2555 (1969).

[2] McElhinny, M. W., Nature, 241, 523 (1973).

[3] McKenzie, D. P., in Nature of the Solid Earth (edit. by Robertson, E. C.) (McGraw Hill, New York, 1972).

[4] Duncan, R. A., Petersen, N., and Hargraves, R. B., Nature, 239, 82 (1972).

[5] Baker, P. E., Gass, I. G., Harris, P. G., and LeMaitre, R. W., Phil. Trans. R. Soc., A256, 439 (1964).

[6] Jackson, E. D., Silver, E. A., and Dalrymple, G. B., Bull. geol. Soc. Am., 83, 601 (1972).

[7] Morgan, W. J., Geol. Soc. Am. Mem., 132, 7 (1972).

[8] Minster, B., Jordan, T., Molnar, P., and Haines, E., Abs. Am. geophys. Un., 54, 238 (1972).

[9] Draper, N. R., and Smith, H., in Applied Regression Analysis (Wiley, New York, 1968).

[10] Grommé, S., and Vine, F. J., Earth planet. Sci. Lett., 17, 159 (1973).

[11] Tarling, D. H., and Gale, N. H., Nature, 218, 1043 (1968).

[12] Carte Géologique de L'Afriquc, ASGA, UNESCO (1963).

[13] Piccoli, G., Bull. Soc. geol. It., 89, 449 (1970).

[14] Wilson, J. T., Can. J. Phys., 41, 863 (1963).

[15] McDougall, I., Geol. Soc. Am. Bull.,
75, 107 (1964).

[16] McDougall, I., Nature phys. Sci.,
231, 141 (1971).

[17] Clague, D. A., and Dalrymple, G. B.,
Earth planet. Sci. Lett., 17,
411 (1973).

[18] LePichon, X., and Heirtzler, J. R.
J. geophys. Res., 73, 201 (1968).

[19] Vogt, P. R., and Conolly, J. R.,
Bull. geol. Soc. Am., 82,
2577 (1971).

[20] McKenzie, D. P., and Sclater, J. G.,
Geophys. Jl R. astr. Soc., 25,
437 (1971).

[21] Heirtzler, J. R., Dickson, G. O.,
Herron, E. M., Pitman, W. C., and
LePichon, X., J. geophys. Res.,
73, 2119 (1968).

[22] Dickson, G. O., Pitman, W. C., and
Heirtzler, J. R., J. geophys.
Res., 73, 2087 (1968).

[23] Tarling, D. H., in Paleogeophysics
(edit. by Runcorn, S. K.), 193
(Academic Press, New York, 1970).

[24] Creer, K. M., Phil. Trans. R. Soc.,
A267, 457 (1970).

# Hot spots must move

## D.D. (1973)

If linear chains of volcanoes in the ocean basins originate as outflows from localised regions near the Earth's surface--hot spots--over which plates have moved during geological times, then these hot spots cannot have formed a fixed frame of reference during the past 40 million years. This is the conclusion of Molnar and Atwater in this issue of Nature (page 288). This result has major implications for the hot spot hypothesis as there has up to now been a hope among geophysicists that hot spots might have been semi-permanent features and thus good markers of absolute motion of plates over the mantle. The implications, however, go beyond the geometry. If hot spots are not stationary relative to each other, there are serious questions to be asked about their physical character and how it permits them to migrate.

Since the idea of very active volcanic sources leaving their trace on plates was revived by Morgan three

years ago, many groups have attempted to use the concept to learn more about past motions. The results have been strangely inconclusive. The one undeniable thing from the geological and geophysical evidence is that the long traces ending in present day volcanoes do indicate something other than pure chance. Many geophysicists would further agree that there is something global to the traces--they are not ten or so entirely unrelated phenomena. But beyond there, opinions differ. Are hot spots simply localised lava sources, or do they have any vertical extension? Are the traces markers of pre-existing fractures or can hot spots punch volcanoes in unfractured plate? Present day geometry of the traces, palaeomagnetic data and a reasonable understanding of plate positions for the past 70 million years are all one has to go by at the moment.

Hargraves and Duncan (Nature, 245, 361; 1973) analysed palaeomagnetic data and concluded that they were inconsistent with hot spots being stationary relative to the Earth's magnetic field.

Since there are reasonably strong grounds now for assuming that the pole does not wander, it was necessary to assume that the hot spots do. Hargraves and Duncan were able, however, to find a rotation of the mantle as a whole which reproduced their results tolerably well without the need for the migration of individual hot spots. There is thus the possibility that not only do plates move, but the material that they slide over is also in motion.

Molnar and Atwater adopt a geometrical approach which does not use palaeomagnetic evidence directly. They use the now generally agreed upon poles of rotation of plates to run the geological clock back 38 million years. They place the 38 million-year-old rocks on the Hawaiian chain over the present position of the Hawaiian hot spot and see what overlay the other hot spots at that time. None of the linear features lies on top of their presumed parental hot spots when this analysis is carried out. This implies that there must have been relative motion among hot spots in the past.

It is now necessary for geophysicists to consider these two papers very carefully. The authors do not come to identical conclusions, but this is perhaps not surprising. What they do point to fairly clearly is some sort of mobility below the lithosphere--a mobility of up to a couple of centimeters a year. How deep this mobility extends there is as yet no idea, and though it is still possible that the Morgan hypothesis that hot spots are fed by deep mantle plumes is tenable, there are now very serious doubts that such a delightfully simple model can apply. Yet more solid ground of nature is eroded.

# Relative motion
# of hot spots in the mantle

**P. Molnar and T. Atwater (1973)**

Carey[1] suggested that many of the
aseismic ridges observed on the ocean
floor are manifestations of hot spots
in the mantle and assumed, as Dietz and
Holden[2] did later, that these aseismic
ridges could be used in reconstructions
of continents earlier in the Cainozoic.
Wilson[3,4] further suggested that they
were generated as the seafloor moved
over localised zones of upwelling that
are fixed with respect to one another
in the mantle beneath the asthenosphere. Morgan[5,6] attributed the hot
spots to convective plumes rising from
the lower mantle. We explore here the
assumption that the hot spots are fixed
with respect to one another and therefore define a reference frame with
respect to which the plates move.

RECONSTRUCTION OF PLATES

One test is to examine the
relative positions of the major plates
at some time in the past and locate one

of them properly over its hot spots.
If all the hot spots are fixed with
respect to one another, then they
should all underlie the linear volcanic
chains that they generated. Minster
et al.[7] used inversion techniques to
obtain a refined version of recent
relative plate motions, and made a
preliminary test by comparing the
trends of linear volcanic chains with
those predicted. They found no
observable relative motion of the hot
spots during the past few million
years. Because changes in spreading
rates occurred between about 5 and 20
m.y. ago at nearly every major spreading centre in the oceans, their
instantaneous velocities are probably
not useful for periods earlier than 10
m.y. ago.

The relative positions of pairs of
plates at different times earlier in
the Cainozoic have been reconstructed
in the North Atlantic Ocean[8], in
various parts of the Indian Ocean[9-11],
and in the South Pacific[12]. Spreading

in the South Atlantic has been
relatively constant, so that
LePichon's[13] original pole and angular
velocity are probably sufficient for
our purposes.  These studies allow one
to interrelate the motions of all the
major plates.

    We reconstructed the plates to
their positions with respect to one
another at different times in the
Cainozoic.  Then, assuming the
Hawaiian-Emperor chain to be the
expression of the motion of the
Pacific Plate over the Hawaiian hot
spot, we rotated the Pacific Plate
along with the reconstruction of the
other plates to their positions with
respect to the Hawaiian hot spot.  The
reasons for constraining the
reconstructions to agree with the
inferred relative positions of the
Pacific over the Hawaiian hot spot
are:  (1) using other island chains in
the Pacific Ocean the pole of relative
motion between the Pacific Plate and
the hot spots can be well determined;
(2) the dating of islands along the
chain is more complete than along any
other; and (3) there is more complete
agreement that the Hawaiian Islands
were generated by a hot spot than for
most of the other suggested ones.  If
the other hot spots did not move with
respect to the Hawaiian spot, then for
each of these reconstructions the
linear volcanic chains should overlie

the hot spot that generated them.

    We considered the hot spots and
linear volcanic chains listed in Table
1.  These were selected because they
are among the more popular ones and
they have a balanced geographical
distribution.  We recognise that there
is no unanimity as to what constitutes
a well defined hot spot.

    We made reconstructions for 21 and
38 m.y. ago.  For earlier times, the
data are not yet adequate to reconstruct the major plates reliably.  For 21
m.y. we used reconstructions to anomaly
6 in the Pacific, South-east Indian and
South Atlantic oceans and interpolated
between earlier and later reconstructions in the North Atlantic and North-
east Indian Ocean.  Assuming the age of
Midway Island to be 18 m.y.[14], the
position of the inferred Hawaiian hot
spot with respect to the Hawaiian chain
21 m.y. ago is relatively well defined
(Fig. 1).  Assuming that the hot spots
are fixed with respect to each other,
the reconstructed positions of the
plates at 21 m.y. show that, except for
the Ninety East Ridge, the other hot
spots do not underlie the aseismic
ridges generated by them (Figs. 2 to
5).  If the 18 m.y. age of Midway
approximately dates the passage of the
Hawaiian hot spot, that spot cannot be
fixed with respect to the others[7].

    In view of the possibility that
the Midway age does not date the forma-

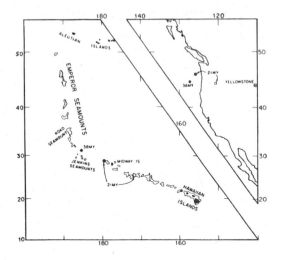

Fig. 1. Left, Hawaiian-Emperor chain and positions of the Hawaiian hot spot at the present and at 21 and 38 m.y. ago. Position at 38 m.y. ago is based on the age of Koko seamount [15]. Two estimates for 21 m.y. ago assume: ●, Jackson *et al.*[14] date of Midway or, ○, constant rate of Pacific motion over hot spot. ⊗, Present position of hot spot. Right, calculated position of Yellowstone hot spot with respect to North America at 21 and 38 m.y. ago.

above (open circles in Figs. 2 to 5), the calculated positions of the other hot spots still do not lie under the traces presumed to be generated by them. Because of the various inter- polations involved in the 21 m.y. reconstruction, this conclusion is, however, not inescapable.

The reconstructions for 38 m.y. ago provide a more definitive test. Dating of Koko seamount[15] constrains the Hawaiian-Emperor bend to be about 42 m.y. old[16]. Anomaly 13, a prominent magnetic anomaly used in the recon- structions of most oceans, was formed approximately 38 m.y. ago. Thus the uncertainties both in the relative positions of the major plates and in the assumed position of the Pacific Plate with respect to the Hawaiian hot spot are likely to be smaller than for other times. The relative positions of the hot spots and their associated linear island chains at 38 m.y. are

tion of the island, we decided to explore the possibility that the rate of motion has been constant for 42 m.y. during the entire formation of the Hawaiian chain (Fig. 1). Although the disagreement is less than for the case

Table 1   Hot Spots and Volcanic Chains

| Hot spot | Linear volcanic chain | Fig. |
|---|---|---|
| Hawaii | Hawaiian Islands | 1 |
| Gough Island and Tristan da Cunha | Walvis Ridge and Rio Grande Rise | 2 |
| Réunion | Mauritius Plateau and Chagos-Laccadive Ridge | 3 |
| Kerguelen | Ninety East Ridge | 4 |
| St Paul's and Amsterdam Island | Ninety East Ridge | 4 |
| Iceland | Greenland-Iceland and Iceland-Faeroes Ridges | 5 |
| Yellowstone | Snake River Basalts | 1 |

shown in Figs. 2 to 5. Except for the
possibility that a hot spot beneath
Kerguelen generated the Ninety East
Ridge, the calculated positions of the
spots are not coincident with the
chains presumed to be generated by
them. If the hot spots formed these
chains, they have moved 300 to 800 km
with respect to the Hawaiian hot spot,
that is, at rates of 0.8 to 2 cm yr$^{-1}$.
These rates are comparable with those
for relative motions of most of the
plates, although they are significantly
slower than those in the Pacific.

UNCERTAINTIES AND ASSUMPTIONS

The uncertainties in the reconstructions result from two sources:

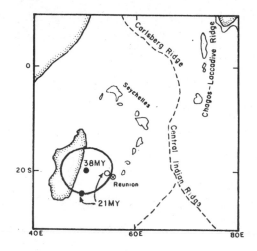

Fig. 3   Relative position of the hot spot under Reunion with respect to the
African Plate. Symbols same as Figs 1 and 2. The Mauritius Plateau and the
Chagos-Laccadive Ridge were continuous 38 m.y. ago[9],[10]. If this hot spot
generated the Mauritius Plateau, and the Chagos-Laccadive Ridge on the
Indian Plate, and fixed with respect to the Hawaiian hot spot, it should lie
beneath the Mauritius Plateau north-east of Reunion. It does not.

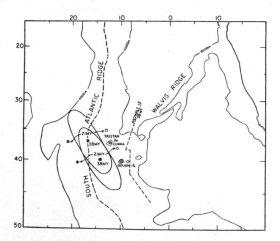

Fig. 2   Relative positions of hot spots under Gough Island and Tristan da
Cunha with respect to the African Plate. Symbols as for Fig. 1. Confidence
oval surrounds position for 38 m.y. ago. If fixed to the Hawaiian hot spot, the
two proposed hot spots should underlie the Walvis Ridge beneath material
older than 21 m.y. and 38 m.y. The calculated positions lie beneath material
that had not yet been formed at these times. Therefore, at these times, they
lay beneath the South American Plate, south of the Rio Grande Risc. They
could not have generated these rises and also been fixed to the Hawaiian
hot spot.

errors in individual reconstructions
and validity of the assumptions. Both
are difficult to evaluate. One may
gain a crude, upper limit for the
reconstruction errors in the following
manner. The uncertainties in the
positions of the poles matter very
little except that they are coupled
to uncertainties in the angles of
rotation. The uncertainties in the
angles are not likely to be greater
than 1° (perhaps 2° for the Pacific hot
spot rotation). Thus at the equator
for each rotation, an error of about
100 km could result. For example, the
position of Gough Island with respect
to Africa requires four rotations
(North America → India → Antarctica →
Pacific → Hawaiian hot spot). The

maximum uncertainty would be about 400 km. As Gough Island does in fact lie near the equator of each of the four rotations, this estimate is a reasonable upper limit. The motion of the Iceland hot spot with respect to the North American Plate involves five rotations (North America → Africa → India → Antarctica → Pacific → Hawaiian hot spot). In three of these (the first one and the last two), the pole of rotation is near Iceland. Thus the error in relative position should be considerably less than 500 km.

Since the above error estimates tend to be excessive, we performed a more realistic calculation. We were not able to treat the uncertainties in a statistically rigorous manner, so we simply estimated the maximum uncertainty in each of the finite rotations and explored the range of possible combinations of them. This procedure allowed determination of ovals describing the largest possible errors in the relative positions of the hot spots with respect to the plates. These ovals are shown in Figs. 2 to 5 for the reconstructions for 38 m.y. ago. They show that the relative motion of any given hot spot with respect to the Hawaiian hot spot can be reduced by about a factor of two. Because the uncertainties are not independent, it is not, however, possible to reduce all of the motions this much.

Apart from assuming the Vine-Mathews hypothesis to be correct so that given anomalies on opposite sides of a spreading centre were formed at the same place, the crucial assumption is that the plates are rigid. There seems to be no reason to doubt this in the oceanic areas. We neglected deformation along the East African Rift, because it is much smaller than the uncertainties discussed above. The possibility of deformation in Antarctica is difficult to eliminate. Motion between East and West Antarctica has occurred since the Jurassic[17-19] and seems to have continued into the Cainozoic[12]. The reconstruction of the South-West Pacific Ocean at 38 m.y. does not require younger deformation[12], so we assume that Antarctica has behaved rigidly since then. In any event, if deformation has occurred since 38 m.y. ago in the same sense as earlier, it seems that the resulting relative positions of the hot spots and associated linear chains would differ even more than shown in Figs. 2 to 5.

Perhaps the strongest justification for the reconstructions used here comes from the estimated relative position of the Pacific and North American plates, for 38 and 29 m.y. (ref. 20). These reconstructions place the extinct Pacific-Farallon spreading centre a few hundred km west of North

America at 38 m.y. ago, and within 200
km of the coast 29 m.y. ago. These
calculated positions agree well with
those inferred from the magnetic
anomales in the Pacific Plate west of
California[21,22] and their differences
are far less than the maximum uncertainties cited above.

   We carried out one other check.
If there was some unknown deformation
within one of the plates or if the
reconstruction between one pair of
plates was grossly in error, and if the
hot spots were fixed, then an additional, arbitrary rotation exists that when

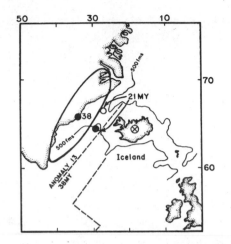

Fig. 5   Relative positions of a hot spot beneath Iceland with respect to the
North American Plate. Symbols as in Figs 1 and 2. If the ridges between
Greenland and Iceland and between Iceland and Europe were produced by
a hot spot beneath Iceland, it would have been centrally located with respect
to Greenland and Europe during the past 38 m.y. When fixed with respect to
Hawaii, it is found to have underlain the North American Plate during this
time, and therefore could not have generated the Iceland-Faeroes Ridge.

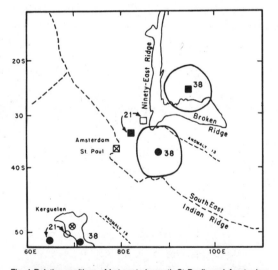

Fig. 4 Relative positions of hot spots beneath St Paul's and Amsterdam
Islands with respect to the Ninety East Ridge on the Indian Plate (squares);
and of Kerguelen hot spot with respect to the Kerguelen Plateau on the
Antarctic Plate (circles) and with respect to the Indian Plate for 38 m.y. ago
(closed circle with confidence oval). Symbols as in Figs 1 and 2. If either of
these hot spots had generated the Ninety East Ridge and were fixed with
respect to the Hawaiian hot spot, it should lie between the Ninety East Ridge
and Kerguelen 21 and 38 m.y. ago. If a hot spot exists beneath Kerguelen
and generated the Kerguelen Plateau, it should have lain beneath it during
the past 38 m.y. The data do not rule out the possibility that a hot spot
beneath Kerguelen, fixed with respect to Hawaii, could have generated both
of these features. No such feature beneath St Paul's and Amsterdam
Islands could have produced the Ninety East Ridge.

added into the sequence of rotations
would allow all of the aseismic ridges
to overlie their respective hot spots.
To carry out such a search rigorously
is expensive and tedious. But a manual
examination using a globe revealed no
such arbitrary rotation. For instance,
rotations that caused the Iceland hot
spot to lie beneath the Iceland-Faeroes
ridge invariably left the Walvis Ridge
far north of the hot spot inferred to
lie at present beneath Gough Island or
Tristan da Cunha.

IMPLICATIONS FOR PLATE MOTION

   If the hot spots are fixed with
respect to each other, what does this
imply for plate motions and the origin

of linear volcanic chains? First[7], it requires that Midway be considerably older than the date given by Jackson et al.[14]. Second, more than one of the pole positions and finite rotations used here must be grossly in error. Third, there must be a large internal deformation within Antarctica since 38 m.y. ago, in some sense different from that calculated to have occurred earlier in geological time[12]. We consider a simpler interpretation to be that the hot spots do not define a fixed reference frame, that the Hawaiian hot spot moves at velocities with respect to the others that are comparable to the velocities of relative plate motions, and that the others also move with respect to one another. Burke et al.[23] reached a similar conclusion concerning motions among the various Atlantic hot spots since the Cretaceous, and McElhinny[24] concluded that hot spots could not be fixed both with respect to one another and to the spin axis of the Earth. Although relative motion of the hot spots of the magnitude suggested here does not prove that plumes do not exist, it does refute one of the most appealing aspects of the idea: a fixed reference frame with which to describe the motions motions of the plates; and it removes a strong impetus for placing their origin in the lower mantle.

We thank J. N. Brune, J. H. Jordan, H. W. Menard, and J. B. Minster for interest and encouragement; R. L. Parker for help in drawing the figures; and the referee for suggestions. This research was supported by grants from the National Science Foundation.

[1] Carey, S. W., Continental Drift, A Symposium, Hobart, 177 (University of Tasmania, 1958).

[2] Dietz, R. S., and Holden, J. C., J. geophys. Res., 75, 4939 (1970).

[3] Wilson, J. Tuzo, Can. J. Phys., 41, 863 (1963).

[4] Wilson, J. Tuzo, Phil. Trans. R. Soc., 258, 145 (1965).

[5] Morgan, W. J., Nature, 230, 42 (1971).

[6] Morgan, W. J., Bull. Am. Ass. Petrol. Geol., 56, 203 (1972).

[7] Minster, J. B., Jordan, T. H., Molnar, P., and Haines, E., Geophys. J. R. astr. Soc. (in the press).

[8] Pitman, III, W. C., and Talwani, M., Geol. Soc. Am. Bull., 83, 619 (1972).

[9] Fisher, R. L., Sclater, J. G., and McKenzie, D., Geol. Soc. Am. Bull., 82, 553 (1971).

[10] McKenzie, D., and Sclater, J. G., Geophys. J. R. astr. Soc., 25, 437 (1971).

[11] Weissel, J. K., and Hayes, D. E., Antarctic Oceanology II: The Australian-New Zealand Sector (edit. by Hayes, D. E.), Antarctic Research Series, 19, 165 (American Geophysical Union, 1972).

[12] Molnar, P., Atwater, T., Mammerickx, J., and Smith, S. M., Geol. Soc. Am. Bull. (in the press).

[13] LePichon, X., J. geophys. Res., 73, 3661 (1968).

[14] Jackson, E. D., Silver, E., and Dalrymple, G. B., Geol. Soc. Am. Bull., 83, 601 (1972).

[15] Clague, D. A., and Jarrard, R. D., Geol. Soc. Am. Bull., 84, 1135 (1973).

[16] Clague, D. A., and Dalrymple, G. B., Earth planet. Sci. Lett., 17, 411 (1973).

[17] Hamilton, W., Tectonophysics, 4, 555 (1967).

[18] Beck, M. E., Geophys. J. R. astr. Soc., 28, 49 (1972).

[19] Hayes, D. E., and Ringis, J., Nature, 243, 454 (1973).

[20] Atwater, T., and Molnar, P., Proc. Conf. Tectonic Problems of the San Andreas Fault (Stanford University, 1973).

[21] Atwater, T., Geol. Soc. Am. Bull., 81, 3513 (1970).

[22] Atwater, T., and Menard, H. W., Earth planet. Sci. Lett., 7, 445 (1970).

[23] Burke, K., Kidd, W. S. F., and Wilson, J. Tuzo, Nature (in the press).

[24] McElhinny, M. W., Nature, 241, 523 (1973).

# Plate Tectonics and the fossil Record

## Section 3.

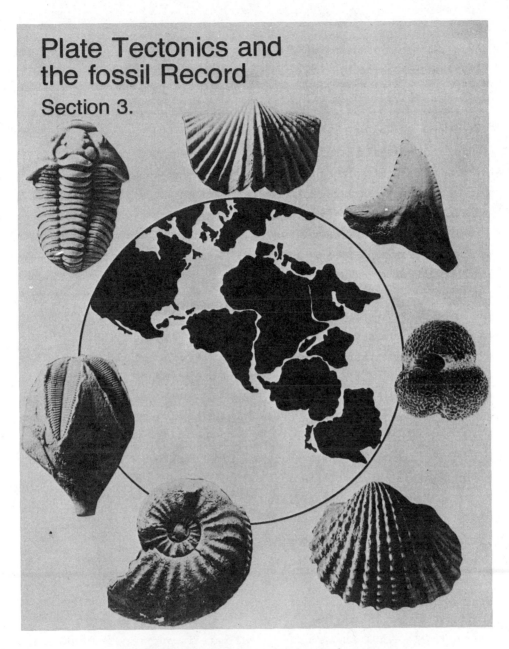

Representative fossils grouped round a reconstructed world.  Montage by Jon W. Branstrator.

It has been known for a long time that the fossil record shows evidence that there have been times of rich faunas, and times when life was not very varied.  Periods of rapid evolution contrast with periods of extinction. Groups of organisms have expanded and then disappeared, and geologists have used these fluctuations to define important reference points in the time scale, such as the boundary between Paleozoic and Mesozoic, or the end of the different periods.

We shall return to the problems of rapid evolutionary radiations and rapid extinctions in a later section; here we will examine controversies about the role played by plate tectonic events in the history of life, and whether in fact the fossil record can be taken at its face value.

If continents and oceans are rifted and drifted about over the Earth's surface as the great crustal plates are created, destroyed and transported, then one would expect that the geography and climate of great areas of the Earth would slowly change through time. If this in fact happened, then one would expect to see the plant and animal life of the planet evolving to adapt to the changing environment. So this poses three questions - is there a clear relation between tectonic and evolutionary events? If so, how is the relation controlled, and can we use paleontological and tectonic evidence together to explain major changes in the rock and fossil record?

The ideas of plate tectonics were first set out in the late 1960's, and although the problem of hot spots is still strongly debated, the outlines of plate-tectonic history were known by the early 1970's. In 1971, and in a longer paper in 1972, VALENTINE and MOORES set out a first explanation of the relationship between plate tectonics and major evolutionary events on Earth. Basically, drifting continents alter climatic regimes, which in turn make it more or less advantageous to be a "specialist" organism at any given time. From time to time, the environment has encouraged specialized animals and plants, and they have evolved to great number and diversity. At other times, the more generalized or opportunistic species have been able to thrive more than specialists, and the latter have suffered large-scale extinctions. How does the correlation work? As continents rift apart, the volume of mid-ocean ridges increases, pushing the oceanic waters out over the edges of the continents, and this increases living space for marine animals and makes for milder climates in several different ways. When continents are joined together, however, the world climate is strongly seasonal, and fewer and more generalized animals and plants can inhabit the world. In other words, if these arguments are correct, then the tectonic events on Earth have largely controlled the biological events, through the intermediate agent of <u>climate</u>.

Naturally, this suggestion was an extremely important one. A major attack on it has come from a paper which disputed the fact that the fossil record was a reliable source of data. Superficially, the data used by Valentine and Moores showed that animals (particularly animals of the shallow seas round the continental margins) had responded in a sensitive way to tectonic events. But if the data they had used happened to be false, then the whole idea would fall apart.

RAUP (1972) pointed out that the discovery of fossils is a process subject to a great deal of uncertainty. Other things being equal, older rocks would have been more eroded, and any fossils they might once have contained would have been destroyed. Raup suggested that the more area of rocks of a given age were preserved in the record, the greater the variety of fossils one would expect to find in them. Raup went on to set up a computer simulation of the chances of finding fossils from rocks subject to destruction, and he finally suggested that the history of life on Earth had been very different from the impression one receives from the fossil record taken at its face value. Indirectly, Raup's suggestion would destroy the correlations which

Valentine and Moores had relied on to show a cause-and-effect relationship between tectonics and the fossil record.

VALENTINE (1973) replied to this, and made some very strong arguments which appear to demolish Raup's suggestion. Although it might be easy to miscount fossil <u>species</u>, argued Valentine, it would be much more difficult to miss a whole biogeographic <u>province</u> in the fossil record. If one can imagine trying to count the number of people on Earth from a satellite, for example, one would probably miscount them. But one would have a good idea how they were distributed through area, because it would be difficult to miss the giant conurbations on the Earth's surface. So, argued Valentine, if we find only a few biogeographic provinces in Triassic rocks, coupled with a low number of fossil species, it looks as if Triassic times were in fact times of low diversity among marine animals.

The controversy rests there for the moment, but since there is no doubt that there are serious flaws in the fossil record, we can look forward to a more thorough examination of these flaws and how important they are, over the next few years.

We include next a paper by HAYS and PITMAN (1973), dealing with the events of the Middle Cretaceous as they

relate to the correlation between
tectonics, climate and diversity of
life.  Hays and Pitman calculated the
volume of the midocean ridges at
various times, and found that this was
greatest at the time between 110 and
85 m.y. ago.  It has been known for a
long time that the Middle Cretaceous
was a time of great marine invasions of
the continental shelves, and that it
was accompanied by an increase in the
variety of marine organisms.  Conversely, at the end of the Cretaceous
there was a slackening in plate
motions, the midocean ridges were
smaller in volume, and sea-level fell
all around the world.  Hays and Pitman
correlate this with a fall in the
organic diversity of the world (an
extinction) at the end of the
Cretaceous.  In effect, then, Hays and
Pitman provide evidence which supports
the ideas of Valentine and Moores, although the details of the processes are
slightly different in the two arguments.

By the middle of 1974 it became
clear that there were doubts about the
validity of Hays and Pitman's ideas.
Their calculations depended critically
on recontructions of the Cretaceous
sea-floor which seemed to show that
there had been very rapid sea-floor
spreading in the Middle Cretaceous.
Baldwin, Coney and Dickinson (1974)
suggested, however, that the Cretaceous
time-scale used by Hays and Pitman was
wrong, and that a new version of the
timing of Cretaceous events might remove altogether the period of alleged
rapid spreading, and therefore the
whole basis for their idea.  It remains
true, however, whatever the outcome of
this argument, that there were important tectonic and paleontological events
in the Middle Cretaceous which require
explanation.

Finally in this section, we reprint a paper which is very exciting
for its own sake, though it does not
try to explain any major geological
process.  The navigational powers of
the green turtle are legendary:  they
navigate over many hundreds of miles
of open ocean, some from the north
shore of Brazil to a volcanic island on
the mid-Atlantic ridge, Ascension
Island, to visit their breeding
grounds.  CARR and COLEMAN (1974)
speculate whether this behavior pattern
could have evolved gradually over
millions of years as the Atlantic
slowly widened and the mid-ocean ridge
thus came to be further and further
from the Brazilian coast.  Carr is one
leading world expert on the biology of
the green turtle, and Coleman is a
well-known Australian geologist.

Such a neat idea was bound to
begin a controversy, and a criticism
appeared later in Nature casting doubt
on its probability.  Nevertheless we

feel that Carr and Coleman have produced a lively contribution with all
kinds of ramifications to do with the
evolution of turtles.  If their idea is
wrong, then we are still looking for a
way to explain the evolution of the
navigational powers of turtles and
their discovery of the tiny pinpoint of
Ascension Island deep in the Atlantic.

FURTHER READINGS

BALDWIN, B., CONEY, P. J. and DICKIN-
    SON, W. R.  1974.  Dilemma of a
    Cretaceous time scale and rates
    of sea-floor spreading.  Geology
    2, 267-270.
BRASIER, M. D.  1974.  Turtle drift.
    Nature 250, 351.
KURTEN, B.  1969.  Continental drift
    and evolution.  Scientific Ameri-
    can, March issue.
VALENTINE, J. W.  1973.  Evolutionary
    paleoecology of the marine bio-
    sphere.  Prentice-Hall, Inc.
    (Chapters 8, 9, 10 particularly)

*

# Global tectonics and the fossil record

## J.W. Valentine and E.M. Moores (1972)

The world ocean at present has rich biotic diversity. Probably it contains about the richest biota of all time in number of species, the greatest variety of ecological communities, and the greatest number of biotic provinces. This present richness permits us to contrast the composition, structure, and function of many different communities and their populations, and to correlate them with parameters of the environments that they inhabit. The chief purpose of this paper is to suggest how past environments have changed with changes in the configuration and location of continents during Phanerozoic time, entailing alterations in the diversity and quality of shelf invertebrates and of their communities and provinces that are reflected in the fossil record.

The general theory of global tectonics holds that the outer part of the earth is composed of a series of rigid

plates passively moving about on a plastic asthenosphere (Le Pichon 1968; Morgan 1968; Vine and Hess 1970). Oceanic crust is created at midoceanic ridges and consumed in subduction zones marginal to continents or island arcs. An important corollary of this theory is that continents are merely passive riders on the much thicker lithospheric plates. The rate of sea-floor creation and consumption is such that little oceanic crust older than Cretaceous exists. Though continental rocks generally are not consumed at oceanic subduction zones, their margins can be altered by attempted subduction, or by collisions with island arcs or with other continents. Their general resistance to subduction enables them to exert important control on the location of plate margins. The presently emerging theory of orogenesis (Dewey and Bird 1970; Dewey and Horsfield 1970; Moores 1970; Vine and Hess 1970) follows from the basic assumption that Alpine-type mountain belts which contain ophiolites sensu stricto are the results of collisions or marginal

interactions of this type.  If this
assumption is correct, then such old
deformed belts now found on the continents represent regions where marginal
interactions have taken place in the
past.  Intracontinental belts such as
the Urals (Hamilton 1970) or the
Appalachian-Caledonides (Dewey 1969;
Bird and Dewey 1970; Naylor 1970) result under this hypothesis, from the
collision of two continental areas
formerly separated by an ocean.  The
dates of opening and closing of oceans
can be inferred from the stratigraphic,
tectonic, plutonic, and metamorphic
history of the belt in question.

The opening of an Atlantic-style
ocean would be accompanied by diabase
intrusion and foundering of the continental margin (ascertainable by
presence of belts of deeper-water
deposits overlying shallow-water sediments, which, in turn, rest upon either
eroded crystalline basements or continental deposits).  Closing of an ocean
basin may be heralded by development of
consuming or Andes-style margins,
marked by batholithic and volcanic
activity and mélange formation.  It is
at any rate, most surely marked by
emplacement on continental margins of
slabs of oceanic crust and mantle
(ophiolites) during the collision of a
continental margin with a seaward-
dipping subduction zone (such as
possibly the Taconic orogeny) (Temple

and Zimmerman 1969; Moores 1970; Coleman 1971; Karig, in press).  Continued
closing would be marked by batholithic
intrusion, regional metamorphism, and
formation and deformation of flysch
basins.  Final suturing of the two continents generally would be accompanied
by massive overthrusting, perhaps even
of one continent over the other as in
the Himalayas and in the Northern
Calcareous Alps, and by the formation
of continental molasse deposits.

Thus, as long as deformation and
metamorphism of major linear belts have
occurred, plate tectonics may have been
operating and producing continental
collisions.  This process may date from
$3 \times 10^9$ years B.P. (Dewey and Horsfield
1970) or since Grenvillian times (Hess
1955; Vine and Hess 1970).  At any
rate, it has probably gone on for a
time considerably greater than that
represented by the Phanerozoic fossil
record.  What patterns of continental
dispersal or assembly can be inferred
from the geological record?

Figure 1 shows the distribution of
ophiolite-bearing Alpine mountain
belts.  They include the late Precambrian Pan-African, Baikalian and
associated systems, the early Paleozoic
Caledonide chain, the late Paleozoic
Appalachian-Hercynian and Uralian
belts, and the Mesozoic-Tethyan system.

From these belts, a model of continental assembly and breakup is in-

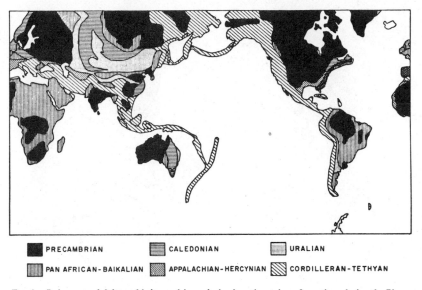

| | | |
|---|---|---|
| ■ PRECAMBRIAN | ▤ CALEDONIAN | ▥ URALIAN |
| ▥ PAN AFRICAN-BAIKALIAN | ▨ APPALACHIAN-HERCYNIAN | ▧ CORDILLERAN-TETHYAN |

Fig. 1.—Index map of deformed belts used in analysis of continental configurations during the Phanerozoic, modified after Dewey and Horsfield (1970, fig. 1).

ferred, employing the ages of metamorphic belts and transgressive marine sedimentation, the relatively well-known timetable of the dispersal of Gondwana and opening of the Atlantic, the reconstruction of the fit of Laurasia (Bullard et al. 1965), Gondwana (Smith and Hallam 1970), and the fit of Gondwana and Laurasia (Bullard et al. 1965). A topological representation of this model is presented in figure 2. It must be emphasized that this figure is topological only and that the relative shapes or positions of the continents in this figure do not possess geographical veri-similitude.

The Pan-African-Baikalian metamorphic belt (Grant 1969; Dewey and Horsfield 1970) is a belt of metamorphic and plutonic rocks which range

in age from 873 to about 450 m.y. B.P. (Grant 1969). Metamorphic rocks of comparable age are found in South America (Dewey and Horsfield 1970), along the Atlantic coast of North America (Long 1969), in western Europe, and in northern Russia and Siberia (Dewey and Horsfield 1970). According to the present orogenic model, these ages represent suturing of formerly separated continents or they represent activity along consuming continental margins. Cratonic nuclei definitely were sutured in Africa, and probably were also in India, Antarctica, and Australia (Grant 1969; Dewey and Horsfield 1970) as shown in figure 1.

The Pan-African metamorphic belt may represent a major convergence of continents culminating in their

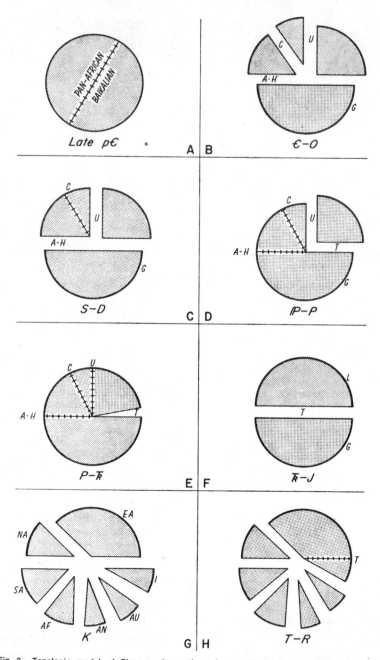

Fig. 2. Topologic model of Phanerozoic continental assembly-dispersal patterns. *A*, Late Precambrian formation of Pangaea I by suturing of Pan-African-Baikalian system. *B*, Cambro-Ordovician fragmentation into four continents separated by pre-Appalachian-Hercynian, Caledonide-Acadion, and pre-Uralian oceans; *B,G* is the Gondwana continent. *C*, Siluro-Devonian suturing of pre-Caledonian-Acadian ocean. *D*, Pennsylvanian-Permian suturing of pre-Appalachian-Hercynian ocean; *D,T* is the Tethyan ocean. *E*, Permo-Triassic suturing of pre-Uralian ocean, resulting in formation of Pangaea II. *F*, Triassic-Jurassic separation of Laurasia (*F,L*) and Gondwana (*F,G*), by extension of Tethys (*F,T*). *G*, Cretaceous opening of Atlantic, fragmentation of Gondwana. Continents are North America (*NA*), Eurasia (*EA*), India (*I*), Australia (*Au*), Anartica (*AN*), Africa (*AF*), and South America (*SA*). *H*, Tertiary-Recent suturing of Alpide belt.

assembly into a super-continent (see fig. 2A). It may not be the first assembly--other, older metamorphic belts (such as, possibly, the Grenville), may represent previous assemblies. Nevertheless, the implication is that near the beginning of Phanerozoic time, nearly or quite all landmasses were assembled. We suggest that just before and during Cambrian time, fragmentation of this supercontinent produced at least four smaller continents, with intervening oceans which may be designated the pre-Caledonian, pre-Appalachian, pre-Hercynian and pre-Uralian oceans (fig. 2B).

The pre-Caledonian ocean was partly open in Early Cambrian time, and deep-water marine conditions were certainly present during Late Cambrian-Early Ordovician times (Dewey 1969; Bird and Dewey 1970). It commenced to close in the Middle Ordovician (as indicated by the emplacement of ophiolites and the Taconic orogeny) and was sutured in Late Silurian to Late Devonian times, first on the Scandinavian end, and last on the New England end (Bird and Dewey 1970; Naylor 1970).

The pre-Appalachian ocean was probably present in late Precambrian time (the Ocoee and Chilhowee sequences possibly represent the continental margin) and was certainly present by Middle to Late Cambrian time (Hatcher 1970). The last major deformation in the Appalachians suggesting oceanic closing, extended from the Pennsylvanian to Early Permian (King 1959; Rodgers 1970). Judging from the presence of Cambrian deep-water sediments in the Hercynian belt, the pre-Hercynian ocean was present by Cambrian time (Rutten 1970). Consuming margins were developed by Devonian time (Dorn 1960; Erentoz 1967; Rutten 1970), and suturing of the Hercynian mountain belt was probably accomplished in Pennsylvanian-Early Permian time (Dorn 1960; Rutten 1970). The Hercynian and Appalachian systems probably represent continental sutures which resulted in the assembly of Europe and North America with Gondwana.

The pre-Uralian ocean probably was in existence by Late Cambrian time and certainly was present by Ordovician time as indicated by the development of a consuming margin (King 1967; Hamilton 1970). The Urals apparently were sutured in Late Permian to Early Triassic time, as indicated by Triassic continental sediments (Hamilton 1970). The Uralian suturing marked the assembly of a single supercontinent, commonly termed Pangaea (see Dietz and Holden 1970).

The stratigraphic sequence in the western Alpine region during the early Mesozoic is compatible with the foundering of a continental margin generally along, but not strictly parallel to, a Hercynian suture, with deep-sea conditions obtaining in the western Alps by Jurassic time (Trumpy 1960). This picture agrees with that developed by Pitman (1970) and Smith (1971) suggesting that a small amount of Triassic and Early Jurassic spreading took place from Gibraltar through the central Atlantic Ocean, extending through into the Caribbean region (see also Dietz and Holden 1970). This opening seaway may not have been very wide, but it probably afforded the opportunity for migration of the Tethyan biota. Dates on emplacement of ophiolites indicate that closure of the Tethyan seasway began by at least Late Jurassic time in the western, and Middle to Late Cretaceous time in the eastern Mediterranean. The breakup of Gondwana began in Triassic with the separation of Australia and climaxed during the Middle Cretaceous (Smith and Hallam 1970).

In summary, the above data from separate orogenic belts lead to a model of the assembly and subsequent breakup of continents. In Early to Late Cambrian time, continental fragmentation occurred (figs. 2A, 2B) giving rise to a pre-Caledonian ocean and probably to pre-Uralian, pre-Appalachian, and pre-Hercynian oceans or seaways (fig. 2B). This fragmentation was accompanied by sea-

level rises which caused seas to transgress the continental platforms. During Ordovician and Silurian times, four major continents may have been in existence: one was composed of the parts of North America and western Europe west of the Acadian-Caledonide suture; one of Europe and North America east of the Caledonide suture and north of the Hercynian suture; one of Asia east of the Urals and generally north of the Alpine suture; and one of Gondwana. Closure of the pre-Caledonide-Arcadian ocean during Late Silurian to Late Devonian time reduced the number of major continents to three (fig. 2C). Suturing of the Appalachian-Hercynian system in Pennsylvanian-Permian time left one very large continent and a single smaller continent (west and central Siberia--see fig. 2D). Finally, the suturing of the Urals in Permo-Triassic time resulted in the final reconstruction of a single supercontinent, Pangaea II (fig. 2E).

Pangaea was short lived, however, and soon parted along the equatorial Tethyan seaway which extended to middle America (fig. 2F). This equatorial belt commenced closing in Jurassic-Cretaceous time, accompanied by the rapid dispersal of fragments of Gondwana and opening of the North Atlantic (fig. 2G, 2H).

## Regulation of Transgressions and Regressions

In an earlier article (Valentine and Moores 1970), we presented a model for transgressions and regressions based upon the topologic model of continental assembly and fragmentation herein presented. Our model, an extension of those presented by Hallam (1963), Bott (1965), Menard and Smith (1966), and Russell (1968), was based on the fact that ridges and trenches are topographic features which contain significant volume, sufficient to cause pronounced fluctuations in the sea level on the continental shelves. Basically, the model holds that as a continent rifts apart, the new oceanic crust appearing at a ridge stands relatively higher than the ocean basin rocks that are being consumed in trenches. Hence, though the relative proportion of continental versus oceanic area remains the same, the volume of ocean basin decreases and transgressions occur, other things being equal. Conversely, when two formerly dispersed continents come together, their relative motion must eventually cease. Hence that portion of spreading which is responsible for their relative motions must cease also,

and either ridges or ridge segments must subside or new subduction zones with accompanying trenches must develop. Either result will increase oceanic volume and cause regression. If crustal shortening and formation of a mountain root takes place this serves to decrease continental area and to increase continental freeboard, again accentuating the regressive tendency. Therefore, continental fragmentations should be marked by transgressions and assemblies by regressions. To a first approximation this seems to be the case. Continental assemblies of the late Precambrian and Permo-Triassic were accompanied by regressions, while their breakups were accompanied by transgressions.

Though the above model explains certain first-order effects, clearly the process is more complex. For example, the Tertiary, a period of rapid spreading, is marked by regression. Kaula (1970) and J. C. Maxwell (personal communication, 1970) observed that one effect of spreading must be depression in the asthenosphere on the flanks of a spreading ridge, as material is supplied to the dilating zone. This effect, similar to that observed in salt domes, would depress the flanks of a ridge relative to "normal" ocean bottom. Such an effect

may be reflected by the deformation of
erosional surfaces in Africa related to
the initiation of the present East
African Rift.  Saggerson and Baker
(1965) and King (1967) indicate that
the early Tertiary (sub-Miocene or
African) erosional surface is deformed
considerably, by a 1,000-2,000-foot
uplift above the present graded erosional surface near the rift valley,
and by as much as 2,000 feet below
grade near the coast.  If these changes
in level are related to the Plio-
Pleistocene (King 1967) initiation of
the present East African rifting, then
they give an idea of the extent of elevation and depression that might be
expected on a continent during the
initiation of a rifting episode.

   The effect of ridges on oceanic
volume may not be as great as formerly
assumed, because forming a ridge is
simply the transfer of material from
the asthenosphere to the ridge area.
The actual bulk volume increase of this
mass transfer is related only to the
phase changes the material undergoes.
The process presumably begins with
asthenospheric material and ends with
depleted mantle plus oceanic crust.
The volume effect of this transfer can
be calculated in the following manner.

   Consider a midoceanic ridge situation as depicted in figure 3.  The
mass of material added to the ridge
must equal that subtracted from the

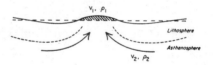

Fig. 3.—Sketch depicting transfer of asthenosphere material to surface at midocean ridge. See
text for explanation.

asthenosphere.  The increase of volume
of the material can be obtained by the
relationship $\underline{V}_1 \cdot \rho_1 = \underline{V}_2 \cdot \rho_2$ where $\underline{V}_1 =$
volume of bulge in ocean floor = volume
of ridges = $118.6 \times 10^6$ km$^3$ (Menard and
Smith 1966); $\rho_1$ = density of material
in the bulge.  As the ridges stand an
average of 1 km above and a maximum of
2.5-3 km above the surrounding oceans
(Menard 1964; Menard and Smith 1966),
$\rho_1$ = average density of diabase or
basalt ~ 2.6 g/cubic centimeter for
layer 2 (Ludwig et al. 1970); $\underline{V}_2 =$
volume of asthenosphere material
removed; and $\rho_2$ = density of asthenosphere material, assumed to be 3.50 g/
cubic centimeter (see Anderson et al.
1971).

   Substituting these values into the
equation yields a volume of asthenosphere removed of $91.0 \cdot 10^6$ km$^3$.  Hence,
the total increase in volume due to the
present ridges is equal to $118.6 \cdot 10^6$ -
$91.0 \cdot 100^6$ km$^3$, or $27.6 \cdot 10^6$ km$^8$.  This
value is the equivalent of a change in
sea level of approximately 74 m over
the oceans or 52 m over the entire
earth's surface.  From this amount one
must subtract the increase of volume

due to trenches. A rough idea of their volumetric effects may be calculated from data presented by Menard and Smith (1966), who report a median depth for island arc and trench provinces of 4 km. The area of these provinces is 1.7% of the world oceans or $6.15 \times 10^6$ $km^2$. Hence their volume is $24.60 \times 10^6$ $km^2$. This volume is probably a maximum figure as Menard and Smith's data do not include areas or elevations of arcs above normal continents.

The conclusion reached from these approximate calculations, however, is that the actual volume effects of a midocean ridge are small and may be almost entirely counterbalanced by trench regions, depending on their location. Transgressions should occur on continents when the latter are rifted apart and become involved in the zone of ridge-flank depression (fig. 4). Ridges which are developed entirely within ocean basins well away from continents, so that the latter do not become involved in the ridge-related changes in elevations, may have little effect upon sea-level changes relative to continents. Transgressions should be limited to those continental margins near enough to spreading centers to become involved in the flank depressions. Present examples of this situation are difficult to find, but may include the Persian Gulf area, the Seychelles Bank, and much of the Arctic

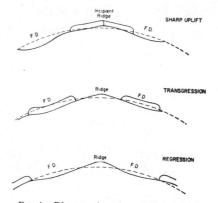

FIG. 4.—Diagram of continental rifting showing transgressive-progressive relations resulting from uplift in ridge area and depression on the flanks. F.D. = zone of flank depression.

Ocean. One problem with these analogies is that there is no area in Africa across the Red Sea rift in a comparable position to the Persian Gulf region, but this lack may result from the complexity of plate interactions involving Africa. Major worldwide transgressions should be correlated with breakup of supercontinents involving most or all of the world's landmasses. Breakup should occur somewhat earlier than the major transgression.

Major regressions should occur when continents are sutured, owing to cessation of spreading the collapse of ridges, the corresponding tendency for areas away from ridges to rise slightly, and to decreases in land surface due to deformation or underthrusting during continent-continent collisions.

A common pattern of a rifting continent's elevation relative to sea level may include: (1) a sharp rise in elevation just prior to initiation of rifting (such as East Africa now or Gondwana during the Jurassic--King (1967) and Russell (1968)); (2) a gradual transgression as the rift opens and the continent moves down to occupy the zone of ridge-flank depression (such as, possibly, the Cretaceous transgression); (3) a regression (such as the Danian) as the continents spread away from the ridge-flank depression areas.

## Provisional Paleozoic Paleogeographic Reconstruction

Using the history of continental dispersal and assembly outlined above together with paleomagnetic data (such as Irving 1964; McIlhenny 1970a, 1970b; and McIlhenny and Briden 1971) and paleoclimatological data (Irving 1964), it is possible to form a provisional reconstruction of Paleozoic continental arrangements.

Figure 5 shows an example of such

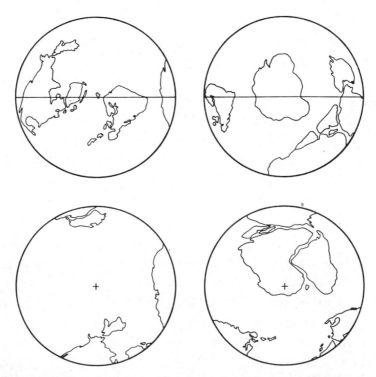

Fig. 5. Preliminary possible cratonic configurations in Ordovician, equatorial and polar views. Topologic equivalent, fig. 2B. The four continents are Gondwana, East Asia, proto-North America, and proto-Europe. East Asia has two possible positions: between Gondwana and North America, or between Europe and Gondwana (shown).

a reconstruction for the Ordovician. The position of Gondwana is fixed by its paleomagnetic pole (McIlhenny 1970b). The relative positions of North America and Europe are suggested by their paleoequators (Irving 1964), but their longitudinal positions are arbitrary. Siberia is shown between Europe and Australia but could just as well be between North America and Australia. Paleobiologic data may eventually resolve this ambiguity. Figure 6 shows a reconstruction of Pangaea II at the Permo-Triassic boundary, and is essentially a replot of that presented by Dietz and Holden (1970).

## CONTINENTAL DRIFT AND EVOLUTION

Continental movements implied by plate tectonic theory must have had profound effects upon many environmental conditions of major importance to living organisms. For marine organisms, the more obvious effects involve

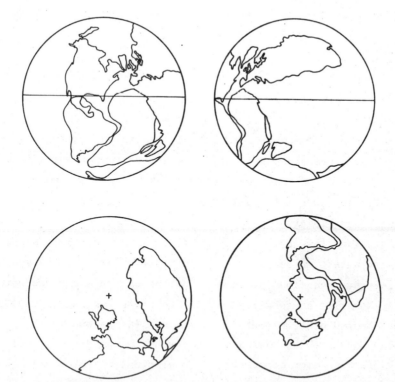

Fig. 6. Possible cratonic configuration for end of Permian using fit of Bullard et al. (1965), after suturing of Urals and formation of Pangaea II. Topologic equivalent, fig. 2E.

changes in dispersal barriers and in marine climates, including changes in temperature regimes, ocean currents, and in vertical stability of oceanic water columns. Regulation of patterns of marine climate across the globe is complex and is not perfectly understood, so that precise climatic predictions for land-sea relations widely different from those of today cannot yet be worked out. Nevertheless, the effects of changing land-sea geographies will be profound, and a few generalities are possible.

First, net movement of continents into high latitudes generally would cause a decrease in average annual temperatures and increase the average seasonality on the shelves, while net movement into lower latitudes would have opposite effects. The exact pattern of continental dispersion is also important, for certain continental patterns can insulate low-latitude oceans from high-latitude influence, or conversely they can isolate cool, high-latitude oceans from low-latitude sources of warm water. Assembly of continents into supercontinents would tend to enhance marine seasonality owing to monsoon and other effects (Valentine and Moores 1970; Robinson 1971), while fragmentation of continents into smaller landmasses and their continuing separation by drift would

tend to raise the average stability of shelf waters, other things being equal. Thus, continental movements change the positions of continents relative to climatic patterns and also change the climatic patterns themselves.

As natural selection is partly an ecological process, and as environmental changes accompanying plate motions must have altered the ecology of marine organisms on an oceanic or global scale, these environmental changes also must have affected pathways of organic evolution. The interrelations between environmental change and biological change are varied and complex; we wish to focus upon only a few aspects of these interrelations that we believe are of special importance in determining the character of the invertebrate fossil record. Chief among these aspects is the ecological regulation of organic diversity--its creation, suppression, and maintenance. This is a complex topic itself but can be simplified by considering two types of diversity. One type is species diversity, by which is meant simply the number of species represented at the time or in the region under consideration. The second is the diversity of "ground plans," hat is, of the basic biological architectures that characterize whole classes or phyla of organisms.

## CONTINENTAL DRIFT AND THE REGULATION OF SPECIES DIVERSITY

Many hypotheses have been formulated to account for species-diversity patterns (Pianka 1966; Sanders 1968). Presently, the hypothesis best supported by empirical data holds that environmental stability is the key factor in species-diversity regulation although the regulation mechanism is still undetermined. Some investigators believe that stability of many or all major environmental parameters can be important (e.g., Sanders 1968, 1969; Bretsky 1969; Bretsky and Lorenz 1969). Others suggest that resource stability is the primary regulator of diversity; this hypothesis exists in several variations (Fisher 1960; Connell and Orias 1964; Paine in Pianka 1966; Margalef 1968; Valentine 1971a).

In the present ocean, the stability patterns of energy resources within ecosystems, represented by sunlight and nutrients for plants and food for animals, correlate highly with the species-diversity pattern (Sanders 1968; Valentine 1971b). These energy supplies are the trophic resources of an ecosystem. Latitudinally, the gradient of seasonality in solar radiation correlates with a gradient on continental shelves. Longitudinally, trends in environmental stability (Sanders 1968), especially of nutrient supplies (Valentine 1971b), correlate well with trends in species diversity. Relatively unstable "continental" marine-shelf climates are associated with a relatively low species diversity, while relatively stable "maritime" shelves have higher species diversities, latitude for latitude.

Many reasons have been suggested for these correlations. At present a likely hypothesis holds that in fluctuating environments, populations that are favored are those best able to endure inclement periods and, during favorable periods, to expand rapidly to take advantage of expanding resources. Characteristics especially useful in these circumstances include: (1) broad food tolerances affording independence from any given food item and adaptability to those items most available at any time; (2) broad habitat tolerances permitting large ubiquitous populations to develop during favorable periods; (3) high reproductive potentials permitting rapid utilization of rising resource levels; and (4) general individual robustness. Other adaptive strategies are possible in fluctuating environments, but a high proportion of species should share a number of these characteristics and thus be rather generalized and opportunistic.

In fluctuating environments, relatively few trophic levels can be supported within an ecosystem--that is, there can be relatively few links in any food chain (see Ryther 1969). When trophic resources expand, there will be a lag before they can be taken up in quantity by organisms on the next higher trophic level, so that there is less time available for population expansion on each higher level before the resources decline again. Thus, fluctuations are greater at higher levels. When decline occurs, there is a "waste" of biomass throughout the ecosystem, which cannot be replaced until the next resource expansion. The ecosystem is inefficient, and a relatively low total number of generalized, opportunistic populations can be accommodated within a fluctuating ecosystem with few trophic levels.

By contrast, populations in stable environments have a steady supply of trophic resources, although in the oceans the supply may be small since high stability is usually correlated with low nutrient levels. Population densities are, therefore, not commonly high, but are rather stable since episodic unfavorable conditions are nearly absent. Hence, successful populations are able to have low reproductive potentials insofar as energy utilization is concerned, although they may require high rates of reproduction if heavily preyed upon or parasitized. Species in stable trophic resource regimes should commonly have: (1) narrow food preferences; (2) narrow habitat preferences; and (3) narrow tolerances for environmental changes. Because of the great trophic stability, the food chains may be long. Clearly, according to this hypothesis the total number of specialized and commonly efficient populations that can be accommodated in ecosystems in stable environments is relatively large.

The qualitative composition of ecosystems is also related to environmental stability. In marine environments with wide fluctuations in primary productivity, such as high latitudes, the amplitude of resources available as living plants such as phytoplankters varies far more than their dead detrital residues. Planktonic blooms are inefficiently exploited and supply much organic detritus to the sea floor. The detritus may remain as a food source long after the bloom has ended, and detritus-feeding species are proportionately more abundant in fluctuating than in stable environments, relative to browsers and planktivores (Odum and De la Cruz 1963). Furthermore, both obligatory and facultative scavengers appear to be proportionately more abundant in high latitudes than in low (Arnaud 1970; Valentine, unpublished data).

Another factor that may affect species diversity is the number of different sorts of habitats available. It is commonly assumed that more habitats and therefore more species will be found in a broad than in a narrow area, although this is not always the case (MacArthur and Wilson 1967). However, during times of widespread shelf seas, that is, of major transgressions, shelf diversity would probably be higher than during regressions, other things being equal.

Provinciality is also an important factor in species diversity, and one in which plate tectonics plays an obvious role through the creation or elimination of dispersal barriers by changing land-sea relations and their climatic consequences. In its simplest terms, the rise of a dispersal barrier creates an opportunity for speciation within all those species that had formerly ranged across the barrier. When barriers appear that endure for long periods of geological time, they permit species diversity to increase by an amount potentially equal to the number of species that are cut into separate populations.

For shelf invertebrates, dispersal barriers are chiefly of two sorts, topographic and climatic. The main topographic barriers are deep-sea basins and landmasses. Whenever continents fragment, the shelves of the daughter continents become progressively more isolated as they drift apart, eventually leading to the appearance of distinctive biotas on each continent, with only those species possessing the greater powers of dispersal being able to maintain their identities across the abyssal barrier. When climates become more varied, as when the latitudinal temperature gradient increases during polar cooling, species ranges become restricted as the climatic regimes to which they are adapted become restricted, and opportunities for new species are created in regions where new climates appear.

Combinations of topographic and climatic barriers can be very effective in increasing provinciality and hence species diversity. Species become restricted to segments of shelves by climatic barriers both to north and south and cannot reach appropriate climatic zones along other coasts owing to topographic barriers. It has been calculated that at present the marine shelves can support more than 10 times more species than would be possible if provinciality were absent owing to climatic amelioration and continental assembly (Valentine 1969, 1970). If any of these major factors that affect species diversity were to change significantly on the scale of a climatic zone or of an ocean, it should cause a meaningful fluctuation in

marine species diversity, while if
these factors were to change in concert
on a global scale it would affect the
diversity and composition of the entire
marine biota in a fundamental manner
that should be easily detectable in the
fossil record.

## CONTINENTAL DRIFT AND THE DIVERSIFICATION OF PHYLA

Each phylum has a distinctive
ground plan, a unique association of
architectural characteristics that set
its members apart from representatives
of other phyla. These characteristics
include the number of tissue layers,
symmetry, presence or absence of a
coelom (body cavity), and similar basic
features. It is generally believed
that wholly new ground plans are
evolved in response to unusual environmental opportunities that become
available to organisms with what is, at
the time, a rather unusual mode of life
(Simpson 1944, 1953; Mayr 1963). The
organisms are able to enter the novel
or uninhabited environment because of
their peculiarities, and as they do
enter it they are subjected to new requirements. In adapting to these requirements, a new ground plan is sometimes developed. Subsequent post-
adaptional consolidation of the ground
plan and evolutionary radiations within
the novel environment establish a new,
higher taxon.

The number of ground plans that
are possible is probably limited in
part by considerations of biological
architecture and engineering, but it
seems that many more are possible than
exist today. Certainly, fewer than
exist today are possible. For example,
if all mollusks were exterminated, a
new phylum would not be expected to
appear in replacement. Instead,
speciation would occur among lineages
of remaining phyla that happened to
function ecologically in ways similar
to the departed mollusks. In other
words, ridding the world of mollusks
would not open up a novel environmental
opportunity; nearly all environments
are already occupied by representative
of numerous phyla.

Thus it seems reasonable that the
number of ground plans in existence at
any time is related to the timing with
which major environments have become
available for habitation. In all
likelihood, major new opportunities
have not been available in the oceans
for a very long time, perhaps since
the early Paleozoic.

Some aspects of diversification of
invertebrate ground plans have been
discussed in detail by Clark (1964).
The function to which ground plans are
chiefly adapted is locomotion, for the
movement of the entire organism
naturally tends to involve the basic
body structure. Locomotion is related

in turn to feeding adaptations of the organisms, for it is the locomotory adaptations that permit an organism to obtain food in a given environment. Also, in the sea, a number of phyla and classes are sessile, or essentially so, and adaptations for this mode of life evidently have had much to do with selection for some ground plans. Sessile adaptations are similar to locomotory ones in that they permit the physical occupation of habitats and they correlate with feeding modes--in this case chiefly with suspension feeding. To call them locomotory adaptations, however, seems inappropriate, and no single term suggests itself. Both locomotory and sessile adaptations are employed for, or coordinated with, many functions, but there is evidence that feeding is the most important (Clark 1964; Barrington 1967). Therefore, it is a reasonable hypothesis that the unusual opportunities that resulted in the development of new invertebrate ground plans were basically trophic ones and that the development of new locomotory or sessile adaptations were required to exploit them. The morphological modifications needed to achieve these new adaptations altered the ground plan of the lineage.

There are two major events in the history of marine invertebrates that we wish to examine in the light of this hypothesis: the development of the coelom in the Precambrian and the appearance of skeletons during the Cambrian. It has been shown that the coelom has most likely evolved as an improvement in burrowing efficiency; it served as a hydrostatic skeleton to permit peristalsis (see Clark 1964 and references therein). An acoelomate organism, such as a flatworm, may insinuate itself between grains in some sediments and so become infaunal in habitat, but it is not well adapted for continuous burrowing in fine sediments since it has little purchase for forcing the sediment aside. With a hydraulic skeleton, however, this becomes easily possible, and coelomate worms may inhabit fine sediments and pursue an actively burrowing mode of life. Clark (1964) and others have suggested that the coelom arose as a burrowing organ that permitted invasion of substrates which were previously poorly exploited and which contained a rich supply of organic detritus and associated bacteria. This idea clearly fits into the hypothesis of ground-plan origin in locomotory modifications to exploit a trophic opportunity.

There seem to have been several major plans of coelomic structure that developed in the late Precambrian; these may represent a radiation from some single primitive coelomate stock or may represent several independent lineages, each of which developed a coelomic cavity. These plans include the metamerous condition found in annelids wherein the coelom is partitioned into a number of similar segments; the oligomerous condition found today in phoronids and brachiopods and probably present in echinoderms or in echinoderm ancestors, wherein the coelom is partitioned into a few (probably originally three) divisions; and the seriated or pseudometamerous condition of some primitive mollusks, wherein partitioning is absent but organs may be repeated serially. These varieties of coelomate architecture are likely to have been evolved as locomotory or sessile adaptations associated with feeding. There is much evidence that the metamerous forms were originally active burrowers, probably detritus feeders; some evidence that the oligomerous forms were first adapted to infaunal suspension feeding; and it is likely that primitive pseudometamerous forms were creeping detritus feeders or browsers.

The skeletonized Cambrian coelomates are chiefly epibenthic. The reason for their sudden appearance has been much debated (see Nicol (1966) for a short review). Cloud (1949, 1968) has argued strongly that the appearance of the skeletons of many of these metazoan phyla represents, more or less, their time of origin. In this event, the Lower Cambrian fauna would represent a radiation of ground plans, chiefly from soft-bodied coelomate worms of several types. Of course a few coelomate phyla (Annelida, Arthropoda) must have had a long Precambrian history, but many of the primitive coelomate worm stocks may have had ground plans that have not survived as such, being modified in living phyla. According to the trophic-opportunity hypothesis, epibenthic radiation of skeletonized forms should involve relatively sudden exploitation of poorly utilized resources, which must at first have been partitioned among the new phyla according to the modes of life to which their ground plans were originally adapted. The skeletons are probably associated with increased locomotory efficiency in the epibenthos where confining effects of sediments are lacking and where rigid skeletons could improve upon hydrostatic ones (L. Fox, personal communication), and also with feeding efficiency in sessile lineages such as brachiopods. Protective functions, while important, may have been secondary, although some subsequent radiations seem to exploit skeletons as protective devices (as among the gastropods).

This is not the place to explore this hypothesis in detail. But it is necessary to develop some sort of historical model in order to show how the environmental changes that follow from plate tectonic processes may be related to diversification on high taxonomic levels by changing the planetary environment in such a way as to make new trophic opportunities available, or to favor their utilization.

The invasion of the substrate by coelomates and their subsequent radiation therein resulted in the rise of infaunal metazoan communities that were chiefly detritus and suspension feeders, with epibenthic forms also chiefly exploiting detritus so far as is known. This sort of community is exactly that which we have described as particularly appropriate to environments wherein trophic resources are highly fluctuating. This factor suggests that late Precambrian continental configurations were such as to cause a fluctuation in trophic resources across most of the world's shelves, perhaps owing to the construction of a supercontinent. Under these conditions, infaunal detritus feeding would have been strongly favored, and perhaps the coelom arose in response to selection for such an advantageous trophic strategy. The coelom proved to be preadaptive to a wide range of locomotory and sessile mechanisms to exploit a variety of trophic resources, and permitted body size increases and many other adaptive

trends. Communities during this time would have been of low species diversity, however. The fossil record of these early coelomates would be restricted to burrows and trails except for unusual circumstances.

Cambrian communities suggest a trend toward stabilization of resources and a concomitant diversification of the epibenthic fauna with the adoption of a variety of new combinations of locomotory or sessile adaptations with various trophic strategies. Thus, the order of appearance of ground plans may have resulted from the sequence of environmental regimes on the shelves, and the number of ground plans may have resulted from an interplay between the constraints on modification of the "super ground plans" of the more primitive lineages and the variety of environments in which new trophic opportunities appeared.

## CONTINENTAL DRIFT AND THE FOSSIL RECORD

Using the suggested hypotheses of the ecological regulation of species diversity and of the diversification of phyla, it is possible to erect a model, for any given sequence of environmental states, of the diversity and, to some extent, the quality expected of marine-shelf ecosystems. The model is testable in many ways; a prima facie test is whether major features of the fossil record are appropriate in terms of the model to the major environmental states that can be inferred from continental positions when these are known. **For** our model, we shall employ the continental patterns inferred in figure 2.

Figure 7 depicts the diversity of benthic marine families of well-skeletonized invertebrate phyla during the Phanerozoic, as described from the fossil record. It would be preferable to use species-diversity patterns, but unfortunately the fragmentary nature of the record and of our knowledge precludes this. Families are inclusive enough so that their record probably reflects their diversities fairly closely, and yet they are close enough to the species level that ecological controls may be inferred for major trends. Species diversities are expected to display much wider swings than do the families.

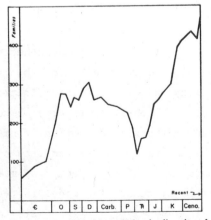

Fig. 7.—Stratigraphic variation in diversity of skeletonized families of shelf benthos.

The early radiation of the infaunal coelomates (see Glaessner 1971) we attribute to unstable conditions associated with continental construction during the late Precambrian. The Cambrian radiation, then, suggests the beginning of an amelioration of those conditions. The spectacular rise in family diversity during the late Cambrian and the Ordovician suggests a continuing stabilization of resources, perhaps resulting from a land-sea pattern such as in figure 5. The communities became more diverse, and as predators appear the number of trophic levels rises, judging by the fossil record. The high diversity levels achieved during the Ordovician are retained throughout the middle Paleozoic periods but tend to fall off during the Carboniferous and the crash to the familiar Permo-Triassic minimum. According to our model, the breakup of a Precambrian supercontinental assembly occurred near the end of the Precambrian, and the Cambro-Ordovician diversifications are thus attributed to the variety of changes, all favoring high diversity, that would accompany or follow such a continental fragmentation. These include the rise of stability of trophic resources throughout the self ecosystems that would accompany the declining continentality. Furthermore, stability would probably be promoted by the subsequent transgressions, for widespread shelf seas would have a damping effect upon diurnal and seasonal heat fluctuations, which would contribute to environmental stability. To the extent that more habitats are opened up to marine organisms, the transgressions would also enhance diversity by increasing the spatial heterogeneity of the shelf environments. And the fragmentation would bring about the genetic isolation of populations and thus create provinciality and raise species diversity. The environment should become more and more closely partitioned by ever more specialized organisms. This trend can be seen in the appearance and development of Paleozoic reef communities to eventually include some of the most specialized benthic organisms known; in the rise and increasing specialization of crinoid and of bryozoan assemblages; and in the elaboration of specialized organisms in many lineages, even including the brachiopods.

During Permian-Triassic time the assembly of a supercontinent (figs. 2E, 6) was completed, and the seas withdrew from most continental platforms in a major regression. Thus, early Triassic environmental conditions should have been roughly analogous to those of the late Precambrian, with many factors favoring low diversity, and with detritus feeders and flexible, generalized populations pre-

dominating. These expectations correspond well with observations. The specialized epifauna of the Paleozoic was hard hit by Permo-Triassic extinctions. The lineages best adapted to Permo-Triassic conditions would not necessarily have been those in the more physically variable environmental regimes, such as in very shallow water, but those adapted to low trophic stability during the late Paleozoic. These might be estuarine forms, opportunistic lineages adapted to exploit the fluctuating fractions of trophic resources, and ubiquitous species adapted to a wide habitat range. The detritus-feeding and browsing gastropods, the ubiquitous bivalves, the linguloid brachiopods, and other groups that were relatively unaffected by the great extinctions appear to share some of these properties.

Mesozoic rediversification accompanied the breakup of the Permo-Triassic supercontinent and must have been roughly analogous to Cambro-Ordovician diversification, with certain important differences. One difference was that no opportunities were presented for the invasion of unexploited kinds of marine environments; all major marine environments contained a variety of metazoans, and thus the rediversification involved the radiation and elaboration of these previously extant taxa, although some new alliances that are now considered as orders first appeared during this event. And additionally, the geographic partitioning of the environment has proceeded farther, so that the re-diversification has achieved levels of diversity greater than those of the Paleozoic among the lower taxonomic levels (Valentine 1969). The continental dispersion pattern that developed in the Mesozoic and that has climaxed during the Cenozoic has created a vast array of shallow marine environments that may well be unparalleled in Phanerozoic history. The great latitudinal spread of continents, the partitioning of oceanic water masses, and the cooling of high latitudes have all combined to raise both latitudinal and longitudinal provinciality to a peak and to provide for shelf environments of both fluctuating regimes in polar regions and relatively stable regimes in tropical latitudes.

The diversity patterns are partially the products of the adaptive strategies followed by populations in different resource regimes; high diversities correlate with stable and low diversities with fluctuating regimes. The proportions of feeding types vary between these types of regime also, resulting in qualitative differences in the faunas. Precambrian and Cambrian radiations from which the higher invertebrate phyla developed may have been primarily adaptive responses to changes in trophic resource regimes brought about by changing land-sea patterns.

## ACKNOWLEDGMENTS

The development of these ideas has especially benefited from discussions with P. W. Bretsky, Linda Fox, W. M. Hamner, J. C. Maxwell, and F. J. Vine. Figures were drafted by R. Darden and L. Valentine.

## REFERENCES CITED

Anderson, D. L.; Sammis, C.; and Jordan, T., 1971, Composition and evolution of the mantle core: Science, v. 171, p. 1103-1112.

Arnaud, P. M., 1970, Frequency and ecological significance of necrophagy among the benthic species of antarctic coastal waters, in Holdgate, M. W., ed., Antarctic ecology: New York, Academic Press, v. 1, p. 259-266.

Barrington, E. J. W., 1967, Invertebrate structure and function: Boston, Houghton Mifflin Co., 549 p.

Bird, J. M., and Dewey, J. F., 1970,
    Lithosphere-plate-continental
    margin tectonics and the evolution
    of the Appalachian orogen:  Geol.
    Soc. America Bull., v. 81, p.
    1031-1060.

Bott, M. H. P., 1965, Formation of
    oceanic ridges:  Nature, v. 207,
    p. 840-843.

Bretsky, P. W., 1969, Evolution of
    Paleozoic benthic marine inverte-
    brate communities:  Palaeo-
    geography, Palaeoclimatology,
    Palaeoecology, v. 6, p. 45-59.

----, and Lorenz, D. M., 1969, Adaptive
    response to environmental stabili-
    ty:  a unifying concept in paleo-
    ecology:  North Am. Paleontology
    Convention, Chicago, Proc., pt. E,
    p. 522-550.

Bullard, E. C.; Everett, J. E.; and
    Smith, A. G., 1965, The fit of the
    continents around the Atlantic, in
    A symposium on continental drift:
    Roy. Soc. (London) Philos. Trans.,
    v. A258, p. 41-51.

Clark, R. B., 1964, Dynamics in meta-
    zoan evolution:  Oxford, Clarendon
    Press, 313 p.

Cloud, P. E., 1949, Some problems and
    patterns of evolution exemplified
    by fossil invertebrates:  Evolu-
    tion, v. 2, p. 322-350.

---- 1968, Pre-metazoan evolution and
    the origin of the metazoa, in

Drake, E. T., ed., Evolution and
    environment:  New Haven, Conn.,
    Yale Univ. Press, p. 1-72.

Coleman, R. G., 1971, Plate tectonic
    emplacement of peridotites at
    continental edges:  Jour. Geophys.
    Res., v. 76, p. 1212-1222.

Connell, J. H., and Orias, E., 1964,
    The ecological regulation of
    species diversity:  Am. Natura-
    list, v. 98, p. 399-414.

Dewey, J. F., 1969, Evolution of the
    Appalachian/Caledonian orogen:
    Nature, v. 222, p. 124-129.

----, and Bird, J. M., 1970, Mountain
    belts and the new global tec-
    tonics:  Jour. Geophys. Res.,
    v. 75, p. 2625-2647.

----, and Horsfield, B., 1970, Plate
    tectonics, orogeny, and conti-
    nental growth:  Nature, v. 225,
    p. 521-525.

Dietz, R. S., and Holden, J., 1970,
    Reconstruction of Pangaea; breakup
    and dispersion of continents,
    Permian to present:  Jour.
    Geophys. Res., v. 75, p. 4939-
    4956.

Dorn, P., 1960, Geologie von Mittel-
    europa:  Stuttgart, F. Schweizer-
    bart, 488 p.

Erentoz, K., 1967, A brief review of
    the geology of Anatolia:  Geo-
    tectonics, no. 2, 1967, p. 85.

Fischer, A. G., 1960, Latitudinal variations in organic diversity: Evolution, v. 14, p. 64-81.

Glaessner, M. F., 1971, Geographic distribution and time range of the Ediacara Precambrian fauna: Geol. Soc. America Bull., v. 82, p. 509-514.

Grant, N. K., 1969, The late Precambrian to Early Paleozoic pan-African orogeny in Ghana, Togo, Dahomey and Nigeria: Geol. Soc. America Bull., v. 80, p. 45-56.

Hallam, A., 1963, Major epeirogenic and eustatic changes since the Cretaceous, and their possible relationship to crustal structure: Am. Jour. Sci., v. 261, p. 397-423.

Hamilton, W., 1970, The Uralides and the motion of the Russian and Siberian platforms: Geol. Soc. America Bull., v. 81, p. 2553-2576.

Hatcher, R. D., Jr., 1970, A working model for the developmental history of the southern Appalachian Blue Ridge and Piedmont: Geol. Soc. America Abs. with Programs, v. 2, no. 7, p. 745-747.

Hess, H. H., 1955, Serpentines, orogeny and epirogeny, in Poldervaart, A., ed., Crust of the earth--a symposium: Geol. Soc. America Spec. Paper 62, p. 391-407.

Irving, E., 1964, Paleomagnetism and its application to geological and geophysical problems: New York, John Wiley & Sons, 399 p.

Karig, D. E., in press, Remnant arcs: Geol. Soc. America Bull.

Kaula, W., 1970, Earth's gravity field: relation to global tectonics: Science, v. 169, p. 982-984.

King, L. C., 1967, Morphology of the earth (2nd ed.): London, Oliver & Boyd, 726 p.

King, P. B., 1959, Evolution of North America: Princeton, N.J., Princeton Univ. Press, 189 p.

Le Pichon, X., 1968, Sea-floor spreading and continental drift: Jour. Geophys. Res., v. 73, p. 3661-3698.

Long, L. E., 1969, Whole rock Rb-Sr age of the Yonkers gneiss, Manhattan prong: Geol. Soc. America Bull., v. 80, p. 2087-2090.

Ludwig, W. J.; Nafe, E.; and Drake, Charles L., 1970, Seismic refraction, in Maxwell, A. E., et al., eds., The seas: New York, John Wiley & Sons, v. 4, pt. 1, p. 53-84.

MacArthur, R. H., and Wilson, E. O., 1967, The theory of island biogeography: Princeton, N.J.; Princeton Univ Press, 203 p.

McIlhenny, M. W., 1970a, Paleomagnetic directions and pole positions X pole numbers 10/1 to 10/200:

Royal Astron. Soc. Geophys. Jour., v. 18, p. 305-327.

---- 1970b, Polar wander curve for Gondwana: Nature, v. 228.

----, and Briden, J. C., 1971, Continental drift during the Paleozoic: Earth and Planetary Sci. Letter, v. 10, p. 407-416.

Margalef, R., 1968, Perspectives in ecological theory: Chicago, Univ. Chicago Press, 111 p.

Mayr, Ernst, 1963, Animal species and evolution: Cambridge, Mass., Harvard Univ. Press, 797 p.

Menard, H. W., 1964, Marine geology of the Pacific: New York, McGraw-Hill Book Co., 271 p.

----, and Smith, S. M., 1966, Hypsometry of ocean basin provinces: Jour. Geophys. Res., v. 71, p. 4305-4325.

Moores, E. M., 1970, Ultramafics and orogeny, with models for the U.S. Cordillera and the Tethys: Nature, v. 228, p. 837-842.

Naylor, R., 1970, Continent collision and the Acadian orogeny: Geol. Soc. America Abs. with Programs, v. 7, p. 634-635.

Nicol, David, 1966, Cope's rule and Precambrian and Cambrian invertebrates: Jour. Paleontology, v. 40, p. 1397-1399.

Odum, E. P., and De la Cruz, A. A., 1963, Detritus as a major component of ecosystems: Am. Inst. Biol. Sci. Bull., v. 13, p. 39-40.

Pianka, E. R., 1966, Latitudinal gradients in species diversity: a review of concepts: Am. Naturalist, v. 100, p. 33-46.

Pitman, W. C. III, 1970, Sea floor spreading in the North Atlantic and its implications regarding the closing of the Tethys: Geol. Soc. America Abs. with Programs, v. 2, no. 7, p. 752-754.

Robinson, P. L., 1971, A problem of faunal replacement on Perm-Triassic continents. Palaeontology, v. 14, p. 131-153.

Rodgers, John, 1970, The tectonics of the Appalachians: New York, John Wiley & Sons, 271 p.

Russell, K. L., 1968, Oceanic ridges and eustatic changes of sea level: Nature, v. 218, p. 861-862.

Rutten, M. G., 1970, Geology of western Europe: Amsterdam, Elsevier Publishing Co., 520 p.

Ryther, J. H., 1969, Photosynthesis and fish production in the sea: Science, v. 166, p. 72-76.

Saggerson, E. P., and Baker, B. H., 1965, Post-Jurassic erosion surfaces in east Kenya: Geol. Soc. London Quart. Jour., v. 21, p. 51-72.

Sanders, H. L., 1968, Marine benthic diversity: a comparative study:

Am. Naturalist, v. 102, p. 243-292.

---- 1969, Benthic marine diversity and the stability-time hypothesis, in Diversity and stability in ecological systems, Brookhaven Symposium in Biology, no. 22, p. 71-81.

Simpson, G. G., 1944, Tempo and mode in evolution: New York, Columbia Univ. Press, 237 p.

---- 1953, The major features of evolution: New York, Columbia Univ. Press, 434 p.

Smith, Alan Gilbert, 1971, Alpine deformation and the oceanic areas of the Tethys, Mediterranean and Atlantic: Geol. Soc. America Bull., v. 82, p. 2039-2070.

Smith, Alan Gilbert, and Hallam, A., 1970, The fit of the southern continents: Nature, vo.225, p.

Smith, Alan Gilbert, and Hallam, A., 1970, The fit of the southern continents: Nature, v. 225, p. 139-144.

Temple, P. G., and Zimmerman, J., 1969, Tectonic significance of Alpine ophiolites in Greece and Turkey: Geol. Soc. America Abs. with Programs, v. 7, no. 2, p. 22.

Trumpy, R., 1960, The paleotectonic evolution of the central and western Alps: Geol. Soc. America Bull., v. 71, p. 843-908.

Valentine, J. W., 1969, Patterns of taxonomic and ecological structure of the shelf benthos during Phanerozoic time: Palaeontology, v. 12, p. 684-709.

---- 1970, How many marine invertebrate fossil species? A new approximation: Jour. Paleontology, v. 44, p. 410-415.

---- 1971a, Resource supply and species diversity patterns. Lethaia, v. 4, p. 51-61.

---- 1971b, Plate tectonics and shallow marine diversity and endemism, an actualistic model: Systematic Zoology, v. 20, p. 253-264.

----, and Moores, E. M., 1970, Plate tectonic regulation of biotic diversity and sea level: a model: Nature, v. 228, p. 657-659.

Vine, F. J., and Hess, H. H., 1970, Sea-floor spreading, in Maxwell, A. E. et al., eds., The seas: New York, John Wiley & Sons, v. 4, pt. 2, p. 587-622.

# Taxonomic diversity during the Phanerozoic

## D.M. Raup (1972)

The evolution of taxonomic diversity is receiving increasing attention among geologists. The immediate reason for this is that diversity data may have a direct bearing on problems of plate tectonics and continental drift. The tantalizing possibility exists that diversity may be a good indicator of past arrangements of continents or climatic belts, or both. Valentine (1,2) has related temporal changes in fossil diversity to changes in climate and to the evolutionary consequences of continental drift. Stehli (3) and others have used spatial differences in diversity to interpret paleoclimates and paleolatitudes for single intervals of time.

Diversity information from the fossil record is also important because of its bearing on general models of organic evolution. Is the evolutionary process one that leads to an equilibrium or steady-state number of taxa, or should diversification be expected to continue almost indefinitely? Has equilibrium (or saturation) been attained in any habitats in the geologic past? If mass extinction has led to a significant reduction in diversity, what are the nature and rate of recovery? The answers to these and comparable questions depend in part on theoretical arguments, but their documentation must come ultimately from the fossil record itself.

The large-scale analysis of taxonomic diversity has been facilitated in the past few years by several important publications. The American Treatise on Invertebrate Paleontology (4) and the Russion Osnovy Paleontologii (5) are particularly valuable in having brought together vast amounts of taxonomic data with a minimum of inconsistency. Also, the British publication The Fossil Record (6) provides a useful synthesis of the geologic ranges of the higher taxa. This new literature, plus advances in data-processing

technology, makes possible a more
sophisticated study of diversity
problems than has been possible heretofore (7).

Valentine (1,2) used the newly
published data to estimate temporal
changes in diversity during the
Phanerozoic, the geologic time since
the end of the Precambrian. His conclusions were not dramatically
different from those of earlier
workers, but the breadth of documentation was far greater.

My purpose in this article is to
investigate the nature of the diversity
data to determine if more can be
learned from it. In particular, I will
examine the proposition that systematic
biases exist in the raw data such that
the actual diversity picture may be
quite different from that afforded by a
direct reading of the raw data. My
study will be limited to the major
groups of readily fossilizable marine
invertebrates (as was Valentine's) and
to changes in their worldwide diversity
through time.

TRADITIONAL VIEW

Figure 1 shows three histograms of
taxonomic diversity for the Phanerozoic. The three sets of data differ
somewhat in scope. Those of Valentine
(1) and Newell (8) are principally tied
to the family level, whereas Müller's
(9) are numbers of genera. All three
are limited mostly to the major groups
of fossilizable marine invertebrates:
Protozoa, Archaeocyatha, Porifera,
Coelenterata, Bryozoa, Brachiopoda,
Arthropoda, Mollusca, and Echinoder-

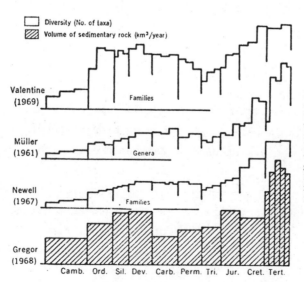

□ Diversity (No. of taxa)
▨ Volume of sedimentary rock (km³/year)

Valentine (1969)   Families

Müller (1961)   Genera

Newell (1967)   Families

Gregor (1968)

Camb. Ord. Sil. Dev. Carb. Perm. Tri. Jur. Cret. Tert.

Fig. 1. Comparison of the number of taxa and the volume of sedimentary rock during the Phanerozoic. The diversity data are based mainly on well-skeletonized marine invertebrates (1, 8, 9, 12).

mata, but Newell's data also include vertebrates. All three sets of data inevitably include some nonmarine and terrestrial taxa, but in none is this influence numerically significant.

The important fact is that all three show essentially the same picture and the one that has constituted the consensus for many years. The overall pattern is one of (i) a rapid rise in the number of taxa during the Cambrian and Early Ordovician, (ii) a maximum at about the Devonian, (iii) a slight but persistent decline to a minimum in the Early Triassic, and (iv) a rapid increase to an all-time high in diversity at the end of the Tertiary. Valentine (1,2) has suggested that the rise in diversity at the species level in Mesozoic to Tertiary time was an exponential one, with the late Tertiary having up to 20 times more species than the average for the mid-Paleozoic. This rise would appear even greater if insects, land plants, and terrestrial vertebrates were considered. These are particularly "noticeable" groups, important to man, and the history of their diversity has influenced thinking on the general subject.

It should be emphasized that the Phanerozoic diversity pattern yielded by the published taxonomic data depends on the choice of taxonomic level. As Valentine has pointed out, diversities at the levels of phylum, class, and

order have behaved very differently from those at the lower levels. The number of phyla has been essentially constant since the Ordovician, for example.

SEDIMENTARY RECORD AND DIVERSITY

It has been established that the general quality of the sedimentary rock record improves with proximity to the Recent (10,11). That is, the younger parts of the record are represented by larger volumes of rock (per unit of time), and the amount of metamorphism, deformation, and cover by overlying rocks is generally less. This is usually interpreted as resulting from the fact that the younger rocks are closer to "the top of the stack" and that, being younger, they have had less chance to be destroyed by erosion, metamorphism, and the like.

Figure 1 includes a graphic display of Gregor's estimate (12) of change in the sedimentary record through the Phanerozoic. The vertical coordinate in the lower graph is what Gregor calls the "survival rate" and is expressed as cubic kilometers of sediment per year now known and dated stratigraphically. This shows, for example, that the Devonian is represented by about twice the volume of sediments as the Cambrian (after adjustment for the relative durations of the periods). Gregor's survival data are comparable to estimates made on quite different bases by others (10, 13).

There is unquestionably a strong similarity between the patterns of taxonomic diversity at the genus and family levels and the pattern of sediment survival rate. This similarity suggests that changes in the quantity of the sedimentary record may cause changes in apparent diversity by introducing a sampling bias.

In spite of the fact that the patterns in Fig. 1 are correlated, a causal relationship is by no means demonstrated. Furthermore, the correspondence is not perfect, and both the diversity and sedimentary data are subject to many errors and uncertainties. The remainder of this article is devoted to a more detailed assessment of these relationships.

Gregor's data (Fig. 1) are estimates of survival rate for all sedimentary rocks, without distinction between marine and nonmarine. This detracts from the comparison with diversity because the biologic data are nearly free of nonmarine elements. Also, with the exception of the interval from the Devonian through the Jurassic, Gregor's numbers are derived from estimates of maximum sediment thickness (14). This part of the data is suspect because of the logical

problems involved in going from the maximum known thickness (in a local section) for a geologic system to the total volume of rock in that system (11). Furthermore, Gregor's rates are all sensitive to errors in estimates of the absolute time durations of the periods.

Thus, although there is little doubt about the general validity of Gregor's pattern, the inherent weaknesses prevent its use in more rigorous analysis.

By far the best data for sediment volumes are those published by Ronov (15). They are based on the results of a 10-year project of compiling lithological-paleogeographic maps and must be considered the most comprehensive data available. They are limited, however, to the Devonian-Jurassic interval. Ronov's data were used by Gregor where possible, but were modified by his calculation of survival rates. Ronov carefully distinguished between continental clastics, marine clastics, evaporites, marine and lagoonal carbonate rocks, and volcanics.

In Fig. 2, the taxonomic diversity data of Newell, Müller, and Valentine are compared with Ronov's estimates for the total volume of marine and lagoonal clastics and carbonates. Absolute time does not enter in because for each stratigraphic series total number of

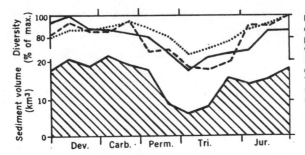

Fig. 2. Apparent taxonomic diversity compared with estimated volume of marine and lagoonal clastic and carbonate sediments. The diversity data are from Fig. 1. (Solid line) Valentine (*l*), (dotted line) Müller (*9*), (dashed line) Newell (*8*).

taxa and total sediment volumes are used. The diagram is thus free of most of the effects of errors in radiometric dating.

The correspondence between diversity and quantity of sediments is much stronger than indicated in Fig. 1. In particular, it should be noted that the Early Triassic diversity minimum coincides with a sediment minimum, which was not the case when Gregor's data were used. This is primarily because Gregor used Ronov's data for all sedimentary facies and because of the effect of Gregor's rate calculation.

It could be argued that the similarity between the patterns in Fig. 1 is due simply to a broad but independent increase in both sediment volume and diversity from the Cambrian through the Tertiary and that similarity in detail is quite accidental. Figure 2 largely denies this interpretation because the Carboniferous-Permian interval shows the reverse trend in both measures. Thus, although a causal relation is not

proved, the empirical relation appears to be strong enough to justify further investigation.

There is no disagreement on the proposition that the number of taxa known from the fossil record is less than the number that actually lived. This stems simply from the fact that some taxa (particularly at the species level) are rarely or never preserved. The effect is most striking when late Tertiary diversity is compared with the diversity of living organisms. There is no evidence for widespread extinction in the late Tertiary yet most groups have much smaller Tertiary records than would be predicted from neontological data. Furthermore, it is agreed that some biologic groups show fossil diversities closer to their actual diversities than do other groups because of inherent differences in preservability. Crustaceans, for example, are clearly underrepresented as fossils when compared with brachiopods or bivalves. The real problem, however, in the present context, is to evaluate relative changes in diversity over time, using the fossil record as the only available measure.

SAMPLING PROBLEMS

Many fossil taxa remain to be discovered. At the species level, this number probably exceeds even the number that have been described, although this would vary greatly from group to group. The diversity problem is thus in the realm of sampling theory and can be attacked from a mathematical viewpoint.

Exploration for fossils is analogous to problems in probability theory known variously as cell occupancy and urn problems. Consider a wooden tray which is divided into small compartments or cells, and assume that small balls are thrown randomly at the tray in such a way that each ball falls into a cell, without being influenced by the position of the cell or whether it is already occupied. The first ball thrown will inevitably result in the occupancy of one cell. The second ball may fall in the same cell and thus not add to the number of cells occupied: The probability of this event will be greatest if the total number of cells is small. At some point, all the cells will be occupied by at least one ball, and the waiting time necessary to accomplish this (measured in number of balls thrown) will depend only on the number of cells in the tray.

As noted above, the waiting time for occupancy of one cell is equal to 1 (one ball thrown). It can be shown (16) that the average additional waiting time for occupancy of a second cell is:

$$\frac{m}{m - 1}$$

where $m$ is the total number of cells in the tray. The additional waiting time for the third occupancy is:

$$\frac{m}{m - 2}$$

and so on. The total waiting time for complete cell occupancy then becomes:

$$m \left( \frac{1}{m} + \frac{1}{m-1} + \frac{1}{m-2} + \ldots + \frac{1}{2} + 1 \right)$$

Calculated curves for the expected waiting time for various values of $m$ are shown in Fig. 3A.

The appropriate paleontological analogy is as follows: Let $m$ be the total number of taxa available for discovery (thus, one cell equals one taxon), and let the balls thrown be the number of fossils found and identified or described. The first fossil discovered inevitably means recognition of one taxon. The second fossil may be the same or it may be from a second taxon (second cell occupied). Groups with fewer subgroups will require less sampling to be completely discovered.

This reasoning can be applied directly to the influence of taxonomic

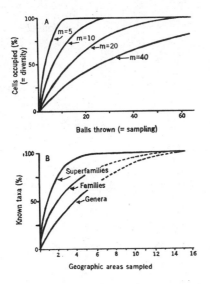

Fig. 3. Diversity as a function of sampling. (A) Illustration of cell occupancy problem. The average waiting time for cell occupancy varies with the number of cells to be occupied (*m*). (B) Effect of sampling on apparent diversity in fossil ammonoids of the *Meekoceras* zone (Triassic).

progresses, the ratio of the numbers of lower to higher taxa (genera per family, for example) steadily increases.

Figure 3 also shows a paleontological analog of the calculated curves. It is based on published data for ammonoids of the Meekoceras zone (Lower Triassic) (17). The data include the known occurrences of 58 genera in 15 geographic assemblages around the world. The ammonoid data (Fig. 3B) show the relationship between sampling and apparent diversity. Sampling is in this case expressed as the number of sites or areas sampled and is analogous to the number of balls thrown in the cell occupancy problem. The number of taxa found at one site in the Meekoceras case depends, of course, on which site is used. China, for example, yields well over half the genera and about three-quarters of the families; at the other extreme, the assemblage from the Caucasus has only two of the genera. The curves in Fig. 3 are therefore based on average expectations. For each taxonomic level, some of the values could be calculated directly; other values were determined by simulation based on a random selection of the published distributional data. The remainder (dashed lines) were extrapolated.

The ammonoid example demonstrates that apparent diversity is severely

level on observed diversity. In any fossiliferous rock unit, the number of families represented is inevitably equal to or greater than the number of phyla, the number of genera is equal to or greater than the number of families, and so on. Thus, much less sampling is required to find all or nearly all the phyla (low *m*) than the families or genera (higher *m* values). At any point in the sampling process, a larger percentage of the phyla will be known than of the lower taxa. In Fig. 3, the curves of low *m* are what would be expected for discovery of high taxa, and the curves of high *m* would be representative of lower taxa. It should be noted that as sampling

controlled by (i) the extent of sampling and (ii) the taxonomic level. At
the order level (Ammonoidea) any one of
the 15 sites is sufficient to yield 100
percent of the known diversity. At the
generic level it requires (in this
case) an average of 5 sites to exceed
50 percent. It should be emphasized
that the leveling off of the curves at
100 percent does not mean that the 15
sites yield all of the ammonoid
diversity in the Meekoceras zone: New
genera and new localities are still
being found.

As noted above, an increase in
sampling is accompanied by predictable
increases in the apparent number of
genera per family, and so forth, and
the effect is seen in the Meekoceras
zone data. That this is a general
phenomenon was noted by Simpson (18) as
follows: "Sampling at few, restricted
localities certainly reveals a much
higher percentage of the genera than of
the species that existed at any one
time."

The sampling problem need not be
analyzed only in the context of geographic extent of collecting. The
sampling axes of Fig. 3 could be
replaced by various measures of the
intensity of collecting or study (such
as number of paleontologists or years
of study) or by measures of the quality
of the fossil or rock record (extent of
outcrops, type of preservation, and

even accessibility of outcrops). The
fact that new taxa are constantly being
defined or discovered means that the
fossil record is still in a relatively
early stage of sampling and thus may be
represented by the steeper parts of the
curves in Fig. 3.

SOURCES OF ERROR IN DIVERSITY DATA

In the following numbered sections
I consider seven major sources of error
that may affect any set of diversity
data. All of them certainly have
influenced published diversity data of
the type shown in Fig. 1.

1) Range charts. When the objective of a diversity study is to estimate how many taxa lived during a given
interval of geologic time, the primary
source of information is usually a
range chart drawn at the appropriate
taxonomic level. If a family has a
range from the base of the Silurian to
the top of the Lower Devonian, for
example, it is assumed that the family
lived throughout the entire range.
Thus, the family is registered for the
Upper Silurian even though the Upper
Silurian fossil record may not actually
contain species of the family.

This procedure is valid biologically as a means of estimating actual
diversity, but it does have the effect
of overestimating "observed" diversity
for relatively unfossiliferous

intervals. In fact, an interval can be completely unfossiliferous yet still be credited with having considerable fossil diversity. This source of error becomes important when one is assessing the biasing effect of low sediment volume, as in the Permian-Triassic of Fig. 2. In this instance, the drop in fossil content of Permian rocks may be greater than it appears from the range chart data.

More important, the use of range charts introduces a systematic, time-related bias, as follows. Many (or most) range charts are incomplete in that the true first and last occurrences have not yet been found. In fact, the fossil record may not even contain the first or last occurrences (due to nonpreservation). Ranges of taxa may be truncated at either end, but truncation at the older end (first occurrence) has a higher probability because the older rocks have a greater chance of nonexposure or destruction by erosion and metamorphism. This means that the Phanerozoic diversity data are inevitably biased toward an increase in observed diversity through time.

2) Influence of "extant" records. Cutbill and Funnel (7) have already noted the biasing effect of the fact that ranges of fossil taxa are generally said to include the Recent if the taxa have living representatives. A not uncommon example would be a living group which has only one fossil occurrence, let us say in the Jurassic. Its range would be listed as Jurassic-Recent which, again, is valid for many purposes but causes problems in the present context. If the group had the same sparse fossil record but had not survived to the Recent, its range would be given as simply Jurassic. Cutbill and Funnel concluded that truncation at the "last occurrence" end of a range through nondiscovery is less likely if the group has living representatives, and since younger rocks contain more extant forms, the late Mesozoic and Cenozoic diversity data are consistently biased toward larger diversity and fewer extinctions than older parts of the column.

3) Durations of geologic time units. Consider the effects of the durations of periods and epochs on the diversity data in Fig. 1. The horizontal axis in the diagram is roughly adjusted for relative durations--albeit with little justification in many cases --but the vertical axes showing numbers of taxa are not. The height of each bar on the histograms indicates the total taxa which are found anywhere in the system or series or which have ranges that include those rocks. All things being equal, a long time interval will show a higher diversity than a short one. The effect of the bias is probably to overestimate diversity in

the early Paleozoic, where period and epoch durations are generally greater (7). This bias thus operates in a direction opposite to that of the two discussed above.

Furthermore, the bias is not easily corrected. The calculation of a simple ratio, such as families per million years, is valid when working with, for example, extinction rates, but only makes matters worse in the present context, where "standing crop" is the objective.

4) _Monographic effects_. The effects of the quality and quantity of taxonomic activity on apparent diversity are well known. It has been noted, for example, that the peak number of brachiopod genera shifted from Devonian to Ordovician largely as a result of the publication of one monograph (19). It is interesting to note that the generic peak has since shifted back to the Devonian.

Some of the monographic effects stem from the stratigraphic distribution of taxonomic specialists and taxonomic and phylogenetic philosophy, and perhaps even from the geographic distribution of taxonomists. Fossiliferous rocks in western Europe and eastern North America are more likely to be fully studied and thus to show higher diversity than rocks in other parts of the world.

If monographic effects are randomly distributed among the major phyla and throughout the stratigraphic column, then the consequences for overall trends in Phanerozoic diversity are minimal. Whether this lack of systematic bias exists is difficult to prove. If more families and superfamilies have been defined in the lower Paleozoic than in other parts of the geologic time table, it is impossible to say whether the difference reflects a tendency of lower Paleozoic paleontologists to be quick to erect such taxa, or whether it results from different kinds of diversity and states of preservation. At the very least, the monographic factors make highly precise studies of diversity impossible.

One special type of monographic effect is surely time dependent. If a group of organisms has many living representatives, and if biologists have subdivided it into many higher taxa, fossil representatives of these higher taxa are more likely to be recognized than if living forms are absent. This says in effect that it is easier to recognize a fossil taxon as distinct if the classification has already been established on the basis of the more complete morphological information afforded by living species. This bias has the effect that diversity is underestimated in extinct groups relative to nonextinct groups. For example, the discovery in Japan of a bivalved gastropod led to the reassignment of its Eocene counterpart from the Bivalvia to the Gastropoda. This greatly extended the stratigraphic range of the gastropod order Sacoglossa and thus increased the apparent gastropod diversity of the Tertiary (20).

5) Lagerstätten. Our knowledge of the history of life would be very different were it not for the occasional instances of spectacular preservation of large assemblages (Lagerstätten). Individual formations such as the Solnhofen, the Burgess shale, and the Baltic Amber as well as unusually fossiliferous groups of rocks such as in Timor and Madagascar have significant effects on diversity curves. In some cases, the lack of Lagerstätten is also significant. For example, the observed diversity of insects during the Cretaceous is essentially zero, but this is presumably only an artifact resulting from the lack of the special conditions required for good insect preservation during that period.

The distribution of Lagerstätten through time does not appear to be systematic although they are probably more common in younger rocks. To the extent that this is true, there will be a bias toward high diversity in younger rocks. The greatest effect, however, is to add "noise" to the diversity data in much the same way that monographic

bursts produce irregularity in diversity trends in the affected groups.

6) Area-diversity relationships. When a new geographic region is opened to exploration, new taxa are almost inevitably discovered. This is due in part to increased sampling, but it also results from the fact that taxa tend to be geographically restricted because of either climatic factors or barriers to dispersal. Also, diversity has been shown empirically to be area dependent (21).

Many instances of geographic effects could be cited. One example comes from Mortensen's tabulation of distributions of living cidarid echinoids, which shows that the 148 species and subspecies of the 27 genera are distributed among 18 geographic regions (22). Only one genus, Eucidaris, is found in as many as half the 18 regions, and 63 percent of the genera are confined to fewer than four regions. No single region contains even one-third of the species. This is in spite of the fact that most cidarids have a free-swimming larval stage.

If the cidarid distribution is looked at in terms of the probable fossil record it will leave, the potential effect of geographic restriction becomes greater. The biogeography of living echinoids is based on a reasonably good sampling of three-quarters of the earth's surface--that is, the oceanic areas. In the fossil record, sampling is limited for all intents and purposes to one-quarter of the earth's surface (the continents and islands), and a significant part of that quarter has remained out of the marine realm by being emergent during most of the Phanerozoic. Thus, the paleontologists can examine only a small fraction of the ocean area for any point in the geologic past. If one were to look at only 5 percent of the present ocean area (or even 5 percent of the present continental shelf area), the apparent diversity in groups such as echinoids would be greatly reduced at all taxonomic levels. This is particularly true since, in most geologic systems, the bulk of the record is usually concentrated in a few areas--rather than being randomly scattered over the world.

The effect of biogeography on diversity is greatest at the species level and decreases upward in the taxonomic hierarchy. Most modern phyla have worldwide distributions but even so are missing in some large regions, mainly due to climatic factors. At the family level, endemism becomes much more common, although this varies greatly from group to group.

The net effect of the biogeographic factor in the present context is to make the observed fossil diversity dependent not only on the area of rock exposure but also on the nature of the world distribution of exposures. Relatively small exposures on several continents are likely to yield a higher overall diversity than the same total exposure concentrated on one continent.

Although Gilluly has demonstrated a clear increase in area of exposure through the Phanerozoic column (10), no studies have been made on the manner in which these rocks are distributed spatially. However, because the probability of finding older Phanerozoic rocks is less than that of finding younger ones (assuming equal time durations) it would seem reasonable that geographic coverage improves toward the Recent. This should produce higher observed diversities in younger rocks.

7) Sediment volume. This article started with the empirical correlation between sediment volume per unit of time and diversity of major marine groups. It is clear from sampling considerations that more sedimentary record should produce more diversity. The correlation shown in Fig. 2 is thus quite plausibly a causal one. But the strength of the resulting bias depends on (i) the taxonomic level and (ii) the kinds of differences in sediment volume from one part of the column to another. A figure for sediment volume for one geologic system (such as used by Ronov (15)), may be higher than the figure

for another geologic system for many reasons. Discontinuous sedimentation may mean that many short-lived taxa are not preserved, but the fossil record of longer-lived taxa, characteristic of families and orders, may not be much affected. Thus, for example, if the Paris Basin had twice the volume of sediments, species diversity would be higher but family diversity little if any different. If sediment volume figures are influenced by differences in area of sedimentation, then the biogeographic relationships discussed above become significant, even at high taxonomic levels.

Postdepositional destruction or covering of sediments is the most widely accepted explanation for the temporal trends in sediment volume. Such losses of record are likely to have a spotty geographic distribution. That is, loss of the sedimentary record from one or more whole regions is more likely than small-scale reductions in all areas. This suggests that loss of biogeographic coverage is the important factor for diversity and that the sediment volume bias is closely tied to the geographic bias discussed earlier.

## MODELS FOR PHANEROZOIC DIVERSITY

Figure 4 shows in generalized form Phanerozoic diversity patterns at several taxonomic levels for shelf invertebrates with well-developed skeletons. The illustration is a composite of several from Valentine (1) and one from Müller (9). Minor irregularities were removed in making the composite, and vertical scales were adjusted. Valentine based the species curve on inference, but all the others were drawn directly from observed diversities.

Valentine concluded that the patterns are a plausible result of a combination of the evolutionary process of diversification and certain events in the physical history of the Phanerozoic. The basic biologic process envisioned requires that diversification take place first at high taxonomic levels (phylum, class, order) and later at successively lower taxonomic levels. The number of phyla (not shown) reached a maximum during or before the Early Ordovician, classes and orders later in the Ordovician, families in the

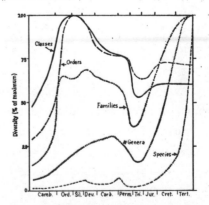

Fig. 4. Variation in apparent taxonomic diversity for several taxonomic levels of well-skeletonized marine invertebrates during the Phanerozoic.

Devonian, and genera and species in the Carboniferous or earliest Permian. According to Valentine, the diversity of the higher taxa (except phyla) declined after the initial peaks because as high taxa became extinct, they were replaced not by equally distinct groups but rather by specialized lower taxa (genera and species) within the surviving groups.

Still following Valentine's interpretation, the Permian-Triassic mass extinctions sharply reduced the diversity at all levels, and this was followed by a dramatic rise in diversity at the family, genus, and species levels, leading to the present-day array. Valentine argues that the driving forces behind this Mesozoic-Cenozoic rediversification were (i) continental drift and (ii) an increase in latitudinal temperature gradients. The diversity increase would presumably have taken place anyway--but to a lesser degree--as a continuation of the trend to specialization that was interrupted by the Permian-Triassic extinctions.

Figure 4 and its interpretation represent, therefore, one model for Phanerozoic diversity. It is an appealing one in that it is based largely on a "face value" use of empirical data and because it is biologically and ecologically plausible.

The foregoing interpretations are subject to several problems. The patterns in Fig. 4 contain elements that are qualitatively those which would be predicted from the biases discussed in this paper, as follows:

1) If the quality or quantity of sampling increases through time, it is inevitable that the ratios of species to genera, genera to families, and so on, will also increase.

2) Time-dependent biases should produce a rise in diversity at the lower taxonomic levels as the Recent is approached. The post-Paleozoic increase in numbers of families, genera, and species seen in Fig. 4 may be due to this factor.

3) Time-dependent biases should also shift any diversity peak toward the Recent (to the right in Fig. 4), and the amount of shift should be greatest at the lowest taxonomic levels. The fact, noted by Valentine (1), that diversities at lower taxonomic levels appear to have peaked after those at higher levels may actually be due to the effects of biases.

The last point deserves more consideration. From an evolutionary viewpoint, it is certainly plausible that diversity maxima for species and genera should occur after those for higher taxa in the same group. The question is whether the time lag is large enough and sufficiently universal to produce distinct offsets when diversities of several major animal groups are plotted together as in Fig. 4. If this were the case, periods of widespread extinction should be followed by recognizable intervals of low diversity, during which rediversification takes place. But the fact is that most major extinctions are not followed by periods of low diversity. Lowered diversity must have occurred at such times, but it evidently did not last long enough to be noticeable on the time scale used here. Valentine points out that the Permian-Triassic extinction is the only one which is followed by a diversity drop. Figure 2 indicates that in that interval the diversity drop may be an artifact of sampling.

An alternate model for Phanerozoic diversity is suggested by the dashed line in Fig. 5 and consists of a diversity maximum followed by a decline to an equilibrium level. The time scale in Fig. 5 is arbitrary, but a mid-Paleozoic position for the maximum is implied: The curve was suggested by the curves in Fig. 4 for classes and orders (where effects of biases should be least). The alternative model makes no distinction between taxonomic levels and thus is meant to apply to all levels below phylum. Thus, the assumption has been made that the

offset of diversity peaks caused by gradual diversification either is not large enough to be observed at this scale or is masked by noise resulting from the fact that many animal groups with different evolutionary histories are plotted together. The proposed model is, of course, valid only if the biases described in this article are quantitatively significant.

The plausibility of the alternative model was checked by a computer simulation. By using random numbers, hypothetical first and last occurrences were generated, and a range chart was constructed showing the distributions in time of 2000 hypothetical species (segregated into 100 genera). The dashed diversity curve in Fig. 5 was computed from the simulated range chart. The curve thus represents a hypothetical diversity pattern before biasing factors are applied.

Next, information was removed from the range chart by a random process designed to simulate the biasing factors. For each species, portions of the record were "destroyed," with the probability of destruction increasing back in time. Record losses occurring only inside a range had no effect. If, however, a loss included the beginning or end of the range, the range was shortened accordingly. In many cases, species were completely removed by this process. The Recent was made immune

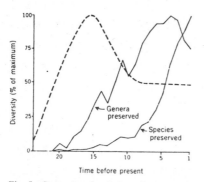

Fig. 5. Computer simulation of taxonomic diversity. The dashed line is a hypothetical diversity distribution before fossilization and is based on simulated ranges of 2000 species constituting 100 genera. The solid lines indicate the diversity trends after biases are applied to the range data.

from these information losses to stimulate the biasing effect of "extant" records.

Finally, a new range chart was constructed from what was left after the information removals. The diversity curves computed from this are also shown in Fig. 5. Species diversity increases sharply toward the Recent whereas generic diversity shows a maximum, offset to the right of the original maximum. When genera are grouped into hypothetical families (not shown), the diversity maximum is offset to the right but not as far.

The simulation demonstrates that diversity patterns such as are observed in the fossil record can be produced by the application of known biases to quite different diversity data. The simulation does not, of course, prove

the alternative model for Phanerozoic diversity because of our present ignorance of the actual impact of the biases. The simulation does suggest, however, that the model proposed in Fig. 5 is a plausible one for the Phanerozoic record of marine invertebrates.

The alternative model cannot be applied literally to land-dwelling forms because the exploitation of terrestrial habitats started much later in geologic time and may be still going on. The fossil record of terrestrial organisms is subject to the same biases, however, and so should be read with caution.

SUMMARY

Apparent taxonomic diversity in the fossil record is influenced by several time-dependent biases. The effects of the biases are most significant at low taxonomic levels and in the younger rocks. It is likely that the apparent rise in numbers of families, genera, and species after the Paleozoic is due to these biases. For well-skeletonized marine invertebrates as a group, the observed diversity patterns are compatible with the proposition that taxonomic diversity was highest in the Paleozoic. There are undoubtedly other plausible models as well, depending on the weight given to each of the

biases. Future research should therefore be concentrated on a quantitative
assessment of the biases so that a
corrected diversity pattern can be
calculated from the fossil data. In
the meantime, it would seem prudent to
attach considerable uncertainty to the
traditional view of Phanerozoic
diversity.

## REFERENCES AND NOTES

1.  J. W. Valentine, Paleontology 12,
    684 (1969).

2.  ----, Bull. Geol. Soc. Amer. 79,
    273 (1968); J. Paleontol. 44, 410
    (1970).

3.  F. G. Stehli, in Evolution and
    Environment, E. T. Drake, Ed. (Yale
    Univ. Press, New Haven, Conn.,
    1968), p. 163.

4.  R. C. Moore et al., Eds., Treatise
    on Invertebrate Paleontology
    (Geological Society of America and
    Univ. of Kansas Press, Lawrence,
    1953-1972).

5.  Y. A. Orlov, Osnovy Paleontologii
    (Akademiia Nauk SSSR, Moscow,
    1958-1964).

6.  W. B. Harland et al., Eds., The
    Fossil Record (Geological Society
    of London, London, 1967).

7.  J. L. Cutbill and B. M. Funnel,
    ibid., p. 791.

8.  N. D. Newell, in Uniformity and
    Simplicity, C. C. Albritton, Jr.,
    Ed. (Geological Society of
    America, New York, 1967), p. 63.

9.  A H. Müller, Grossabläufe der
    Stammesgeschichte (G. Fischer,
    Jena, Germany, 1961).

10. J. Gilluly, Bull. Geol. Soc.
    Amer. 60, 561 (1949).

11. J. D. Hudson, in The Phaenerozoic
    Time-Scale, W. B. Harland et al.,
    Eds. (Geological Society of
    London, London, 1964), p. 37.

12. C. B. Gregor, Proc. Kon. Ned.
    Akad. Wetensch. 71, 22 (1968).

13. See also the general discussion by
    R. M. Garrels and F. T. Mackenzie,
    Evolution of Sedimentary Rocks
    (Norton, New York, 1971).

14. The data on maximum thickness are
    mostly from A. Holmes, Trans.
    Edinburgh Geol. Soc. 17, 117
    (1959); M. Kay, in Crust of the
    Earth, A. Poldervaart, Ed.
    (Geological Society of America,
    New York, 1955), p. 665.

15. A. B. Ronov, Geokhimiya 1959,
    397 (1959), translated in
    Geochemistry USSR 1959, 493
    (1951).

16. M. Dwass, Probability (Benjamin,
    New York, 1970).

17. B. Kummel and G. Steele, J.
    Paleontol. 36, 638 (1962). A few
    of the ammonoid occurrences were
    designated as doubtful; these were

eliminated for the present purpose
with the effect that one of the
original 16 localities was eliminated.

18. G. G. Simpson, The Major Features
    of Evolution (Columbia Univ.
    Press, New York, 1953), p. 31.
    See also the excellent discussions
    of sampling problems and biases by
    J. W. Durham, J. Paleontol. 41,
    559 (1967) and by G. G. Simpson,
    in Evolution After Darwin, S. Tax,
    Ed. (Univ. of Chicago Press,
    Chicago, 1960), vol. 1, pp. 117-
    180.

19. G. A. Cooper, J. Paleontol. 32,
    1010 (1958); see also the general
    discussion of monographic effects
    in A. Williams, Geol. Mag. 94, 201
    (1957).

20. L. R. Cox and W. J. Rees, Nature
    185, 749 (1960).

21. F. E. Preston, Ecology 43, 185
    (1962); N. D. Newell, Amer. Mus.
    Nov. 2465 (1971).

22. T. Mortensen, A Monograph of the
    Echinodea (Reitzel, Copenhagen,
    1928), vol. 1.

# Phanerozoic taxonomic diversity —
# a test of alternate models

## J.W. Valentine (1973)

Although the fossil record forms
our only direct evidence of the course
of evolutionary and ecological history,
it is notoriously incomplete (1).  Many
of our historical interpretations must
be based on interpolations between
scattered datum points; in effect we
construct historical models that
explain the data at hand and that are
tested as new data appear.  Two such
models are available to describe the
course of taxonomic diversity of marine
biota during the Phanerozoic (2,3).
The purpose of this comment is to show
that the fossil data are adequate to
falsify one of them.

1) <u>Empirical model</u>.  Although the
processes of evolution and ecology
operate chiefly on species, the fossil
record of species is far too incomplete to serve as an adequate basis for
the interpretation of many paleo-
ecological patterns.  Taxa in

Reprinted from Science 180, 1078-1079
by permission of the author and publisher.  Copyright 1973 by the American
Association for the Advancement of
Science.

progressively higher categories,
however, are represented by
progressively more individuals over
progressively broader geographical and
temporal ranges and thus have increasingly better chances of being discovered in the record.  For diversity
estimation the family level is commonly
employed.  As diversity regulators
apparently operate on species rather
than directly on higher taxa, however,
it is important to estimate the species
diversities associated with the family
data.

Figure 1 depicts the Phanerozoic
diversity trends of well-skeletonized
marine benthic phyla, classes, orders,
and families as known from the fossil
record (2); note that each category has
a separate vertical scale.  The
diversity of taxa in increasingly lower
categories is increasingly volatile.
Below the phylum level, late lower
Paleozoic to early middle Paleozoic
diversity levels were high, but they
declined in late Paleozoic to a low at
the beginning of the Mesozoic.  Classes

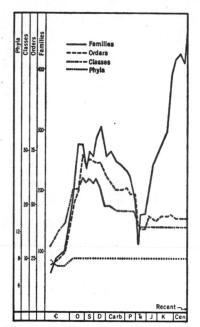

**Fig. 1.** Diversities of higher taxa of well-skeletonized benthic marine invertebrates as actually described from the Phanerozoic fossil record, plotted by period from Cambrian to Recent.

have remained at this low level, but orders increased somewhat during the Mesozoic and families underwent a great increase during the Mesozoic and Cenozoic. Genera of the best-known higher taxa that have contributed most to the post-Paleozoic rise in family diversity show an increase even more spectacular than that of the families. From such data it has been inferred that marine species diversity (Fig. 2B) rose to a mid-Paleozoic high, declined to a low at the close of the Paleozoic, and then underwent a Mesozoic-Cenozoic rise that raised species diversity by at least an order of magnitude over the

early Mesozoic level (2,4).

2) <u>Bias simulation model</u>. Our knowledge of diversity patterns and levels for living species far exceeds our knowledge of these factors for any time in the past. In general it is expected that preservation of ancient biotas would become successively poorer in successively older rocks, since the chances of destruction of fossils should increase with the time available. Raup (3) examined the main sources of bias in the fossil record in some detail, and while some of his points are arguable, it certainly seems clear that time-dependent biases do exist. Since higher taxa have a better chance of being recorded than lower taxa, higher categories should be proportionately better represented than lower ones at times when the record is poor. Therefore as the record improves through time the taxa in successively lower categories should display proportionately larger gains in diversity, even if diversities in all categories were temporally constant.

From such considerations, Raup (3) erected a model of Phanerozoic species diversity trends that is quite different from the empirical one (Fig. 2). He assumed an early species diversity maximum, presumably to correspond with the Ordovician to Devonian peaks in higher categories displayed in Fig. 1, and then a decrease to an intermediate species diversity plateau. He then employed a time-dependent bias to determine by computer simulation the diversities of genera and of species that would be registered in the fossil record. These resultant diversities rise toward the present, naturally, and the genera are proportionately better preserved than the species in progressively older rocks.

These two models imply radically different species diversity levels at certain times in the past (Fig. 2), so that if there were a way to obtain an estimate of actual diversity at one of these times it should be possible to falsify at least one of the models. In fact there is a way, and although it is indirect and does not involve actual species counting it nevertheless provides a strong test of these hypotheses.

The test revolves around our knowledge of how species diversity is accommodated in the marine biosphere at present. The regulators of diversity within habitats are still uncertain, though environmental stability is commonly considered to be a major factor. However, there is no question as to the way in which marine benthic diversity is chiefly accommodated on a planetary scale; the world's shelves,

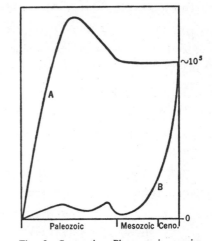

Fig. 2. Contrasting Phanerozoic species diversity trends as predicted by the bias simulation model (A) and the empirical model (B). The point where the curves meet on the right represents about 100,-000 species of well-skeletonized benthic marine invertebrates.

which contain more than 90 percent of the world's benthic marine species, are partitioned into provinces by major dispersal barriers--primarily by changes in thermal regimes in latitudinal directions, and by deep-sea or land barriers in longitudinal directions. The average species difference between latitudinally contiguous provinces with a common boundary is well over 50 percent (5), and between provinces separated by longitudinal barriers it is much higher. As a result, the more provinces that exist, the more species that are present. Allowing for diversity gradients, it has been conservatively calculated that marine species diversity levels associated with the present degree of shelf provincialization (over 30 marine provinces) is over ten times the level that would be accommodated in a single extensive tropical province, even one of high intraprovincial diversity (5).

The empirical model predicts that diversity in the early Mesozoic following Permian-Triassic extinctions was low, while the bias simulation model predicts that it was approximately the same as today but that the fossil record is so poor, especially for Permian-Triassic time, that the fauna appears to have been depauperate. Now, if the species diversity of Permian-Triassic time was anything like that of today, there must have been numerous marine provinces then. The amount of species packing required to accommodate such a diversity within the communities of a single province cannot be justified, especially in view of the sorts of invertebrate species that are known to have existed at the time. If, on the other hand, early Mesozoic diversity was quite low, then numerous provinces are not required and indeed would be difficult to account for.

The detection of a whole biotic province in the fossil record is far more likely than the detection of a fossil lineage or of a fossil community (6) for the same general reason that higher taxa are more easily detected than lower. Even for times when the record is poor, it is difficult to miss an entire province, considering the geographic extent of provinces and the density of fossil sampling even at worst. For early Mesozoic times, even well into the Jurassic, provinciality appears to have been quite low, for at times many fossil species are found on numerous continents while the number of species endemic to local regions is relatively low considering the nature of the fossil record. A good example is documented for the Lower Jurassic, for which sampling is reasonably abundant and is widespread; provinciality is nearly absent (7). In Late Jurassic and Early Cretaceous times provinciality increased, but compared with today it was still very low (8). It is conceivable that a province or two has been overlooked at times in the Mesozoic, but one or two provinces would not much affect the general picture. It is difficult indeed to escape the conclusion that early Mesozoic diversity was very low, and that as provincialization increased, diversity rose, eventually to its present level. In fact, these sorts of biogeographic considerations were implicit in the establishment of the empirical model (2,4).

Most other objections to the bias simulation model require extended discussions and add little to the conclusions. Two objections that require no preamble are: (i) the fossil associations that are found in the fossil record during times of low recorded diversity, such as the Early Triassic, have consistently low diversities themselves; and (ii) the time-dependent bias does not appear to operate during the Paleozoic, although this era encompasses the first 400 million years of Phanerozoic time.

Finally, we must ask why time-related biases, which must exist (3), do not play a larger role in the fossil record than is accorded them by the evidence reviewed here. Probably a major reason is that at the family level the record is good enough so that the temporal biases are greatly reduced, and in fact the empirical model was based chiefly on the family level for just this reason. Coupled with this is the possibility that short-term natural fluctuations in diversity together with episodic non-temporal biases in the record frequently outweigh the time-related biases on the family level. And finally, the progressive provincialization of late Mesozoic and Cenozoic times must have resulted in a progressive shrinkage of the average geographic range of species (and of genera and families to a lesser extent). Thus, the chances of discovery of a certain proportion of taxa in these categories were decreasing. And while Upper Cretaceous seas were widespread on continental platforms, Cenozoic seas were much more restricted, which further lessened the chances of discovery of taxa. How such decreases compare with an imputed increase in chances of discovery due to better sampling in younger rocks is not known, but they might well mask it completely.

It is concluded that the diversity trends suggested for the bias simulation model are not historically correct. Probably the general trends suggested by the empirical model are real. Nevertheless, it would be a mistake to suggest that the species diversity levels inferred from the empirical model are any more than very rough estimates (indeed it is probable that the Paleozoic levels were underestimated by a factor of 2 or so, owing to an underestimate of Paleozoic provinciality). Clearly, paleontologists should work to develop improved estimates of biases and incompletenesses of all types in the record, as Raup has done.

REFERENCES

1. C. Darwin, On the Origin of Species by Means of Natural Selection (Murray, London, 1859).
2. J. W. Valentine, Palaeontology 12, 684 (1969).

3.  D. M. Raup, Science 177, 1065
    (1972).

4.  J. W. Valentine, J. Paleontol. 44,
    410 (1970).

5.  ----, Geol. Soc. Amer. Bull. 79,
    273 (1968).

6.  ----, J. Paleontol. 42, 253 (1968).

7.  W. J. Arkell, Jurassic Geology of
    the World (Oliver and Boyd, London,
    1956); R. W. Imlay, J. Paleontol.
    39, 1023 (1965); A. Hallam, Amer.
    Ass. Petrol. Geol. Bull. 49, 1485
    (1965).

8.  N. F. Sohl, Proc. North Amer.
    Paleontol. Conv. (1969), p. 1610;
    G. R. Stevens, J. Roy. Soc. N.Z.
    1, 145 (1971).

# Lithospheric plate motions, sea level changes and climatic and ecological consequences

## J.D. Hays and W.C. Pitman (1973)

The global nature of the great marine transgressions and regressions such as occurred in the Upper Cretaceous has been recognised for nearly a century[1]. These fluctuations of sea level have been variously attributed to gradual filling of ocean basins by detritus displacing water onto the continents, down faulting of ocean basins to cause regressions[1] and simultaneous vertical movements of both continents and ocean basins[2,3]. The transgressions and regressions have been linked to orogenic cycles; it has been argued[2] that during intervals of low orogenic activity reduced horizontal compression caused continental subsidence relative to sea level, hence transgressions, and that conversely orogenic episodes produced increased horizontal compression, increased continental elevation and marine regression[4]. Late Cainozoic sea level changes are certainly attributable to glaciation, but this cause is not applicable to the Upper Mesozoic.

Changes in the volume of ocean basins could explain flooding of portions of continental surfaces. It has been proposed[5-7] that such changes occur due to alterations in the volume of the mid-oceanic ridges. Valentine and Moores[5] linked these volumetric changes to the assembly and breakup of super continents. Hallam[8] speculated that the Upper Cretaceous transgression and regression may have been caused by a contemporaneous pulse of rapid spreading which substantially increased the volume of the world ridge system. We show that this latter proposal is correct.

CAUSE OF UPPER CRETACEOUS TRANSGRESSION

The lithosphere formed at a spread spreading ridge axis is initially hot and therefore elevated; as it moves away from the axis, it cools and subsides[9,10]. This cooling and subsidence is time dependent[11]; the depth to which any portion of flanking crust has subsided is essentially a function of age only. So one empirical age-depth relation fits most ridges regardless of spreading rate[11] (Table 1). There-

fore, the volume of any ridge is a function of its spreading rate history and changes in the spreading rate cause, in time, changes in ridge volume. (It has been suggested that the axial portion of fast spreading ridges is deeper than that of slow spreading ridges[11]; however, subsequent analysis has shown that this is not systematically true[12].) Larson and Pitman[13] correlated anomaly lineations of Middle and Upper Mesozoic age in the Atlantic and Pacific. They calibrated

Table 1    Empirical Values showing the Relationship between Crustal Age and Ridge Elevation[11]

| Age m.y. | Mean depth corrected (m) | Standard deviation (m) | No. of values | Mean depth at 5 m.y. intervals (m) | Elevation above 5.5 base level (m) |
|---|---|---|---|---|---|
| 0 | 2,644 | $\underline{+109}$ | 40 | 2,644 | 2,856 |
| 2 | 2,948 | $\overline{+93}$ | 46 | | |
| 4 | 3,178 | $\overline{+96}$ | 46 | | |
| 5 | | | | 3,243 | 2,257 |
| 6 | 3,309 | +94 | 33 | | |
| 8 | 3,480 | $\overline{+81}$ | 32 | | |
| 10 | 3,545 | $+\overline{155}$ | 51 | 3,545 | 1,955 |
| 15 | | | | 3,788 | 1,712 |
| 20 | | | | 4,029 | 1,471 |
| 21 | 4,078 | $\underline{+33}$ | 33 | | |
| 25 | | | | 4,205 | 1,295 |
| 29 | 4,332 | $\underline{+17}$ | 17 | | |
| 30 | | | | 4,371 | 1,129 |
| 33.5 | 4,507 | $\underline{+77}$ | 13 | | |
| 35 | | | | 4,556 | 944 |
| 38 | 4,655 | $\underline{+46}$ | 27 | | |
| 40 | | | | 4,718 | 782 |
| 45 | | | | 4,879 | 621 |
| 50 | | | | 5,041 | 459 |
| 53 | 5,137 | $+47$ | 20 | | |
| 64 | 5,432 | $+\overline{100}$ | 15 | | |
| 69.3 | | | | 5,500 | 0 |
| 77 | 5,600 | $\underline{+63}$ | 13 | | |

these lineations with Deep-Sea Drilling Project data, thereby extending the magnetic polarity time scale to -160 m.y. From the geometry of these lineations they showed that during the Upper Cretaceous (-110 m.y. to -85 m.y.) there was an episode of rapid spreading in the central and south Atlantic and the Pacific. The geological evidence suggests an initial rise of sea level that began at or before the boundary between Upper and Lower Cretaceous (-100 m.y.), the cresting of this rise some time between the Turonian and Lower Maastrichtian (-90 and -70 m.y.) and a withdrawal that was most pronounced in the Maastrichtian but continued into the Cainozoic[14-17].

We have computed the volume of most of the mid-oceanic ridge system at several times, beginning in the Cretaceous and extending to the upper Tertiary (Table 2). For each time and for each ridge segment an average cross-sectional area was computed using a standard age-depth curve[11] (Table 1). The data are referred to present day sea surface. Since sea level varies with time it is more meaningful to consider ridge volume in terms of elevation above an abyssal reference plain of 5,500 m. A history of spreading rates for the ridge segments is adapted from Larson and Pitman[13] (Table 3).

We use the Upper Cretaceous system of ridges proposed by Larson and Pitman[13] (Fig. 2a, Table 3). In the Atlantic this system has not altered significantly since -180 m.y. except to extend both northwards and southwards. The Pacific ridges, however, are today different from the Cretaceous configuration[13] (Fig. 2, a and c). For example, the lineations of Mesozoic age form a double magnetic bight indicating that a ridge-ridge-ridge triple junction must have existed at that time

Table 2    The Calculated Volume of the Ridge Segments at Various Times is Given as are the Changes in Total Volume and Sea Level

|         | -110 | -100 | -90 | -85 | -80 | -70 | -60 | -50 | -40 | -30 | -20 | -10 |                    |
|---------|------|------|------|------|------|------|------|------|------|------|------|------|--------------------|
| PAC-PH  | 24.2 | 44.3 | 49.0 | 64.9 | 57.5 | 47.5 | 39.1 | 32.6 | 26.3 | 22.0 | 20.0 | 19.4 |                    |
| PH-FAR  | 21.8 | 35.7 | 45.9 | 50.0 | 42.9 | 33.0 | 24.9 | 18.6 | 13.1 | 9.4  | 7.6  | 7.3  |                    |
| PAC-FAR | 26.7 | 43.7 | 56.1 | 61.1 | 57.7 | 53.7 | 48.2 | 44.6 | 39.1 | 47.1 | 35.8 | 35.6 | Volume             |
| KUL-FAR | 36.4 | 59.5 | 76.4 | 83.3 | 78.7 | 71.9 | 65.8 | 60.9 | 53.3 | 50.6 | 48.8 | 48.5 | $\times 10^6 (km^3)$ |
| PAC-KUL | 51.8 | 49.4 | 44.7 | 43.4 | 35.6 | 23.6 | 14.6 | 7.9  | 3.8  | 1.3  | 0.1  | 0    |                    |
| A-NA    | 6.5  | 10.6 | 13.6 | 14.8 | 14.8 | 14.7 | 14.5 | 14.3 | 13.6 | 13.2 | 13.0 | 12.9 |                    |
| A-SA    | 0.0  | 11.5 | 20.0 | 23.4 | 22.6 | 21.5 | 20.1 | 19.3 | 16.7 | 15.3 | 14.6 | 14.6 |                    |
| E-NA    |      |      |      |      | 2.0  | 5.0  | 7.3  | 9.0  | 10.2 | 10.9 | 11.2 | 11.3 |                    |

| | -110 | -100 | -90 | -85 | -80 | -70 | -60 | -50 | -40 | -30 | -20 | -10 |
|---|---|---|---|---|---|---|---|---|---|---|---|---|
| $\Delta V \times 10^6 \ km^3$ | | 85.4 | 148.2 | 173.7 | 144.9 | 102.8 | 67.5 | 40.0 | 9.7 | -5.6 | -14.3 | -15.7 |
| $\bar{h}$ - dm | | 212 | 362 | 421 | 354 | 254 | 168 | 100 | 24.0 | -14.0 | -33.0 | -40.0 |

at points A and B of Fig. 2a. This system of ridges evolved into the present system (Fig. 2c). The Phoenix plate is now the Antarctic plate and Juan de Fuca, Cocos and Nasca plates are all that remain of the Farallon plate; the Kula plate is gone. The Galapagos Ridge (between the Cocos and Nasca plates) is no more than 10 m.y. old and does not affect our computations. Beginning at -75 m.y. there was a phase of spreading at about 3 cm yr$^{-1}$ between the small remaining eastern fragment of the Kula plate and the Pacific plate. (This event yielded the "Great Magnetic Bight of the Northeast Pacific"[18].) By -38 m.y. at the latest this spreading had ceased and the entire Kula plate with its bordering ridges had been subducted[19]. The volumetric increase due to this brief episode would have been small.

In calculating the volumetric changes for each interval, we have kept the length of each ridge segment fixed. In reality, the length of the ridges changes with time, but we assume that as some ridges have grown in length an equivalent volume of other ridges has been subducted.

The Indian Ocean and Tethys Sea have been left out of our calculation. The spreading of the Indian Ocean is known in sufficient detail[10] but since the Tethys is gone its history cannot be determined. As a first approximation we assume that as Tethys closed due to the drift of various Gondwana fragments into Eurasia, any ridge system that existed in Tethys gradually subsided, ridges and ridge axes were subducted. We also assume that this consequent volumetric decrease was about equal to the simultaneous

---

Table 3    Lengths and Half Spreading Rates of Ridge Segments from -110 m.y. to -9 m.y.

| | Length of ridge (km) | Half spreading rates (cm yr$^{-1}$) | | |
| --- | --- | --- | --- | --- |
| | | Prior to 110 m.y. | 110-85 m.y. | 85-10 m.y. |
| Pacific-Phoenix (PAC-PH) | 3,300 | 5 | 18 | 4 |
| Phoenix-Farallon (PH-FAR) | 4,950 | 3 | 9 | 1 |
| Pacific-Farallon (PAC-FAR) | 6,050 | 3 | 9 | 4 |
| Kula-Farallon (KUL-FAR) | 8,250 | 3 | 9 | 4 |
| Pacific-Kula (PAC-KUL) | 8,800 | 4 | 3 | 0 |
| Africa-North America (A-NA) | 4,400 | 1 | 3 | 2 |
| Africa-South America (A-SA) | 4,950 | 0 | 5 | 2 |
| Europe-North America (E-NA) | 3,850 | 0 | 0 | 2 |
| Total | 45,550 | | | |

volumetric increase in the ridges of the Indian Ocean.

To convert changes of ridge volume into changes of continental freeboard, we calculated isostatic adjustment due to increased water depth. If water depth increases by a thickness $h$, oceans subside a distance $d$. Assuming 3.3 g cm$^{-3}$ for the density of the upper part of the mantle, $h = 3.3d$ and the change in freeboard will be $(h - d) = 0.7h$.

A second correction arises because the boundaries of the oceans are not vertical. As the oceans rise, more surface area is covered (Fig. 1). Approximately 1/6 of the Earth's surface (85 x 10$^6$ km$^2$) lies at elevations between 0 and +500 m (ref. 20). We assume that as sea level rises the area covered by the sea increases linearly (0.170 x 10$^6$ km$^2$ for each 0.001 km rise in sea level). Thus, the

change in continental freeboard $(h - d)$ may be calculated from the equation

$$V = hA_0 + (0.7h)^2 170/2$$

where $A_0$ is the surface area of the ocean at -110 m.y. and $(0.7h)^2 170/2$ is the volume occupied by the sea encroaching on the continent. Hallam[8], using the data of Termier and Termier[21], shows that 33 x 10$^6$ km$^2$ of presently emergent crust was covered by seas at -110 m.y. The present area occupied by the oceans is 359 x 10$^6$ km$^2$; thus the oceanic area at -110 m.y. was 392 x 10$^6$ km$^2$. The freeboard change $(h - d)$ has been calculated from the equation for each cumulative total change in volume.

The values of $(h - d)$ given in Table 2 have been computed relative to sea level taken as zero at -110 m.y. But since 33 x 10$^6$ km$^2$ of the present day land area was covered by the seas[8] at that time we estimate (using a standard hypometric curve[20]) that sea level was about 100 m higher than today. The curve in Fig. 3 has been adjusted to give sea levels relative to the present.

Estimates of the late Mesozoic-Cainozoic changes in sea level computed from our model agree well with other estimates based on outcrops of marine sediments (Fig. 3). All except Termier and Termier[21] place the time of the maximum transgression at about -85 m.y.

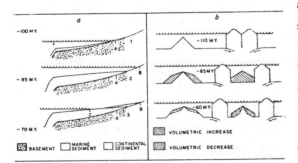

**Fig. 1** *a*, Schematic Upper Cretaceous two ocean ridge system. As spreading rates increase at −110 m.y. ridges expand horizontally causing a transgression by −85 m.y. Decrease in spreading rates after −85 m.y. causes ridges to contract and the seas to withdraw. *b*, As sea level rises from −100 m.y. (A) to −85 m.y. (B), sediments fill and level the inland seafloors. In contrast, as sea level lowers an equivalent amount between −85 and −70 m.y., the area of continent exposed is much greater (curve B to C).

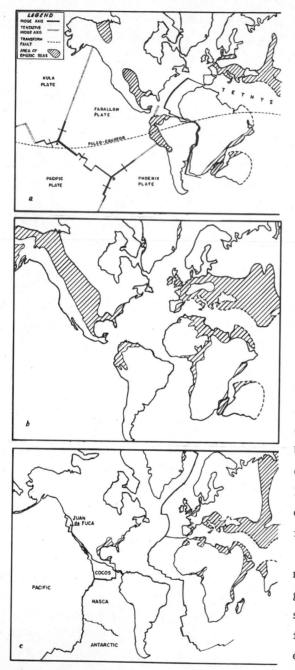

Fig. 2. The extent of the seas for three geological epochs 14,
16,17,21; *a*, Cenomanian (— 100 to — 94 m.y.); *b*, Santonian-
Campanian (— 70 to — 85 m.y.); *c*, Eocene (— 50 to — 40 m.y.).
The arrangement of the ridge axes at present is shown in *c*,
where they are plotted with respect to North America.

Our curve indicates that sea level 10 m.y. ago was 40 m above the present stand. The total present continental ice volume, most of which accumulated during the past 10 to 20 m.y., is equivalent to 60 m of sea level chang change[23]. Therefore, our model predicts a present day sea level within 20 m of that observed.

Our estimates differ from those inferred geologically by others[21-23] in two respects (Fig. 3). First, we obtain higher sea levels during the maximum transgression. Our calculation may contain errors inherent in our assumptions but the corresponding geological estimates of past marine transgressions are based on the extent of Upper Cretaceous marine sediments, and hence are low, since 70 m.y. of erosion must have removed some of the Upper Cretaceous deposits[7]. Second, our model predicts that the rise of sea level during the transgression will occur at about the same rate as the fall during the regression.

Some geological data point to a more rapid regression than trans-gression. But it is likely that sedimentation led to the continuous filling and levelling of the epicontinental seafloors. Therefore, depths of the Cretaceous epicontinental seas in general would have been less than 500 m, perhaps only 100 to 200 m

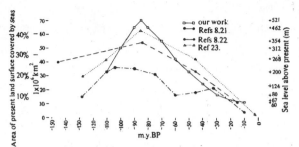

Fig. 3. Area of present day land covered by former seas plus the height of these seas above present levels plotted against age (adapted from ref. 8).

as suggested by the limited palaeontological evidence[24]. Consequently, the area drained by the first 200 m drop of sea level would have been very large relative to the area flooded by the latest 200 m rise of the transgression (Fig. 1b). We estimate that between -85 m.y. and -60 m.y. sea level would have fallen by more than 250 m. This would have been sufficient to drain most continental areas and is in agreement with geological evidence.

## CONSEQUENCES OF OUR MODEL

What is the relationship between tectonism, climate change and fluctuations in faunal diversity? We believe that our model provides a working hypothesis to connect these phenomena. The effects of the Upper Cretaceous transgression and regression on climatic, stability, faunal diversity and ocean circulation (Fig. 4) are not necessarily newly discovered relationships. But we discuss them

briefly to emphasize that all are related to changes in the rate of plate motion.

World climate: The uniformity and relative mildness of climate at the height of the Upper Cretaceous transgression are well documented[25-27]. An increase in the ratio of oceanic to continental area moderates and stabilizes climate[4]. This is because the heat capacity of water is very large compared with continental rocks or the atmosphere. With more than 40% of the continents covered by several hundred metres of water, this moderating effect must have been large. Also, the expanse of the epicontinental seas provided new corridors along which heat could be advectively transferred between low and high latitudes. Conversely, as the seas withdrew beginning in the Upper Cretaceous, continentality increased probably producing the well documented high latitude cooling and increased seasonal

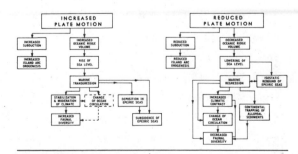

Fig. 4. Proposed sequence of events causally related to an Upper Cretaceous worldwide increase in the velocity of plate motion and subsequent decrease.

contrasts that continued through the Tertiary. This cooling trend leads to polar glaciation in the mid-Tertiary[28] and mid-latitude glaciation in the Quaternary. Palaeomagnetic data[36] demonstrate that the Arctic Ocean and possibly Antarctica had drifted to a polar position by the Upper Cretaceous[30,36]. We therefore attribute these Late Cainozoic glaciations to continued fall of sea level (which caused increasing emergence of all continents, including Antarctica) and increased constriction of oceanic thermal advection between low latitudes and the Arctic Ocean. Post-Cretaceous opening of the North Atlantic did not significantly improve circulation with the Arctic because, as the Atlantic opened, routes to the Pacific closed[31].

Ocean circulation: All available evidence suggests that the Upper Cretaceous oceans and their epeiric seas were well mixed and productive in spite of the low thermal gradient between poles and equator. The pattern of circulation is not known, but it was probably quite different from that of the present.

Today the deep circulation of the oceans is maintained by high latitude formation of dense water in restricted seas or shallow shelf areas. We expect that a similar mechanism operated in the Upper Cretaceous and the source areas for deep and bottom waters lay in some of the epeiric seas. With the withdrawal of the seas at the close of the Cretaceous some or all of these source areas may have been drained. New source areas would develop as the polar regions cooled but there may have a short but important interval of stagnation as one regime of deep circulation was replaced by another. If a brief period of stagnation did occur there is no evidence that it led to the development of anaerobic conditions in any of the ocean basins.

Mesozoic-Cainozoic faunal crisis: The Upper Cretaceous stable and uniform climates allowed significant increases in diversity of many groups[32,33]. The adaptation to these conditions produced an inherent vulnerability to change[34]. We conclude that the marine withdrawal was most rapid between -70 m.y. and -60 m.y. because of isostatic rebound and sediment infilling of epeiric seas (Fig. 1). Thus, in latest Cretaceous, the land area would have nearly doubled. The consequent environmental changes (increased thermal gradients and seasonal contrast[33,35] and marked alteration of the ocean circulation and thermal regime) were probably equally rapid, producing stresses on species adapted to stable Upper Cretaceous conditions.

We acknowledge discussions with R. Batten, W. Broecker, A. Gordon, D. Hayes, M. Langseth, M. McKenna, and N. Newell.

[1] Seuss, E., in The Face of the Earth II (Oxford University Press, 1906).

[2] Stille, H., Einführung in don Bau Amerikas (Oebruder Borntraeger, Berlin, 1940).

[3] Umgrove, J. H. F., in The Pulse of the Earth, second ed. (Nijhoff, The Hague, 1949).

[4] Brooks, C. E. P., in Climate Through the Ages (Ernest Benn, London, 1949).

[5] Valentine, J. W., and Moores, E. M., Nature, 228, 657 (1970).

[6] Menard, H. W., in Marine Geology of the Pacific (McGraw-Hill, New York, 1964).

[7] Hallam, A., Am. J. Sci., 201, 397 (1963).

[8] Hallam, A., Nature, 232, 180 (1971).

[9] Menard, H. W., Earth planet. Sci. Lett., 6, 275 (1969).

[10] McKenzie, D., and Sclater, J. G., Geophys. J. R. astr. Soc., 25, 437 (1971).

[11] Sclater, J., Anderson, R., and Bell, M. L., J. geophys. Res., 32, 7888 (1971).

[12] Anderson, R. N., McKenzie, D. P., and Sclater, J. G., Earth planet. Sci. Lett., 18, 391 (1973).

[13] Larson, R. L., and Pitman, W. C., III, Bull. geol. Soc. Am., 83, 3645 (1972).

[14] Gignoux, M., Stratigraphic Geology: Translation of fourth French Edition 1950, 1-16 (Freeman, San Francisco, 1955).

[15] Naidin, D. P., in Stockholm Contrib. Geol. (edit. by Gavelin, S., and Hessland, I.), 3, 127 (1960).

[16] Schuckert, C., in Atlas of Paleo-geographic Maps of North America (Wiley, New York, 1955).

[17] Brinkman, R., The Geological Evolution of Europe (Hafner, New York, 1960).

[18] Elvers, D. J., Mathewson, C. C., Kohler, R. E., and Moses, R. L., in Operational Data Report C and GSDR-1 (1967).

[19] Atwater, T., Bull. geol. Soc. Am., 81, 351 (1970).

[20] Sverdrup, H. U., Johnson, M. W., and Fleming, R. H., in The Oceans (Prentice Hall, Englewood Cliffs, New Jersey, 1942).

[21] Termier, H., and Termier, G., in Atlas de Paleogeographie (Masson, Paris, 1960).

[22] Strakhov, N. M., Stages in the development of external geospheres and the formation of deposits in the history of the Earth (Akad. Nauk, SSSR, Geol. Ser., 12: 3-22).

[23] Ronoff, A. B., _Sedimentology_, _10_, 25 (1968).

[24] Ollson, R. K., _Bull. Am. Ass. Petrol. Geol._, _47_, 654 (1963).

[25] Lowenstam, H. A., and Epstein, S., _Proc. Int. geol. Congr. twentieth Symp._, _1_, 65 (1959).

[26] Dorf, E., _Am. Scient._, _48_, 364 (1960).

[27] Newell, N., _Geol. Soc. Am. Spec. Pap._, _89_, 63 (1967).

[28] Denton, G., Armstrong, R. L., and Stuiver, M., in _Late Cenozoic Glacial Ages_ (edit. by Turekian, K. K.)(Yale University Press, 1971).

[29] Donn, W. L., and Ewing, M., _Science, N. Y._, _152_, 3730 (1966).

[30] Pitman, W. C., III, and Talwani, M., _Bull. geol. Soc. Am._, _83_, 619 (1972).

[31] Talwani, M., and Edholm, O., _Bull. geol. Soc. Am._, _83_, 3575 (1972).

[32] Tappan, H., _Palaeogeogr., Palaeoclim., Palaeoecol._, _4_, 187 (1968).

[33] Newell, N., _Am. Mus. Novitates_, 2465 (1971).

[34] Bretsky, P. W., and Lorenz, D. M., _Bull. geol. Soc. Am._, _81_, 2449 (1970).

[35] Axelrod, D. F., and Bailey, H. P., _Evolution_, _22_, 595 (1968).

[36] McElhinny, M. W., _Paleomagnetism and Plate Tectonics_ (Cambridge University Press, 1973).

# Seafloor spreading theory and the odyssey of the green turtle

## A. Carr and P.J. Coleman (1974)

A subpopulation of the green turtle (Chelonia mydas) lives on the coast of Brazil but breeds and nests 2,000 km away on Ascension Island in the central equatorial Atlantic[1-6]. The feat of navigation involved in this odyssey has not been explained[5,7]. A puzzle of equal stature, and of more fundamental importance, arises in trying to account for the initial stages in the evolution of the adaptation[3,4]. Populations of the genus Chelonia have distinctive characteristics--enormous shoulder musculature together with its support, heavy fat deposits, and a peculiar jaw structure. The ecological regimen responsible for these characteristics is evidently the herbivorous feeding habit[1,8] and the need that this imposed for travel between a protected, shallow water pasture ground and the exposed shores on which suitable nesting beaches develop. Offshore and oceanic islands

are likely to have good surf-built beaches and they almost always offer relative freedom from the nest-predators which plague mainland nesting beaches. In spite of these advantages, however, the Ascension colony has established the adaptations in the face of huge difficulties, inherent in the initial stages, which seem to make impossible demands on the process of natural selection. Seafloor spreading theory may bear directly upon this aspect, and it also offers clues to the navigation problem.

TURTLES AND CENTRAL ATLANTIC OPENING

Marine turtles of Chelonia type inhabited the seas between 'North America' and northwestern Gondwanaland by the beginning of the late Cretaceous, about 100 million years (Myr) ago. Because of the cheloniid fossil record in this general area (refs 9-12 with the references provided there), we postulate that these ancient turtles included the

ancestors of <u>Chelonia</u> <u>mydas</u>, even
though its phylogeny cannot at present
be reconstructed. (<u>Chelonia</u> has been
found in Miocene sediments[9,10].) The
northern coast of South America was a
suitable habitat for these early
turtles, providing a tropical to sub-
tropical environment. There is fossil
evidence that even then the herbivorous
habit had evolved and that there was a
separation of residence-pasture and
breeding grounds[12]. Mammalian egg-
predators no doubt acted as a limiting
factor in turtle ecology at this time,
augmenting the advantage in island
nesting.

The opening of the equatorial
Atlantic[13-16] (Fig. 1<u>a</u> and <u>b</u>), marking
the final separation of South America
and Africa at about 80 Myr ago, took
place in steps: first, rift valley
formation at about 110 Myr ago, or even
earlier; second, sporadic but
progressive ocean flooding with a

of a ridge system which linked the
spreading ridges of the North and South
Atlantic and made of it a single ocean
by 80 Myr ago.

Volcanic piles are a frequent
feature of midoceanic ridges, and some
may grow sufficiently to emerge as
islands. As spreading continues, they
become inactive and are carried outwards and downwards and become seamounts[17], the deepest being furthest
from the ridge axis. They are sporadi-
cally replaced by new volcanoes on the
ridge so that for the continental
observer the volcanoes march seawards
through time (Fig. 2). The ancestral
turtles could thus migrate outwards
with the volcanic islands, without the
demands on the process of natural
selection becoming excessive.

## TURTLE MOVEMENT INTO THE BRAZIL-WEST AFRICAN CHANNEL

Before 100 Myr ago, turtle populations travelled between residence-
pasture and breeding grounds along the
shores of northern South America. By
90 Myr ago, a corridor had opened at
the head of the gulf between Brazil and
West Africa in line with the establish-
longshore travel paths. By 80 Myr ago,
the corridor was complete, much new
coastline had been added, and there was

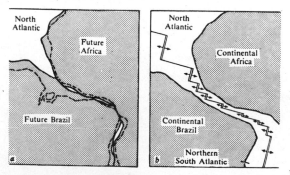

Fig. 1. *a*, Predrift reconstruction of equatorial 'South America'
and 'West Africa', as parts of north-western Gondwanaland, based
on the computer-tested bathymetric fit of Bullard *et al.* (ref. 23).
*b*, Reconstruction for the time of about 80 Myr ago. A 'Red Sea'
connects the northern Atlantic with the younger southern Atlantic
and its youthful spreading ridge. Spreading ridge detail is
schematic.

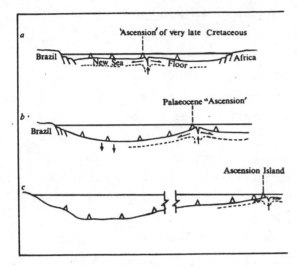

Fig. 2. The progressive appearance, then disappearance, of spreading ridge volcanes as part of seafloor spreading mechanism. *a*, Approximately 70 Myr BP; *b*, approximately 60 Myr BP; *c*, approximately 1 Myr BP.

a string of offshore islands. The exploitation of this oceanic channel by the turtles simply required repetitive extension of previous travel paths (Fig. 3). The bearings were WNW-ESE. This is close to a latitudinal course, and could be navigated relatively easily. Locating a particular beach on an island within this narrow sea or along its coasts would involve a longitudinal fix: local currents, rivers, even surfbeat, would offer individual signatures. It seems reasonable to assume that this pattern of travel, constant over a long period of time would become an established, heritable part of the turtle's behaviour.

By 80 Myr ago the Brazil-West Africa sea channel had connected with the northern South Atlantic and its

active spreading ridge (Fig. 1b). The ridge volcanoes which had formed earliest, were close to those which had arisen or were still arising at the eastern end of the sea channel. For these new ridge islands to become included in the range of breeding grounds it was necessary only that some turtles extend their travel path, that is, simply stay on course, assuming either that the turtle was distracted from its prime target; or that the target had become submerged and unusable. On return trips the turtles' would be guided to the final landfall by chemical cues, by surfbeat or even by direct vision. A degree of elasticity in the establishment and recognition of breeding grounds is necessary, but not much more than seems consistent with that found in the

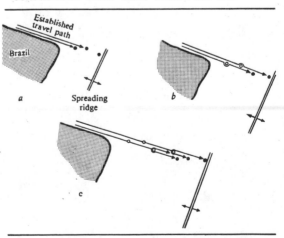

Fig. 3. As seafloor spreading proceeds, through stages *a*, *b* and *c*, new volcanic islands are generated near the ridge crest (double line). Older ones have moved out with the flanks of the ridge and gradually sink until they are submarine (open circle). As a former breeding ground becomes unusable, the turtle swims further on the same, or nearly the same, course.

modern green turtle[18].

We envisage that by 70 Myr ago, the ancestors of the Ascension Island colony were making seaward breeding migrations of up to 300 km--a figure consistent with the spreading rate of 2 cm yr$^{-1}$ (refs 14, 19). This process continued (Fig. 3) assuming there was sporadic creation of volcanic islands; Ascension Island is the last and youngest of them (less than 7 Myr old[19,20]). What was probably the penultimate island is now a seamount 15 km away, which is submerged to a depth of 1,500 m (ref 21).

We cannot establish the existence of a narrow swathe of seamounts connecting north-east Brazil with Ascension Island, because the detailed bathymetry is not well known. There is, however, a suggestion of a line of seamounts running south-eastwards from near Recife to about 30° W and then swinging east to Ascension (Fig. 4). This last limb is close to the trace of the Ascension Fracture Zone and its ridge[19]. The early Tertiary migration may have been controlled by volcanic islands that arose along the ridge.

## BRAZIL TO ASCENSION

We have argued that, since early in the Cainozoic era, the turtle has had an inherited tendency to swim a particular travel path, roughly WNW-ESE. This path can be followed by using the rising sun as a beacon to stay on latitude, an operation made simpler by the fact that the path is equatorial and the migration period is seasonal; turtles arrive at Ascension from February until May. Thus, for a December departure from the coast of northern Brazil the turtle heads directly into the rising sun (that is, ESE) and does the same every day. At night it rests below the equatorial current or even drifts eastwards with the counter-equatorial current[5]. The journey takes perhaps eight weeks. Because of the northerly drift of the sun at this time of the year, the general path followed from Brazil towards Ascension will describe a gentle arc, convex to the south, of the sort shown in Fig. 4. The required target is the intersection of the travel path with a plume of sensory clues arising from the island. Because of the equatorial current, which at this latitude has a southerly

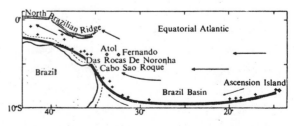

**Fig. 4** The western equatorial Atlantic. The arrows indicate the generalised direction of the Equatorial Current (for the northern hemisphere summer). A possible line of seamounts (there are others) is indicated by the vertical crosses. Linking these is a hypothetical migration path (heavy pattern, still to be tested) taken by the Ascension turtles from Brazil to Ascension Island.

component, the stream line of solubles released from the island is favourably disposed to make the intersection and provide the turtle with a pathway to landfall; the stream line gives a longitudinal fix. The equatorial current also serves another important function, that of carrying the hatchlings, imprinted to Ascension, away from the ecologically unsuitable island[5]. The same requirements probably existed for the early Cainozoic populations, when the separation between Africa and South America was insufficient for development of an equatorial current. Reconstructions of late Cretaceous current systems[22] allow us to postulate an equally favourable smaller current system which would perform the same functions during the early stages of the migration to the islands.

This navigatory scheme is simpler than that of Koch et al. (ref 7). In particular, it calls for a smaller plume from Ascension, in keeping with the size and nature of the island, and allows for recognition of Ascension emanations, as a new element, against the African background or 'noise' with which the equatorial current is loaded (the Congo contribution alone is massive). If the sun, in combination with chemoreceptive detection of the Ascension plume, is used as a navigational beacon, there would seem to

be a corollary which can be examined: turtles from northern parts of the resident grounds would depart first and those from southern parts (south of Recif) would depart later. This is a necessary consequence of the northerly swing of the sun during the visitation period.

INHERITANCE OF BEHAVIOUR

We suggest that the process of racial learning is of the repetitive, stepping-stone type, which requires no radical change in behaviour at any point. The hypothesis puts the migration to Ascension Island in an evolutionary framework by allowing selective and adaptive processes to operate over a great period of time. Indeed, the period of time may be too long to be acceptable to some. If we are right it means that an extant species has inherited a vital behaviour pattern from ancestors that lived 40 Myr ago and more. In purely taxonomic terms these ancestral species were different, and perhaps very different, from Chelonia mydas. In the face of this somewhat unpalatable notion, we take comfort in the realisation that species are more than flesh and bones, especially bones. For the Ascension Island colony, we accept the seaward migration drive as an inherited feature that predates other morphological features conventionally used in systematics to define taxa at species and genus levels. We also accept the far reaching implications which lie behind this statement and its preamble, but we do not have space to discuss them here.

This article is the result, and therefore an acknowledgment, of the scholarly opportunities given to one of us (P.J.C.) during a year as Visiting Professor at the Hawaii Institute of Geophysics, University of Hawaii, where the Director, George Wollard, gave special help. A.C. received support from the National Science Foundation.

[1] Carr, A., and Hirth, H., Am. Mus. Novit., 2091 (1962).

[2] Carr, A., Scient. Am., 212 (1965).

[3] Carr, A., in Animal Orientation and Navigation (edit. by Storm, R. M.), 35 (Proceedings of 27th Annual Biological Colloquium, Oregon State University Press, Corvallis, 1967).

[4] Carr, A., So excellent a fishe (Natural History Press, Doubleday, New York, 1967).

[5] Carr, A., in Animal Orientation and Navigation (edit. by Galler, S. R. et al.), 469 (National Aeronautics and Space Administration, SP-262, Washington DC, 1972).

[6] Da Costa, R. S., Bol. Estud. Pesca (Braz.), 9, 21 (1969).

[7] Koch, A. L., Carr, A., and Ehrenfeld, D. W., J. theor. Biol., 22, 163 (1969).

[8] Ferreira, M. M., Arg. Estud. Biol. mar., Univ. Fed. Ceara, 8, 85 (1968).

[9] Bergounioux, F. M., in Traite de Paleontologie (edit. by Piveteau, J.), 5, 487 (Masson, Paris, 1955).

[10] Hay, O. P., Carnegie Inst., Washington, Spec. Publ. 75 (1908).

[11] Zangerl, R., Fieldiana: Geol. Mem., 3, 279 (1960).

[12] Zangerl, R., and Sloan, R. E., Fieldiana: Geol., 14(2), 7 (1960).

[13] Dietz, R. S., and Holden, J. C., J. geophys. Res., 75, 4939 (1970).

[14] Larson, R. L., and Pitman, W. C., Bull. geol. Soc. Am., 83, 3645 (1972).

[15] Le Pichon, X., and Hayes, E. D., J. geophys. Res., 76, 6283 (1971).

[16] Maack, R., Kontinentaldrift und Geologie des Sudatlantischen Ozeans (Walter de Gruyter, Berlin, 1969).

[17] Menard, H. W., *J. geophys. Res.*, **74**, 4827 (1969).

[18] Carr, A., and Carr, M. H., *Ecology*, **53**, 425 (1972).

[19] van Andel, Tj. H., Rea, D. K., von Herzen, R. P., and Hoskins, H., *Bull. geol. Soc. Am.*, **84**, 1527 (1973).

[20] Bell, J. D., Atkins, F. B., Baker, P. E., and Smith, D. G. W., *Eos (Trans. Am. geophys. Un.)*, **53**, 168 (1972).

[21] Daly, R. A., *Proc. Am. Acad. Arts Sci.*, **60** (1925).

[22] Gordon, W. A., *J. Geol.*, **81**, 269 (1973).

[23] Bullard, E., Everett, J. E., and Gilbert-Smith, A., *Phil. Trans. R. Soc.*, **A 1088**, 41 (1965).

*

# Origin of Life
## Section 4.

Life may have originated in tide-pools.
Photograph by Jere H. Lipps.

Man is naturally curious about his origins, and this curiosity seems to have focussed on two questions: where did life come from in the first place, and what was the origin of man himself? The first now seems to be the more difficult question, and geologists, paleontologists, biologists, chemists and astronomers have all contributed to its discussion. Needless to say, there is no accord yet.

In the broad sense, these discuss-

ions have dealt with three general cat-
egories--the philosophical and religi-
ous aspects of life; a terrestrial
origin of life resulting from processes
known to occur on the earth; and an
extraterrestrial origin of life. Some
of the explanations come from ancient
times; Aristotle, for example, thought
that life originated spontaneously from
items such as morning dew. Others
followed him, proposing that mice, for
example, could arise from a mixture of
the squeezings of dirty underwear and
wheat grains. All of this was finally
laid to rest by Louis Pasteur only
slightly more than a hundred years ago.

A little before Pasteur did his
experiments, Charles Darwin was writing
about warm little ponds somewhere on
the earth, full of various chemicals,
affected by light, heat and electricity
to give rise to complex molecules that
later evolved into living creatures.
As with so many other things, Darwin
had great insight into future developments, for it has since been demonstrated time after time that various
mixtures of chemicals, when subjected
to a source of energy, can give rise to
the organic molecules required for
life.

Much has been written on the possibilities for a terrestrial origin of
life by people such as Thomas Huxley,
Alexander Oparin, Stanley Miller, and
others. These arguments are summarized

in our readings by a final chapter from
a book by Leslie ORGEL (1973). After
sketching out the main requirements for
the origin of life, he concludes that
it would indeed be a rare event in the
universe at large.

Another group of scientists,
however, has been studying the possibility of the existence of life, perhaps
even civilizations, on other planets or
planetary systems in the universe. The
discipline, termed "exobiology", has
yet to discover its subject matter, as
George Gaylord Simpson has acidly
pointed out. The study is nevertheless
important. If life does exist elsewhere, then there is the intriguing
possibility that earth might have been
infested with life from an extra-terrestrial source. This view is examined
in the paper by CRICK and ORGEL (1973).
Crick is a Nobel Prize winner for his
work on nucleic acids, which carry the
genetic code, and Orgel is a leading
worker on the chemical processes which
might have led to life. BALL (1973)
examines the idea that another civilization might be maintaining the earth
as a sort of wild-life preserve or zoo.
Ball finds the idea unpleasant and
grotesque, and out of the realm of
science. But his paper is a good example of the way scientists look at ideas
that might otherwise be left to laymen
to consider--in this case, science
fiction writers and film-makers.

This whole problem will surely result in the next few years in one of the

The idea of Crick and Orgel has already been assailed by other workers. For example, it was suggested that the "directed panspermia" hypothesis could not work, because the abundance of molybdenum on the earth was of the wrong magnitude. Orgel (1974) has replied that his idea and Crick's is not affected by this criticism. We have probably not heard the last of this debate, and the search for extra-terrestrial life will continue. The United States has a tremendous financial commitment to this very search. Millions of dollars have already been spent or committed to a systematic program of searching the skies with a radio telescope in West Virginia, in part to try to detect signals from other civilizations, and a program to send a space probe to Mars in the late 1970's with a "life detector" on board.

Any positive discoveries would be startling, but they would still not answer our question of where life came from in the first place--they would only remove its origin from the earth. Certainly, somewhere in the universe there was, and maybe still is, life being created from inorganic materials. most exciting scientific adventures ever undertaken.

FURTHER READINGS

CHAPPELL, W. R., MEGLEN, R. R., and RUNNELLS, D. D. 1974. Comments on "Directed Panspermia". Icarus 21, 513-515.

JUKES, T. H. 1974. Sea-water and the origins of life. Icarus 21, 516-517.

ORGEL, L. E. 1973. The origins of life: molecules and natural selection. John Wiley and Sons. New York. 237pp.

ORGEL, L. E. 1974. Reply: "Comments on 'Directed Panspermia'" and "Seawater and the Origins of Life". Icarus 21, 518.

PONNAMPERUMA, C. 1972. The origins of life. Thames and Hudson. London. 215pp.

*

# Summary of the main argument

**L.E. Orgel (1973)**

The earth was formed from a cloud of dust and gases about four and a half billion years ago. The cloud was made up very largely of hydrogen and helium, but these gases, together with most other volatile material, escaped during the condensation process. Our atmosphere and oceans are derived from gases that were expelled from the interior of the earth at a later date.

The primitive atmosphere was reducing. It contained water, methane, carbon dioxide and nitrogen; in addition, some ammonia and hydrogen may have been present. We know that the atmosphere was still reducing a billion and a half years later, by which time organisms similar to algae or bacteria had evolved on the surface of the earth. When we talk about the origins of life, we mean the series of processes which occurred in the atmosphere, oceans, and lakes of the primitive earth and led, more than three billion years ago, to the

Reprinted from The Origins of Life, pp. 229-232 by permission of the author and publisher. Copyright 1973 by John Wiley and Sons Inc.

appearance of the first living organisms.

We have no geological record of the events that occurred so long ago. However, it seems very likely that the first step in the origins of life occurred when very simple organic molecules were formed in the earth's reducing atmosphere by the action of lightning, ultraviolet energy from the sun, shock-waves, and so forth. These compounds were washed into oceans and lakes where they reacted to form a complex mixture of organic substances, including many that were to form part of the most primitive living organisms.

The mixture of chemical compounds that was formed on the primitive earth is referred to as the prebiotic soup. Prebiotic chemistry is concerned with laboratory experiments that simulate the processes believed to have been involved in the formation of the prebiotic soup. The most striking success achieved so far in the study of the origins of life is the demonstration that the components of living organisms are particularly abundant in the mixture of organic chemicals formed in prebiotic reactions. Most of the important components of modern organisms, for example, amino acids, sugars, and nucleic-acid bases, have been synthesized in the laboratory under conditions that could have prevailed on the primitive earth.

The next stage in the origins of life must have been the concentration or "thickening" of the prebiotic soup. Evaporation was almost surely important, but freezing and adsorption on the surface of rocks or colloidal particles may also have played a part. Once the prebiotic soup was sufficiently concentrated, polymers resembling proteins and nucleic acids could be formed. Not much success has been attained in simulating these polymerization reactions in the laboratory, but some progress has been made.

The major intellectual problem presented by the origins of life is concerned with the next stage, the evolution of biological organization. How did a complex self-replicating organism evolve from an unorganized mixture of polymeric molecules? Little experimental evidence is available, so one is forced to attempt a speculative reconstruction of this phase in the origins of life.

The key to the understanding of the evolution of biological organization is the theory of natural selection. Before the evolution of complicated self-replicating organisms, natural selection must have acted on something much simpler, probably on polymeric molecules resembling nucleic acids. It is believed that nucleic acid-like molecules were formed in the prebiotic soup and were able to reproduce without the help of enzymes. The theory of natural selection then shows that those molecules that could replicate fastest would have become dominant in the prebiotic soup.

As the competition became fiercer, the more successful families of self-replicating molecules must have "learned" to make use of small molecules in their environment to help them to replicate even faster. The most important of these adaptations involved the amino acids; ultimately a family of self-replicating nucleic acids evolved to the point where they could begin to control the synthesis of polypeptide sequences that had useful catalytic properties. This adaptation led ultimately to the evolution of protein synthesis and the genetic code.

The evolution of protein synthesis is not understood in detail. One of the great challenges of the problem of the origins of life is to demonstrate in the laboratory how polynucleotides, without the help of preformed enzymes, could have replicated and begun to control the synthesis of peptides with determined sequences. Once this has been done we shall be well on our way to understanding the origins of the first living cells.

Until recently, theories of the origin of the prebiotic soup were based entirely on laboratory experiments. Within the last few years dramatic developments in astronomy have changed this situation. A number of small organic molecules have been detected in interstellar dust clouds far from the earth. The most abundant of these include formaldehyde, hydrogen cyanide, and cyanoacetylene. These three molecules had been proposed previously as three of the most important precursors of the prebiotic soup. Thus radioastronomy has provided evidence supporting strongly our general ideas about prebiotic synthesis.

The discovery of many of the twenty naturally occurring amino acids in the Murchison meteorite, a meteorite that could not have been contaminated with terrestrial organic material, is equally important. In fact the mixture of amino acids in the meteorite matches almost exactly the mixture of amino acids formed by the action of an electric discharge on a reducing gas mixture. Clearly, amino acids are formed in large amounts somewhere in the solar system far from the earth.

These new findings suggest that the range of organic compounds that can be formed under prebiotic conditions is small. It is not unlikely, therefore, that there are other planets on which rich prebiotic soups have accumulated. We do not know how likely it is that life would evolve on such planets, but we can see no good reason why earth-like life should not exist elsewhere in the universe. It is also possible that totally alien forms of life have evolved on other planets, but we are not as yet in a position to say much about this possibility.

Studies of the origins of life have reached a particularly exciting point. We hope within a few decades to understand how life on earth evolved from a random mixture of organic compounds. It may take a little longer to create novel self-replicating systems, but it is likely that this will be achieved within, say, a hundred years. Once we understand more about the evolution of biological organization, we should be able to say something quantitative about the probability that life exists elsewhere in the universe. Perhaps we shall not have to wait so long--a systematic search for signs of intelligent life elsewhere in the universe is likely to begin in the near future.  If we succeed in making contact with an extraterrestrial civilization, our view of man's place in the universe will certainly change.

# Directed panspermia

## F.H.C. Crick and L.E. Orgel (1973)

INTRODUCTION

It was not until the middle of
the nineteenth century that Pasteur
and Tyndall completed the demonstration
that spontaneous generation is not
occurring on the Earth nowadays.
Darwin and a number of other
biologists concluded that life must
have evolved here long ago when
conditions were more favourable. A
number of scientists, however, drew a
quite different conclusion. They
supposed that if life does not
evolve from terrestrial nonliving
matter nowadays, it may never have
done so. Hence, they argued, life
reached the earth as an "infection"
from another planet (Oparin, 1957).

Arrhenius (1908) proposed that
spores had been driven here by the
pressure of the light from the
central star of another planetary
system. His theory is known as

Reprinted from Icarus 19, 341-346 by
permission of the authors and publisher. Copyright 1973 Academic Press Inc.

Panspermia. Kelvin suggested that
the first organisms reached the
Earth in a meteorite. Neither of
these theories is absurd, but both
can be subjected to severe criticism.
Sagan (Shklovski and Sagan, 1966;
Sagan and Whitehall, 1973) has shown
that any known type of radiation-
resistant spore would receive so
large a dose of radiation during its
journey to the Earth from another
Solar System that it would be
extremely unlikely to remain viable.
The probability that sufficiently
massive objects escape from a
Solar System and arrive on the planet
of another one is considered to be so
small that it is unlikely that a
single meteorite of extrasolar
origin has ever reached the surface
of the Earth (Sagan, private
communication). These arguments may
not be conclusive, but they argue
against the "infective" theories
of the origins of life that were
proposed in the nineteenth century.

It has also been argued that
"infective" theories of the origins

of terrestrial life should be
rejected because they do no more than
transfer the problem of origins to
another planet. This view is
mistaken; the historical facts are
important in their own right. For
all we know there may be other
types of planet on which the origin
of life ab initio is greatly more
probable than on our own. For
example, such a planet may possess
a mineral, or compound, of crucial
catalytic importance, which is rare
on Earth. It is thus important to
know whether primitive organisms
evolved here or whether they
arrived here from somewhere else.
Here we reexamine this problem in
the light of more recent biological
and astronomical information.

## Our Present Knowledge of the Galaxy

The local galactic system is
estimated to be about $13 \times 10^9$ yr
old (See Metz, 1972). The first
generation of stars, because they were
formed from light elements, are
unlikely to have been accompanied
by planets. However, some second
generation stars not unlike the Sun
must have formed within $2 \times 10^9$ yr
of the origin of the galaxy (Blaauw
and Schmidt, 1965). Thus it is quite
probable that planets not unlike the
Earth existed as much as $6.5 \times$

$10^9$ yr before the formation of our
own Solar System.

We know that not much more than
$4 \times 10^9$ yr elapsed between the
appearance of life on the Earth
(wherever it came from) and the
development of our own technological
society. The time available makes it
possible, therefore, that technological
societies existed elsewhere in the
galaxy even before the formation of
the Earth. We should, therefore,
consider a new "infective" theory,
namely that a primitive form of life
was deliberately planted on the Earth
by a technologically advanced society
on another planet.

Are there many planets which
could be infected with some chance
of success? It is believed, though
the evidence is weak and indirect,
that in the galaxy many stars, of
a size not dissimilar to our Sun,
have planets, on a fair fraction of
which temperatures are suitable for
a form of life based on carbon
chemistry and liquid water, as ours
is. Experimental studies of the
production of organic chemicals
under prebiotic conditions make it
seem likely that a rich prebiotic
soup accumulates on a high proportion
of such Earthlike planets. Unfortunately, we know next to
nothing about the probability that
life evolves within a few billion

years in such a soup, either on our own special Earth, or still less on other Earthlike planets.

If the probability that life evolves in a suitable environment is low we may be able to prove that we are likely to be alone in the galaxy (Universe). If it is high the galaxy may be pullulating with life of many different forms. At the moment we have no means at all of knowing which of these alternatives is correct. We are thus free to postulate that there have been (and still are) many places in the galaxy where life could exist but that, in at least a fraction of them, after several billion years the chemical systems had not evolved to the point of self-replication and natural selection. Such planets, if they do exist, would form an excellent breeding ground for external microorganisms. Note that because many if not all such planets would have a reducing atmosphere they would not be very hospitable to the higher forms of life as we know them on Earth.

## Our Proposal

The possibility that terrestrial life derives from the deliberate activity of an extraterrestrial society has often been considered in science fiction and more or less light-heartedly in a number of scientific papers. For example, Gold (1960) has suggested that we might have evolved from the micro-organisms inadvertently left behind by some previous visitors from another planet (for example, in their garbage). Here we wish to examine a very specific form of Directed Panspermia. Could life have started on Earth as a result of infection by microorganisms sent here deliberately by a technological society on another planet, by means of a special long-range unmanned spaceship? To show that this is not totally implausible we shall use the theorem of detailed cosmic reversibility; if we are capable of infecting an as yet lifeless extrasolar planet, then, given that the time was available, another technological society might well have infected our planet when it was still lifeless.

## The Proposed Spaceship

The spaceship would carry large samples of a number of microorganisms, each having different but simple nutritional requirements, for example blue-green algae, which could grow on $CO_2$ and water in "sunlight." A payload of 1000 kg might be made up of 10 samples each containing $10^{16}$ microorganisms, or 100 samples each of $10^{15}$ microorganisms.

It would not be necessary to accelerate the spaceship to extremely high velocities, since its time of arrival would not be important. The radius of our galaxy is about $10^5$ light years, so we could infect most planets in the galaxy within $10^8$ yr by means of a spaceship traveling at only one-thousandth of the velocity of light. Several thousand stars are within a hundred light years of the Earth and could be reached within as little as a million years by a spaceship travelling at only 60,000 mph, or within 10,000 yr if a speed of one-hundredth of that of light were possible.

The technology required to carry out such an act of interstellar pollution is not available at the present time. However, it seems likely that the improvements in astronomical techniques will permit the location of extrasolar planets within the next few decades. Similarly, the problem of sending spaceships to other stars, at velocities low compared with that of light, should not prove insoluble once workable nuclear engines are available. This again is likely to be within a few decades. The most difficult problem would be

presented by the long flight times; it is not clear how long it will be before we can build components that would survive in space for periods of thousands or millions of years.

Although there are some technological problems associated with the distribution of the microorganisms in viable form after a long journey through space, none of them seems insuperable. Some radiation protection could be provided during the journey. Suitable packaging should guarantee that small samples, including some viable organisms, would be widely distributed. The question of how long microorganisms, and in particular bacterial spores, could survive in a spaceship has been considered in a preliminary way by Sneath (1962). He concludes "that life could probably be preserved for periods of more than a million years if suitably protected and maintained at temperatures close to absolute zero." Sagan (1960) has given a comparable estimate of the effects of radiation damage. We conclude that within the foreseeable future we could, if we wished, infect another planet, and hence that it is not out of the question that our planet was infected.

We can in fact go further than this. It may be possible in the future to send either mice or men or elaborate instruments to the planets of other Solar Systems (as so often described in science fiction) but a rocket carrying microorganisms will always have a much greater effective range and so be advantageous if the sole aim is to spread life. This is true for several reasons. The conditions on many planets are likely to favour microorganisms rather than higher organisms. Because of their extremely small size vast numbers of microorganisms can be carried, so much more wastage can be accepted. The ability of microorganisms to survive, without special equipment, both storage for very long periods at low temperatures and also an abrupt change back to room temperatures is also a great advantage. Whatever the potential range for infection by other organisms, microorganisms can almost certainly be sent further and probably much further.

It should be noted that most of the earliest "fossils" so far recognized are somewhat similar to our present bacteria or blue-green algae. They occur in cherts of various kinds and are estimated to be up to $3 \times 10^9$ yr old. This makes it improbable that the Earth was ever infected merely by higher organisms.

## Motivation

Next we must ask what motive we might have for polluting other planets. Since we would not derive any direct advantage from such a programme, presumably it would be carried through either as a demonstration of technological capability or, more probably, through some form of missionary zeal.

It seems unlikely that we would deliberately send terrestrial organisms to planets that we believed might already be inhabited. However, in view of the precarious situation of Earth, we might well be tempted to infect other planets if we became convinced that we were alone in the galaxy (Universe).[1] As we have already explained we cannot at the moment estimate the probability of this. The hypothetical senders on another planet may have been able to prove that they were likely to be alone, and to remain so, or they may have reached this conclusion mistakenly. In either

[1] In a somewhat different context the seeding of Venus and other solar planets has been suggested by C. Sagan (1961), and T. Gold, private communication.

case, if they resembled us psychologically, their motivation for polluting the galaxy would be strong, if they believed that all or even the great majority of inhabitable planets could be given life by Directed Panspermia.

The psychology of extra-terrestrial societies is no better understood than terrestrial psychology. It is entirely possible that extraterrestrial societies might infect other planets for quite different reasons than those we have suggested. Alternatively, they might be less tempted than we would be, even if they thought that they were alone. The arguments given above, together with the principle of cosmic reversibility, demonstrate the possibility that we have been infected, but do not enable us to estimate the probability.

## Possible Biological Evidence

Infective theories of the origins of terrestrial life could be taken more seriously if they explained aspects of biochemistry or biology that are otherwise difficult to understand. We do not have any strong arguments of this kind, but there are two weak facts that could be relevant.

The chemical composition of living organisms must reflect to some extent the composition of the environment in which they evolved. Thus the presence in living organisms of elements that are extremely rare on the Earth might indicate that life is extraterrestrial in origin. Molybdenum is an essential trace element that plays an important role in many enzymatic reactions, while chromium and nickel are relatively unimportant in biochemistry. The abundance of chromium, nickel, and molybdenum on the Earth are 0.20, 3.16, and 0.02%, respectively. We cannot conclude anything from this single example, since molybdenum may be irreplaceable in some essential reaction--nitrogen fixation, for example. However, if it could be shown that the elements represented in terrestrial living organisms correlate closely with those that are abundant in some class of star--molybdenum stars, for example--we might look more sympathetically at "infective" theories.

Our second example is the genetic code. Several orthodox explanations of the universality of the genetic code can be suggested, but none is generally accepted to be completely convincing. It is a little surprising that organisms with somewhat different codes do not coexist. The universality of the code follows naturally from an "infective" theory of the origins of life. Life on Earth would represent a clone derived from a single extraterrestrial organism. Even if many codes were represented at the primary site where life began, only a single one might have operated in the organisms used to infect the Earth.

CONCLUSION

In summary, there is adequate time for technological society to have evolved twice in succession. The places in the galaxy where life could start, if seeded, are probably very numerous. We can foresee that we ourselves will be able to construct rockets with sufficient range, delivery ability, and surviving payload if microorganisms are used. Thus the idea of Directed Panspermia cannot at the moment be rejected by an simple argument. It is radically different from the idea that life started here ab initio without infection from elsewhere. We have thus two sharply different theories of the origin of life on Earth. Can we choose between them?

At the moment it seems that the experimental evidence is too feeble to make this discrimination. It is difficult to avoid a personal prejudice, one way or the other, but such prejudices find no scientific support of any weight. It is thus important that both theories should be followed up. Work on the supposed terrestrial origin of life is in progress in many laboratories. As far as Directed Panspermia is concerned we can suggest several rather diverse lines of research.

The arguments we have employed here are, of necessity, somewhat sketchy. Thus the detailed design of a long-range spaceship would be worth a careful feasibility study. The spaceship must clearly be able to home on a star, for an object with any appreciable velocity, if despatched in a random direction, would in almost all cases pass right through the galaxy and out the other side. It must probably have to decelerate as it approached the star, in order to allow the safe delivery of the payload. The packets of microorganisms must be made and dispersed in such a way that they can survive the entry at high velocity into the atmosphere of the planet, and yet be able to dissolve in the oceans. Many useful feasibility studies could be carried out on the engineering points involved.

On the biological side we lack precise information concerning the life-time of microorganisms held at very low temperatures while traveling through space at relatively high velocities. The rocket would presumably be coasting most of the time so the convenient temperature might approximate to that of space. How serious is radiation damage, given a certain degree of shielding? How many distinct types of organism should be sent and which should they be? Should they collectively be capable of nitrogen fixation, oxidative phosphorylation and photosynthesis? Although many "soups" have been produced artificially in the laboratory, following the pioneer experiments of Miller, as far as we know no careful study has been made to determine which present-day organisms would grow well in them under primitive Earth conditions.

At the same time present-day organisms should be carefully scrutinized to see if they still bear any vestigial traces of extraterrestrial origin. We have already mentioned the uniformity of the genetic code and the anomalous abundance of molybdenum. These facts amount to very little by themselves but as already stated there may be other as yet unsuspected features which, taken together, might point to a special type of planet as the home of our ancestors.

These enquiries are not trivial, for if successful they could lead to others which would touch us more closely. Are the **senders** or their descendants still alive? Or have the hazards of 4 billion years been too much for them? Has their star inexorably warmed up and frizzled them, or were they able to colonise a different Solar System with a short-range spaceship? Have they perhaps destroyed themselves, either by too much aggression or too little? The difficulties of placing any form of life on another planetary system are so great that we are unlikely to be their sole descendants. Presumably they would have made many attempts to infect the galaxy. If the range of their rockets were small this might suggest that we have cousins on planets which are not too distant. Perhaps the galaxy is lifeless except for a local village, of which we are one member.

One further point deserves emphasis. We feel strongly that under no circumstances should we risk infecting other planets at the present time. It would be wise to wait until we know far more about the probability of the development of life on extrasolar planets before causing terrestrial organisms to escape from the solar system.

## ACKNOWLEDGMENTS

We are indebted to the organisers of a meeting on Communication with Extraterrestrial Intelligence, held at Byurakan Observatory in Soviet Armenia in September 1971, which crystallized our ideas about Panspermia. We thank Drs. Freeman Dyson, Tommy Gold, and Carl Sagan for discussion and important comments on our argument.

## REFERENCES

Arrhenius, S. (1908). "Worlds in the Making." Harper and Row, New York.

Blaauw, A., and Schmidt, M. (1965). "Galactic Structure." University of Chicago Press, Chicago.

Gold, T. (May 1960). "Cosmic Garbage" "Air Force and Space Digest", p. 65.

Metz, W. D. (1972). Reporting a speech by Allan Sandage. Science 178, 600.

Oparin, A. I. (1957). "The Origins of Life on Earth." Academic Press, New York, New York (gives a general discussion on Panspermia).

Sagan, C. (1960). Biological contamination of the Moon. Proc. Nat. Acad. Sci. 46, 396.

Sagan, C. (1961). The planet Venus. Science 133, 849.

Sagan, C., and Whitehall, L. (1973). To be submitted to Icarus.

Shklovskii, I. S., and Sagan, C. (1966 and 1967). "Intelligent Life in the Universe," p. 207. Holden-Day, San Francisco and Dell Publishing Co., New York.

Sneath, P. H. A. (1962). Longevity of microorganisms. Nature (London) 195, 643.

*

# The zoo hypothesis

## J.A. Ball (1973)

INTRODUCTION

The most interesting scientific
problem of our age involves the
question of the existence of extra-
terrestrial intelligent life.
Arguments summarized below make it
likely that intelligence exists on
many planets throughout our galaxy
and that most of these civilizations
are much older than our own.  This
problem has been the subject of
considerable work both theoretical
and experimental (see Oparin and
Fesenkov, 1960; Cameron, 1963;
Shklovskii and Sagan, 1966; Sagan,
1973; and other references therein)
and our understanding of the subject
has certainly progressed rapidly
in the last decade or so.  However,
this problem has proved to be
extremely difficult, in part because
it involves understanding what a
civilization much older than ours
might be like.  It is difficult

Reprinted from _Icarus_ 19, 347-349 by
permission of the author and publisher.
Copyright 1973 Academic Press Inc.

enough to predict our own development for a few decades hence, but we
need to know about other civilizations
that may be older than ours not by
decades but by eons.

Among currently popular ideas
about extraterrestrial intelligence,
the idea that "they" are trying to
talk to us has many adherents (see,
e.g., Drake, 1963).  This idea seems
to me to be unlikely to be correct
and the zoo hypothesis is in fact
the antithesis of this idea.

Starting Premises

Three working hypotheses or
starting premises are used in most
discussions of the problem of
extraterrestrial life.  These
premises are stated below with a
discussion of their origin and
references to the literature.  Although this discussion is brief,
these premises are in fact crucial
and if any of them proves to be
incorrect, then the zoo hypothesis
falls.

A.  Whenever the conditions are

such that life can exist and evolve, it will. Life is to be understood as a chemical reaction that occurs whenever the necessary reactants are present under the appropriate conditions for a sufficient time. This statement represents a considerable extrapolation of our present knowledge. In fact the opposite hypothesis, that life is statistically unlikely even in ideal conditions, has been expressed (e.g., by Townes, 1971). **Discovery of primitive** life on Mars or Venus would probably settle this question. Our current understanding of biochemistry seems to support premise A (Shklovskii and Sagan, 1966, Chapter 14; and Calvin, 1963).

B. There are many places where life can exist. Planets are probably quite common in the universe. As many as 20% of all stars may have planets and as many as 10% of these planets may have surfaces on which life can form. (However Oparin and Fesenkov, 1960, think that only one star in $10^5$ or $10^6$ has a planet with a surface suitable for life. See also von Hoerner, 1963). This statement also represents more than we know at present; no star other than our sun is definitely known to have planets comparable to the earth. Objects that may be planets have been detected around a few

other stars (see Shklovskii and Sagan, 1966, Chapter 11; Huang, 1963; and van de Kamp, 1969), however these objects are much more massive than the earth. Planets comparable to the earth around almost any other star would go undetected with present techniques. The opposite hypothesis, that the solar system is unique, was believed by Jeans, 1929, Chapter XVI, but is now discredited (see, e.g., Levin, 1964, for a summary of current thinking).

C. We are unaware of "them."

Who is Out There?

It is statistically unlikely that there exists anywhere in our whole galaxy any other civilization whose level of development is at all comparable to ours. We would expect to find either primitive life forms, perhaps comparable to those on the earth a few million years ago, or very advanced life forms, perhaps comparable to what will be on earth a few million years hence(!)

There are three general categories of possibilities defining the technological evolution of a civilization:

(1) Destruction (from within or without).

(2) Technological stagnation.

(3) Quasi-continuous technological progress.

Also there are many other mixed possibilities such as partial destruction and rebuilding, and the surprisingly popular finite-lifetime idea. These possibilities are sketched diagramatically in Fig. 1 with specific reference to our own extrapolated future. It is likely that some fraction of all civilizations follow each of these possibilities. However, analogy with civilizations on earth indicates that most of those civilizations that are behind in technological development would eventually be engulfed and destroyed, tamed, or perhaps assimilated. So, generally speaking, we need consider only the most technologically advanced civilizations because they will be, in some sense, in control of the universe.

Technological progress may be defined as increasing ability to control one's environment. Already at our level of technology we affect almost everything on earth from elephants to viruses. But we do not always exert the power we possess. Occasionally we set aside wilderness areas, wildlife sanctuaries, or zoos in which other species (or other civilizations) are allowed to develop naturally, i.e., inter-acting very little with man. The perfect zoo (or wilderness area or sanctuary) would be one in which the fauna inside do not inter-act with, and are unaware of, their zookeepers.

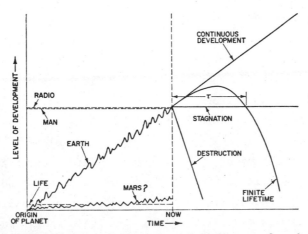

FIG. 1. This is a sketch of the top level of development, defined in terms of complexity, versatility, and ability to control the environment, either of the organism itself or of the civilization to which it belongs. The various possible extrapolations for our future are discussed in the text.

## The Zoo Hypothesis

Premise C above now seems to me to be extremely significant. I believe that the only way that we can understand the apparent non-interaction between "them" and us is to hypothesize that they are deliberately avoiding interaction and that they have set aside the area in which we live as a zoo.

The zoo hypothesis predicts that we shall never find them because they do not want to be found and they have the technological ability to insure this. Thus this hypothesis is falsifiable, but not, in principle, confirmable by future observations.

## CONCLUSIONS

The zoo hypothesis as given here is probably flawed and incomplete. I hope that it can provide some sort of inspiration for further work. Among other hypotheses that one might consider, the laboratory hypothesis is one of the more morbid and grotesque. We may be in an artificial laboratory situation. However, this hypothesis is outside the purview of science because it leads nowhere, it immediately calls into question the premises on which it is based, and it makes no predictions. Or one might suppose that extraterrestrial civilizations have not yet found us or that they know we are here but they are uninterested in us. These latter two hypotheses are probably incompatible with the high level of technological sophistication they undoubtedly possess.

The zoo hypothesis seems to me to be pessimistic and psychologically unpleasant. It would be more pleasant to believe that they want to talk with us, or that they would want to talk with us if they knew that we are here. However the history of science contains numerous examples of psychologically unpleasant hypotheses that turned out to be correct.

## ACKNOWLEDGMENTS

Although the ideas in this paper are not in the mainstream of current scientific thought about the problem, they are also not new. Science-fiction authors, in particular, have toyed with similar notions for many years. And at least a few previous writers have suggested such ideas as a serious possibility.

I thank Sebastian von Hoerner and Mrs. Lyle G. Boyd for pointing out relevant background material and for stimulating discussions. I am grateful to Prof. A. E. Lilley for his encouragement.

## REFERENCES

Calvin, Melvin (1963). Chemical evolution. In "Interstellar Communication" (A. G. W. Cameron, ed.), Chapt. 5. W. A. Benjamin, Inc., New York.

Cameron, A. G. W., ed. (1963). "Interstellar Communication." W. A. Benjamin, Inc., New York.

Drake, Frank D. (1963). How can we detect radio transmissions from distant planetary systems, Project Ozma. In "Interstellar Communication" (A. G. W. Cameron, ed.), Chapts. 16 and 17. W. A. Benjamin, Inc., New York.

Huang, Su-Shu (1963). The problem of life in the universe and the mode of star formation. In "Interstellar Communication" (A. G. W. Cameron, ed.), Chapt. 7. W. A. Benjamin, Inc., New York.

Jeans, James H. (1929). "Astronomy and Cosmogony" Cambridge University Press; also Dover (1961).

Levin, Boris (1964). "The Origin of the Earth and Planets," 3rd ed. Foreign Languages Publishing House, Moscow.

Oparin, A., and Fesenkov, F. (1960). "The Universe," 2nd ed. Foreign Languages Publishing House, Moscow.

Sagan, Carl, ed. (1973). "Communication with Extraterrestrial Intelligence." MIT Press, to be published.

Shklovskii, I. S., and Sagan, Carl (1966). "Intelligent Life in the Universe," Holden-Day and Delta-Dell, San Francisco and New York.

Townes, C. H. (1971). In the 1971 Jansky Lecture at the National Radio Astronomy Observatory, Charlottesville, Virginia, October 4, 1971.

van de Kamp, Peter (1969). Alternate dynamical analysis of Barnard's Star, Astron. J. 74, 757-759.

von Hoerner, Sebastian (1963). The search for signals from other civilizations. In "Interstellar Communication" (A. G. W. Cameron, ed.), Chapt. 27. W. A. Benjamin, Inc., New York.

*

# Evolution and Extinction
## Section 5.

A Cambrian trilobite and a living serolid Isopod, showing
convergent evolution.
Photographs by William Krebs and Jon W. Branstrator.

## THE FOSSIL RECORD

The fossil record, taken at its
face value, shows periods of rapid
evolution among plants and animals, and
periods of rapid extinction. We saw in
Section 4 that there is controversy
about accepting the fossil record at
its face value, but until that argument
is cleared up we must assume that rapid
evolution and rapid extinction are real
features of the history of life that
require explanation.

To illustrate arguments about
rapid evolution, we have chosen readings about the appearance of animals
with hard skeletons in the fossil
record. Other episodes of rapid evolution have similar controversies associated with them. About 570 million
years ago, many quite unrelated groups
of animals all evolved hard skeletons,
which were made of very different
chemical compounds - calcite, aragon-
ite, chitin, phosphate, silica, and
combinations of these materials. The
skeletons were sometimes internal,
sometimes external, and were obviously
adapted for very different functions in
different animals.

The development of hard parts meant that animals became numerous in the fossil record as their chances of preservation increased. The fossil record changed so dramatically that this time period was chosen to mark the beginning of the Paleozoic era, in contrast to the preceding Precambrian era which is almost devoid of any trace of fossils.

There are two main reactions among scientists to the relatively sudden appearance of fossils at the base of the Cambrian. One reaction is to say that it does not necessarily mark a really extraordinary burst of evolution, but was simply the end point of a long and slow evolutionary process among soft-bodied Precambrian animals. Some quite subtle triggering mechanism, or some coincidence, could have led to the apparently sudden appearance of hard part fossils in the early Cambrian. Implicit in this argument is that the early stages of mineralization of skeletons have for some reason not been preserved. This reaction basically says, "There is no scientific problem here which requires a special explanation: let's stop worrying about it."

The other main reaction is to say that the appearance of so many animals with skeletons at the base of the Cambrian was a real burst of evolution, and the question implicit in this line of argument is, "What caused the sudden evolution?" In this line of argument, scientists seeking explanations must put some hypothesis on the line for testing by others, and this is what makes for controversy. To explain a feature like the evolution of skeletons in animals as diverse as sponges, brachiopods, molluscs, arthropods and echinoderms, all at about the same time, it seems inevitable that one must search for a cause which affects the whole of the life in the sea--in other words, a world-scale event. One fascinating part of examining this controversy is in admiring the ingenuity of the scientists who have proposed and criticized so many different hypotheses, all based on the same data.

THE READINGS

We begin with an article by RUDWICK (1964) who is first anxious to show that there is a real problem to be examined. He then offers a double-barreled hypothesis--there are very few Precambrian fossils known because they were small and had very weak skeletons if they had skeletons at all; therefore there must have been a triggering mechanism which encouraged extremely rapid evolution towards larger size and stronger skeleton. Rudwick believes that a world-wide glaciation, well documented from several continents, would have caused conditions difficult for life in the late Precambrian; however, the end of the glaciation just before Cambrian times would have caused milder climates that encouraged rapid evolution. In other words, Rudwick is suggesting an external physical control and trigger to influence rates of evolution.

FISCHER (1965) suggested an alternative trigger. Berkner and Marshall had proposed that oxygen had accumulated slowly through Precambrian time as the Earth's originally anoxic atmosphere and ocean were oxygenated by photosynthesis, mainly from algae. Although Fischer did not agree with the timing that Berkner and Marshall had proposed, he agreed with them that increasing oxygen level could have been the trigger that set off the evolution of multicellular animals. Fischer tried to avoid the idea that the Cambrian was a time of explosive evolution, however, and he envisaged slow evolution in "oxygen oases" in patches of algae during much of the Precambrian. He suggested that we see the rapid development of skeletons in the Cambrian because animals developed shields against ultra-violet radiation. This is a rather complex argument, but it does depend ultimately on an oxygen trigger for an evolutionary event which must have been rapid even if it was not explosive.

NICOL (1966) summarized previous theories, and proposed that a simple effect of size increase would be an increase in the complexity of body organization, and an increase in structural rigidity (skeletonization, in other words). Thus Nicol relates the development of hard parts to increased size. He appeals to "Cope's Rule", an observation which says that animals tend to increase in size through time, and this implies that the Cambrian size increase is thus a part of normal evolution, not something extraordinary (see Stanley 1973 for a more recent discussion of Cope's Rule). Nicol is implying that the problem has been somewhat overstated, and that one might have expected a sudden "burst" of skeletonization of some point in the history of life. It is part of the structure of Nicol's approach that no external, extraordinary trigger is required to explain the data, so he does not discuss one.

TOWE (1970) turns to biochemistry in search of an explanation. He points out that a minimum oxygen concentration is needed by living animals if they are to form collagen--the "glue and tape" used in constructing the anatomy of very many animals. If this requirement also applied in the late Precambrian, then animals would not have been able to form complex body tissues until the ocean reached a critical oxygen level.

In other words, Towe is using the same kind of argument as Berkner and Marshall, and Fischer, but extending it to include biochemical aspects of body organization. In this argument, an external trigger is necessary to explain the data.

The paper by VALENTINE and MOORES (1972; Section 4) points out that the Cambrian was a time at which a supercontinent was breaking up. This would automatically mean a milder climate, an increase in stability among food resources in the sea, and an increase in area and diversity of available habitats. So Valentine and Moores would envisage an external trigger which encouraged the increase in diversity among Cambrian organisms, but a trigger connected with plate tectonics and climate rather than atmospheric and oceanic oxygen levels.

In 1973 SCHOPF, HAUGH, MOLNAR and SATTERTHWAIT published a paper in which they appealed to ecological triggers rather than physical ones. J. William Schopf is a paleobotanist, and even the title of the paper suggests that the authors felt botanical aspects of late Precambrian evolution had been ignored --the title stresses metaphytes (complex plants) ahead of metazoans (comples animals). The original idea expressed in the paper is that the development of sexual reproduction among single-celled algae at or before

1 billion years ago led to the rapid evolution of complex multicellular plants. An ecology based on these could lead to the rapid diversification of complex multicellular animals at the same time. In other words, Schopf et al. see no reason to postulate an external environmental trigger for the Cambrian radiation, but they view it as part of the process of evolution on Earth, once sexual reproduction had been developed.

The same year STANLEY (1973), partly in response to Schopf et al., proposed a different ecological hypothesis. He argued that among living animals, the removal of a predator often results in the disappearance of other members of a natural community. In experimental systems, the introduction of a predator may have a stabilizing effect on population levels among prey species. In other words, predators encourage diversity and stability. This is not always true (compare, for example, the introduction of rats to oceanic islands), so that it cannot be called a "principle", but it is a viewpoint with data to support it. Stanley made a large conceptual jump and tried to apply the idea to the origin of multicellular animals at the base of the Cambrian. He argued that the evolution of "cropping" animals (either herbivores or carnivores) would have encouraged diversity among late

Precambrian prey species (both plants and animals), and that this would have led to the development of a suddenly visible complexity as skeletons were evolved. Thus Stanley's idea also dispenses with external environmental controls and concentrates on biological factors within the organisms and communities of the time as triggering mechanisms.

Thus the controversy stands at the moment, with competing hypotheses based on the same data, but with different suggestions. In summary, they are

NO-PROBLEM HYPOTHESES--there is nothing particularly worrying about Cambrian faunas; for example, they are a natural result of size increase (NICOL). The fact that we have only one reading with this point of view is not significant, because controversy arises where people try to explain what they think is a problem.

EXPLANATORY HYPOTHESES

1.  External environment

    A.  Triggered by the end of glaciation (RUDWICK).

    B.  Triggered by splitting a supercontinent (VALENTINE and MOORES, Section 4).

    C.  Triggered by reaching a critical oxygen level (FISCHER, TOWE).

2.  Biological factors within organisms

    A.  Triggered by a key innovation--development of sexual reproduction (SCHOPF et al.).

    B.  Triggered by a key innovation--development of a "cropping" capability (STANLEY).

This controversy is not likely to go away. The more evidence that accumulates, the moreiit becomes clear that the evolution of skeletons was a predominantly Lower Cambrian set of events, and was therefore really sudden in geological terms (see, for example, Cisne 1974 and Runnegar and Pojeta 1974).

EXTINCTIONS

For extinction as for evolution, we can understand the process on the population level from observational and experimental data on living animals. But it is much more difficult to explain times in the fossil record when either evolution or extinction has been very rapid. The extinction of many groups of marine organisms at the end of the Permian appears to have been a major event in Earth history (see, for example, Newell 1963). Just as for evolution, there are two approaches: one is to explain apparent disastrous extinctions as understandable in comparatively ordinary terms, the other is to try to find some reasonable special circumstances to account for the data at their face value.

Valentine and Moores (1972;

Section 4) associate the dramatic Permian extinction with the re-assembly of a supercontinent, and in a hypothesis that is essentially the reverse of their idea for promoting diversity, they attribute extinctions to more unstable climatic and food resource conditions, triggered ultimately by plate tectonic movements.

As we saw in Section 4, Raup (1972) regards the Permian "extinction" as simply an artefact of the fossil record, whereas Valentine (1973) argued that it is a real phenomenon.

We reprint an extract from a longer paper by RHODES (1967) who has made a careful analysis of the Permian extinction. He comes to the conclusion that it was not so much a rapid extinction at the end of the Permian as it was a failure to replace those animals that became extinct naturally. Rhodes' opinion is that we should change our question, to ask Why no new evolution? rather than Why extinction? Note that this conclusion does not affect the arguments of Valentine and Moores, but could be accommodated along with theirs.

Other contributions to the question of extinction at the end of the Permian have been more "catastrophist", and Rhodes summarizes many of these.

As a new contribution to the continuing controversy, we reprint a recent paper by JOHNSON (1974). He notes that extinctions have been known for a long time to be associated with retreats of sealevel, but that such retreats do not always cause extinction. He suggests that a long period of high sea-level, followed by a rapid retreat, may be a trigger for widespread extinction, and he quotes examples of such events. Note that the idea meshes well with the plate-tectonic ideas of Valentine and Moores, and Hays and Pitman (Section 4).

FURTHER READINGS

BERKNER, L. V. and MARSHALL, L. C. 1965. History of major atmospheric components. Proc. nat. Acad. Sci. 53, 1215-1225.

CISNE, J. L. 1974. Trilobites and the origin of arthropods. Science 186, 13-18.

CLOUD, P. E. 1968. Pre-metazoan evolution and the origins of the Metazoa. In DRAKE, E. T. (ed.), Evolution and Environment, 1-72. Yale University Press.

---- 1972. A working model of the primitive Earth. Am. J. Sci. 272, 537-548.

---- 1974. Evolution of ecosystems. Amer. Scientist 62, 54-66.

NEWELL, N. D. 1963. Crises in the history of life. Sci. Amer. 216 (March issue).

RUNNEGAR, B. and POJETA, J.   1974.
    Molluscan phylogeny:  the pale-
    ontological viewpoint.  Science
    186, 311-317.
STANLEY, S. M.   1973.   An explanation
    for Cope's Rule.  Evolution 27,
    1-26.

*

# The infra-Cambrian glaciation and the origin of the Cambrian fauna

## M.J.S. Rudwick (1964)

REALITY OF THE PROBLEM

The origin of the Cambrian fauna is a problem that has been reviewed very frequently during the present century. But the discussion has tended to become repetitive. New light on the problem is now provided by the recognition of a world-wide glacial period not long before the first appearance of the Cambrian fauna (Harland, 1963). A causal connexion between these events has been suggested briefly by several writers (Harland and Wilson, 1956; Kobayashi, 1956; Termier and Termier, 1956, 1957), but it deserved further discussion, especially because recent reviews (e.g. Glaessner, 1962; Simpson, 1960) have failed even to consider it.

The known contrast between the almost unfossiliferous Precambrian and the highly fossiliferous Cambrian, far

Reprinted from Problems in Palaeoclimatology (A.E.M.Nairn, ed.), p. 150-155 by permission of author and publisher.

from being less striking than in Darwin's time (Simpson, 1960), has been accentuated by the progress of geology. The faunas of the Cambrian are far richer than those known to Darwin; and the discovery of unmetamorphosed Precambrian sediments, lying conformably below Cambrian strata, has made untenable two of the explanations most popular in his day, namely that the Late Precambrian fossil record was missing either through non-deposition or through destruction by metamorphism. Many recent authors have avoided the full force of the problem by underrating the magnitude of the contrast. An evident anxiety to preclude any causes of an extra-scientific or even extra-terrestrial nature has led them to underestimate both the sudden appearance and the "advanced" character of the Cambrian fauna.

Even modern authors continue to assert that the early Cambrian metazoa were "primitive" (Glaessner, 1962) or "markedly simple and generalized"

(Simpson, 1950), although this could justifiably be denied long ago (Brooks, 1894), and although it is doubtful whether it would find much support among competent Cambrian palaeontologists at the present day. These Lower Cambrian fossils are "primitive" in the strict and obvious sense that most of them are the earliest known members of their respective phyla. They are certainly not "primitive" or "simple" or "generalized" in any morphological sense, except by judgement a posteriori.* Nor are they notably "small", though it is true that many are "relatively rare" (Glaessner, 1962) by comparison with the Middle Cambrian faunas.

Their relatively sudden appearance has also been underestimated, especially through the use of charts which fail to distinguish clearly between factual and hypothetical ranges. In fact, most of the alleged Precambrian metazoa are now agreed to be very doubtfully metazoan or very doubtfully Precambrian (Glaessner, 1962; Schindewolf, 1956). Fossils of plant origin--stromatolites, hystrichospheres and, more dubiously, spores--are of course well known in the

*Compare the phylogenetic problems raised by the structural peculiarities of the olenellid trilobites and brachiopods such as Matutella in the Lower Cambrian.

Precambrian; but the only substantial record of possibly Precambrian metazoa is now the soft-bodied fauna from Ediacara in South Australia (Glaessner, 1958a, b, 1960). This fauna lies about 150 m below the earliest Lower Cambrian fauna known in that area; but it may be contemparaneous with earlier Lower Cambrian faunas in other regions, for in South Africa two of its members are found with archaeocyathids, which are generally taken to be of Lower Cambrian age (Haughton, 1959). In any case, even if it is called "Precambrian", the Ediacara fauna is stratigraphically so close to the earliest Cambrian faunas, and zoologically so unlike them, that it can have little bearing on the problem of the origin of the Cambrian metazoa (Sdzuy, 1960). On the time-scale of the whole fossil record, it must clearly be grouped with the Lower Cambrian faunas and regarded as a fauna which lived in a different (littoral) environment at about the same time or shortly before. The problem is not solved by the discovery of a single fauna a few tens of metres below those previously known: the real problem lies in the many thousands of metres of barren strata below that. If, then, the Ediacara fauna is omitted from the "Precambrian" column of a recent tabular view of the problem (Glaessner, 1962: Table 1), the phyla represented in the Precambrian are reduced to very doubtful protozoa and sponge spicules, and a very few trace-fossils (tracks, burrows, etc.) possibly made by "worms" or arthropods. After so much intensive searching for organic traces of any kind in Precambrian rocks, it seems extremely unlikely that Precambrian trace-fossils have been overlooked simply because "they were not thought to be important stratigraphically or palaeontologically" (Glaessner, 1962). Seilacher described a few Precambrian trace-fossils, but concluded that they are very poor in numbers and in variety compared to those of even the Lower Cambrian (Seilacher, 1955, 1956). Several authors (e.g. Simpson, 1960) have asserted that the various phyla appeared gradually throughout Cambrian time, which may have lasted as much as 100 My or one-sixth of all time since the Precambrian (Holmes, 1959). In fact most of the major phyla are now known from Lower Cambrian strata: of phyla liable to fossilization under normal circumstances, only the Polyzoa, Graptolithina and Chordata seem to be missing.

The stratigraphy and palaeontology of the Lower Cambrian itself is not yet known in sufficient detail to determine whether the various phyla really appeared gradually during the 25-30 My (Cloud, 1948) of Lower Cambrian time, or whether their scattered occurrences are due to chances of preservation or ecological or biographical factors. Even if the phyla did appear one by one (Sdzuy, 1960; Termier and Termier, 1957), these 25-30 My must be compared with the subsequent 500-600 My in which their fossil record is almost unbroken, and with the still longer period of Precambrian time in which their record is virtually non-existent (Fig. 1). On the total geological time-scale, it is therefore accurate to characterize the appearance of the metazoan phyla as "sudden".

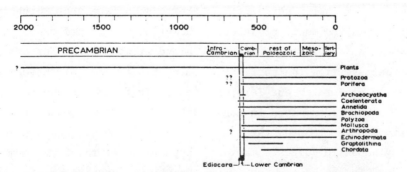

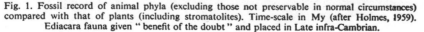

Fig. 1. Fossil record of animal phyla (excluding those not preservable in normal circumstances) compared with that of plants (including stromatolites). Time-scale in My (after Holmes, 1959). Ediacara fauna given " benefit of the doubt " and placed in Late infra-Cambrian.

The problem must therefore be accepted as real; the evidence must be explained, and not merely explained away.

## EXPLANATION OF THE PROBLEM

Three main classes of explanation have been proposed:  either (a) the immediate ancestors of the Cambrian metazoa lived in environments which are not represented among Precambrian sediments (e.g. Axelrod, 1958; Termier and Termier, 1949; Walcott, 1910); or (b) they were not provided with skeletal structures suitable for preservation in the environments that are represented there (e.g. Glaessner, 1960, 1962; Snyder, 1947); or (c) the Cambrian fauna was genuinely novel, being the result of rapid and radical evolutionary change (e.g. Brooks, 1894; Cloud, 1948; Nursall, 1959; Schindewolf, 1956; Seilacher, 1956).

Infra-Cambrian strata are remarkable for the variety of sediments represented and for their "normality" when compared with sediments of later date.  There is no convincing evidence to support the assertion (Walcott, 1914; Leith, 1934; Termier and Termier, 1949) that they were deposited in nonmarine environments.  Indeed, the presence of glauconite suggests marine deposition; and in any case a nonmarine origin is difficult to prove, even in later strata, if fossils are absent (Seilacher, 1956).  It is conceivable that Precambrian metazoa were confined to the littoral zone (Axelrod, 1958); but there are many infra-Cambrian strata with the normal marks of littoral deposition.  Even if there were not, it would be surprising (by analogy with later periods) if the remains of littoral organisms were never drifted into deeper-water environments and preserved there.  Conversely, it is conceivable (though perhaps less likely) that infra-Cambrian metazoa were confined to deep water in geosynclinal belts (Simon, 1958); but not all geosynclinal deposits in the infra-Cambrian have been metamorphosed too strongly for fossils to be preserved (Sdzuy, 1960).

It is more plausible to suppose that the groups represented in the Lower Cambrian fauna evolved first in small confined regions, and only gradually extended their range.  Then, considering the small proportion of the Earth's surface over which infra-Cambrian strata are exposed, chance alone might account for our failure to find the remains of these localized earliest faunas (Sdzuy, 1960).  But this still fails to account for the relatively sudden and widespread appearance of the Lower Cambrian fauna, for Lower Cambrian strata are also exposed only in relatively small areas.

If, then, we do not find the ancestors of the Cambrian fauna preserved in infra-Cambrian strata, it might be due to their lack of preservable skeletal structures. Infra-Cambrian seas might have contained an abundant fauna of large, "advanced" but wholly soft-bodied metazoa (Glaessner, 1960; Snyder, 1947). Then the novelty of the Lower Cambrian fauna would lie simply in the independent acquisition of "hard parts" by several groups during Lower Cambrian time (Glaessner, 1962). Some causal explanations proposed for this development are no longer tenable (Raymond, 1935; Seilacher, 1956; Whittard, 1953). There is no evidence that infra-Cambrian sea-water was poor in calcium, or acidic, or that it differed chemically or physically in any other significant way from the sea-water of later periods. There may have been no carnivores and therefore no need for skeletal protection before the Cambrian (Evans, 1912; Raymond, 1935; Whittard, 1953); but the skeletons of the Lower Cambrian animals appear to have been as much supporting structures, necessary for the functioning of the organism, as protective structures (Seilacher, 1956; Sollas, 1912), and therefore might be expected to precede the first appearance of predators. In any case there is no positive evidence that an abundant Precambrian fauna of large, "advanced" but soft-bodied meta-

zoa ever existed. Apart from the Ediacara fauna from the highest Precambrian (if not the Lower Cambrian), there is nothing but some very sparse trace-fossils. In Cambrian and later strata, traces of the activities of soft-bodied benthonic animals are much more common than actual impressions of the organisms themselves. The scarcity of infra-Cambrian trace-fossils suggests that the origin of the Cambrian fauna involved more than the mere acquisition of preservable skeletal structures (Seilacher, 1956).

In any case, it is not easy to conceive that any immediate ancestor of the Lower Cambrian trilobites or brachiopods, for example, could have been wholly soft-bodied, since rigid skeletal elements seem to be inherently necessary for the functional organization of these groups (Cloud, 1948; Sdzuy, 1960). However, if the ancestral skeletal structures were of unthickened, unmineralized "chitin" (as the rarely preserved limbs of trilobites seem to have been), the ancestral animals would have been skeleton-bearing from the functional point of view, yet soft-bodied from the point of view of their chance of preservation (Glaessner, 1962). But it seems doubtful whether such thin and weak skeletal elements would permit a viable organization unless the animals were small in size. Thus, most infra-

Cambrian metazoa may have been small (Axelrod, 1958; Nursall, 1959), perhaps planktonic (Brooks, 1894; Schindewolf, 1956), and effectively soft-bodied, at least in relation to their chance of preservation under normal conditions (Glaessner, 1962). This hypothesis may yet be confirmed by an intensive search of infra-Cambrian sediments, using modern micropalaeontological techniques. If the infra-Cambrian fauna did indeed have this character, the scarcity of normal macrofossils and trace-fossils in Precambrian strata would be adequately explained (a small and sparse benthonic fauna could have produced the few trace-fossils known).

The hypothesis involves the assumption that major evolutionary change occurred rapidly at the beginning of the Cambrian. This still requires causal explanation. The gradual availability of free oxygen for metabolism (Nursall, 1959), the colonization of the sea-floor by a previously planktonic fauna (Brooks, 1894) or a previously littoral fauna (Axelrod, 1958), and the original inefficiency of chemical change as a source of energy (Haldane, 1944), are all hypotheses on which only a gradual faunal change would be expected. They are therefore inadequate: some "trigger mechanism" must be postulated (Seilacher, 1956). An extra-terrestrial cosmic event of unknown kind, producing typostrophic

evolution (Schindewolf, 1956; Seilacher, 1956), is possible but seems to be an untestable hypothesis.* The evidence of at least one world-wide glacial period in late Precambrian time (Harland, 1963), itself produced perhaps by an extra-terrestrial event, suggests that the latter hypothesis may be nearer the truth than the gradualistic hypotheses which attempt to explain away the sudden appearance of the Cambrian fauna.

A long-continued world-wide glaciation would be unfavourable for the development of life for two reasons: (a) it would create adverse climatic conditions on a world-wide scale; (b) by withdrawing much oceanic water into ice-sheets it would cause an eustatic fall in sea level and hence reduce the total area of shallow seas on the continental shelves, which have probably always contained the most favourable environments for marine life. At the end of the glacial period these factors would have been reversed. The climate would have ameliorated and the melting ice-sheets would have caus-

*Obviously, typostrophism should not be ruled out merely because it happens to be "incompatible with the most widely held evolutionary theories" (Simpson, 1960; cf. Axelrod, 1958 and Glaessner, 1962): even widely held theories may be ultimately proved incorrect or at least inadequate.

ed an eustatic rise in sea level and hence the well-known Cambrian transgression of many of the continental areas. These two factors together could have "triggered off" a phase of evolutionary change on a scale never again repeated. They would have laid open a wide variety of favourable but "empty" marine environments, in which radical adaptive radiation could have occurred (cf. Kobayashi, 1956; Termier and Termier, 1956). This might have been aided by a parallel development in plant life, which would have provided a much enlarged food supply for the evolving metazoa and hence permitted a rapid increase in their body size (most Cambrian metazoa seem likely to have been filter-feeders or deposit-feeders). In some groups, such an increase in size would also have been facilitated by the acquisition of mineralized skeletons. This probably occurred independently in different groups during Lower Cambrian time; it would have increased greatly their chance of preservation, although groups without such skeletons would also have become liable to preservation at least occasionally (i.e. as trace-fossils or in unusual sedimentary environments as at Ediacara or in the Burgess shale). The "pioneer" status of the faunas produced by these changes would account for their initial poverty in species and in numbers, and also perhaps for their apparent differentiation into faunal provinces (Sdzuy, 1960; Termier and Termier, 1957).

Thus the end of the lengthy and world-wide infra-Cambrian glaciation could have provided the "trigger" necessary to induce relatively rapid evolution from an infra-Cambrian fauna of diversified but small and soft-bodied metazoa into the variety of much larger animals, many of them with robust skeletal elements and most of them liable to preservation, which constitute the fauna of the Lower Cambrian. This hypothesis is perfectly actualistic, even though it is not strictly uniformitarian. It does not postulate any cause differing in _kind_ from those believed to be operating at the present day; but it does postulate a causal connexion between climatic and biological events which differ in _degree_ from any similar events that have since occurred on the Earth.

*

# Fossils, early life, and atmospheric history

## A.G. Fischer (1965)

THE FOSSIL RECORD

From the paleontological stand-point, earth history is sharply divided into two parts--the vast Pre-Cambrian, with a sparse fossil record, and the Phanerozoic, beginning with the Cambrian period some 600 million years ago, with its rich documentation of life (Fig. 1).

The only fossils which have been found widespread and abundant in the Pre-Cambrian are stromatolites: head-shaped or branched, laminated structures, generally composed of calcite or dolomite, but in some cases siliceous, and generally attributed to the Cyanophyta (also known as Myxophyta or blue-green "algae"). They appear to be present in most Pre-Cambrian lime-stones and include the oldest fossils known to date, from the African Bulawayan,[1] about 2.7 billion years old. Like most fossil stromatolites,

Reprinted from Proc.Nat.Acad.Sci.USA 53, 1205-1215 (1965) by permission of the author and publisher.

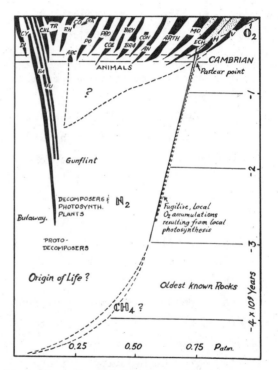

Fig. 1. The fossil record, plotted on a model of the evolving atmosphere. DI, Dinoflagellates (incl. Hystrichospheres); CY, Cyanophyta (blue-green algae); BA, bacteria; FU, fungi; CHL, Chlorophyta; TR, Tracheophyta (the higher plants); RH, Rhodophyta (red algae); CO, Coccolithophorida; DIA, diatoms; ARC, Archaeocyathida; PO, Porifera; PRO, Protozoa, COE, Annelida; ARTH, Arthropoda; MO, Mollusca; ECH, Echinodermata; H, Hemichordata (incl. Graptolites); V, Vertebrata.

those from the Bulawayan show no cellular detail; their interpretation as fossils rests on three lines of evidence: we know of no inorganic processes which form such structures;

cyanophyte colonies produce structures of this type today;[2,14] and well-preserved stromatolites of various ages have yielded microscopic cellular detail which confirms their organic origin. The oldest of these cell and filament-retaining stromatolites studied to date are from the Pre-Cambrian Gunflint chert, about 1.9 billion years old.[3,4] In addition to algae or cyanophyte structures they seem to contain remains of bacteria and perhaps aquatic fungi.

The presence of life in the Pre-Cambrian is also attested by the widespread occurrence of large organic molecules, and microscopic work is turning up filaments, globules, etc., which are of organic origin but of uncertain affinities. But we are only beginning to learn about these.

The animal record is almost non-existent. While a fair number of rock structures have been hopefully interpreted as animal fossils of Pre-Cambrian age, the majority of these are either clearly of inorganic origin, or are dubious as to origin or age.[5] To very skeptical people, such as Cloud,[4] the existence of Pre-Cambrian animal fossils has yet to be established, while Seilacher[6] accepts some bottom markings in late Pre-Cambrian rocks as evidence of animals. The Ediacara fauna deserves special consideration. In several parts of the world, sand-stones occurring below Lower Cambrian rocks with normal Lower Cambrian faunas have yielded imprints of soft-bodied organisms. The best assemblage stems from Ediacara in Australia,[7,8] and contains pennatulid and medusoid coelenterates, annelids, a possible echinoderm, and problematica. Elements of this fauna occur in Africa, Europe, and North America. A Lowermost Cambrian age cannot be entirely ruled out, but a latest Pre-Cambrian age seems more likely.

The base of the Cambrian period is marked in marine sediments round the world by the appearance of abundant animal life. In detrital sediments worm burrows generally appear first, while in carbonate sequences the problematic archaeocyathids pioneer. Brachiopods, trilobites, and other arthropods, sponges, mollusks, and echinoderms tend to appear somewhat later. All of the animal phyla likely to leave a fossil record, and nearly all of the classes thus endowed, appeared by mid-Ordovician time, i.e., over an interval of about 120 million years (Fig. 1).

## THE BERKNER-MARSHALL THEORY

This sudden appearance of the diverse animal stocks has been the most vexing riddle in paleontology, and a voluminous literature has accumulated around it.[4-6,8-10] It has always been tempting to look to some important environmental change, which might or might not have been accompanied by an evolutionary outburst. But neither acid oceans, nor alkaline oceans, nor cosmic events provided plausible and coherent models which fit into earth history as a whole.

This brings us to the Berkner-Marshall hypothesis. Some geologists and some biochemists have long suspected that the earth's atmosphere has evolved through time.[11] In recent years it has become almost certain that the early atmosphere lacked free oxygen,[12] and that the land surface and the upper layers of the water masses were subjected to rather intense ultra-violet radiation. Life originated under these conditions and played a large role in subsequent changes: photosynthetic plants freed oxygen, which accumulated in the atmosphere; and this, in turn, provided the ozone shield which today screens the earth's surface from most of the UV radiation received.[13] Berkner and Marshall[16] suggest that the beginning of Cambrian time marks an important threshold in atmospheric evolution: the attainment of about 1 per cent of present oxygen pressure (Pasteur point), at which level oxygen respiration becomes profitable to organisms and permits

animal life as we know it. They further suggest that the colonization of the continents, which occurred later in Silurian and Devonian times, marked a stage when atmospheric oxygen pressure had risen to 10 per cent of present level, and had thereby reduced UV radiation at the land surface to tolerable levels. In Figure 1, the fossil record is shown superimposed on an evolutionary model of the atmosphere following the general lines (but not the details) of the Berkner-Marshall theory.

The question at hand is whether the fossil record supports the concept of atmospheric evolution in general, and the model proposed by Berkner and Marshall. We may look for evidence of radiation intensity; the evidence is not definite, but it is favorable. We shall also examine Berkner and Marshall's views on animal origins and the history of continental coloniza-tion, and shall formulate alternative views.

UV RADIATION

According to Berkner and Marshall, dangerous quantities of UV radiation penetrated the waters to a depth of about 10 meters during most of the Pre-Cambrian, but decreased to a few centimeters in latest Pre-Cambrian to earliest Cambrian time, and continued to decrease, so that tolerable sub-aerial levels were attained in the Silurian.

This is, of course, not a one-sided matter; not all organisms are equally tolerant of this and other forms of radiation.[15] Tolerance may involve the structure of the proto-plasm, as well as the presence of shielding devices. To most of us, a day at the beach occasionally demon-strates the inadequacy of our own shielding against present-day UV radiation levels. In any case, we might expect the earlier organisms of the shallow waters to have been especially adapted to cope with UV radiation.

## STROMATOLITES

The abundance of Pre-Cambrian stromatolites may be an expression of this: Pérès[17] has found that cyano-phytes are more resistant to UV than are bacteria. Cyanophytes are particularly successful today in such radiation-exposed environments as tropical tidal flats. And cyanophyte colonies of various kinds cover them-selves with mineral particles--either by entrapping and binding extraneous grains, or by secreting calcium carbon-ate or silica. It is this tendency which provides us with a fossil record of stromatolites, in the form of

stromatolite heads, and of horizontally laminated or minutely crinkled calcare-ous "algal mats" and algal mat lime-stones.[18] I suggest that this mineral-ization serves the purpose of shading the cells and the colony as a whole.

Presumably, many Pre-Cambrian stromatolites lived below tidal range.[14] Clearly, they extended into very shallow water, for in the Belt series[19] they are interbedded with mud-cracked sediments. But did they range into the intertidal? This might shed doubt on the theory of intense UV radiation. The Fentons[19] have described some shrinkage cracks in laminated limestones peripheral to stromatolite reefs, but they attributed the cracks to subaqueous syneresis.

Mud-cracked, laminated "bird's-eye" limestones of the type generally attributed to intertidal algal-mat origin occur at least from the Mid-Ordovician on (Lowville limestone, New York),[18] but finite proof of algae is wanting.

In the history of fossil cyano-phytes, stromatolites were dominant up to Middle Ordovician times, and then declined through the Paleozoic to play a very minor role in Mesozoic marine limestones and to virtually disappear in the Cenozoic. Algal mat limestones characterize the span from the Mid-Ordovician on. It is tempting to think that the stromatolites were dominantly subtidal and characterize an early stage on high radiation intensity; that in the next episode of lower radiation intensity cyanophytes flourished primarily in the intertidal, while in the subtidal realm they declined due to competition with other plants, or became less mineralized and therefore less readily preserved.

## EXOSKELETONS

The fossil outburst in the Cambrian and Ordovician is really a double puzzle: one problem is the appearance of abundant and diversified life. Another is the simultaneous appearance of exoskeletons in so many different kinds of plants and animals. Such skeletons now serve a variety of functions such as predator protection and tissue support, but we may well raise the question whether their primary function may not have been (and in some cases still is) one of radiation protection. If so, then we might expect to see a tendency to decrease this surficial armor through geologic time. A number of groups show such a trend.

Among the dasycladacean algae, the most completely sheathed forms known are the Ordovician Cyclocrinus[20] and the Ordovician to Devonian Ischadites.[21]

The Bryozoa, which made their appearance in Ordovician time, included at that time many very heavily calcified forms; some such persist through the Paleozoic, but they contrast markedly with the delicate present-day bryozoans.

The cephalopod mollusks of the early Paleozoic were covered by exoskeletons, which required complex clumsy modifications for a swimming mode of life. Naked forms, in which the tissues overgrew the reduced skeleton, appeared in Devonian (?) or Mississippian time, and in the Cenozoic the squids became the dominant cephalopods.

Even the fishes, in the broad sense, made their Ordovician debut encased in a clumsy, boxlike armor, which in later periods gave way to less massive protection.

Surely these examples do not prove the point, but they are in harmony with the theory of decreasing radiation levels. We shall return below to the problem of skeletal origins.

## COLONIZATION OF THE LANDS

Berkner and Marshall chose the Silurian as the time of oxygen accumulation to the 10 per cent present pressure level and associated UV screening, because this system has yielded the earliest land animals (scorpions) and, at its upper boundary, the earliest generally accepted land plants (vascular plants). But their model strikes me as too simple, and I visualize a much longer period of colonization, as follows:

(1)  Colonization of the soil:

Algae, especially cyanophytes, play a large role in present-day soils; some live at the surface, others below the surface but within the range of light penetration, and others at still greater depth.  It seems possible that, even at times of intense UV radiation, there remained a zone in the soil which could harbor photosynthetic organisms. This possibility should be investigated by experiment.  Cyanophytes lead in the colonization of present-day soils,[28] and if they colonized Pre-Cambrian or earliest Paleozoic soils, they could have formed the base of an ecologic pyramid which came to include bacteria and fungi, as well as animals--worms, palpigrades, collembolids, etc.[22]  The maturity of many Pre-Cambrian sediments[24] furnishes a measure of evidence for development of mature soils at that time, and thus provides support for the concept of an early colonization of soil by organisms.  If this concept is correct, the continents may have been an important source of oxygen in pre-Silurian times.

(2)  Colonization of shade oases:

The lands offer many shady habitats, especially in ruggedly dissected upland regions of the middle latitudes, where no direct sunlight is received,

and where UV radiation levels must have been below those in sunny locations.  I suggest these "shade oases" were the places where surficial land plants developed, derived in part perhaps from soil plants, in part from aquatic algae.  This may have happened in the Cambrian or earlier.  The isolated nature of these shade oases would have favored the development of highly divergent floras of higher plants, before Devonian time.

(3)  Colonization of the intertidal:

The invasion of marine cyanophytes and possibly algae into tidal flats, probably in Ordovician time, has been dealt with above.

(4)  Spread of land plants:

The great spread of land plants in early Devonian time, which Berkner and Marshall regard as an evolutionary explosion, thus becomes mainly a spreading out and commingling of previously isolated floras.  The trigger is the same:  reduction of radiation over the lands as a whole to levels which could be tolerated by a tree flora. This commingling of older floras which is here proposed no doubt would have led to intense evolution, but it would also have led to much extermination.

OXYGEN

Berkner and Marshall propose that the beginning of Cambrian time marks the attainment of the "Pasteur point" in the atmosphere, i.e., 1 per cent of present oxygen pressure, at which point oxidiation respiration becomes advantageous (Cloud[4] places this event somewhere between 1.2 and 0.6 billion years). They believe that animal life before this time was of a very low and primitive sort (though they are not specific), and visualize a great "quantum jump" in evolution, which produced the various animal phyla at this time. They challenge "the idea of a long Precambrian history (i.e., long compared to a few tens of millions of years) of advanced evolution of organisms to account for the diversification found at the base of the Cambrian," an idea which is "deeply embedded in the whole literature of geology and paleontology." They propose that the "geological record should be read exactly as presented in nature." Cloud[4] feels that there may have been a somewhat longer lag between the attainment of the Pasteur point and the evolutionary outburst of phyla.

The suggestion that the widespread appearance of sea animals at the beginning of Cambrian time reflects the attainment of a critical oxygen level in the atmosphere seems to me the first plausible explanation of this milepost of the fossil record, and as a brilliant contribution. Their appeal to an evolutionary explosion seems quite unnecessary, and I am proposing the following alternative model of Pre-Cambrian animal evolution.

Photosynthesis probably began more than 2.7 billion years ago. As Cloud[4] has pointed out, it provided local, fugitive supplies of free oxygen in the water during a large part of Pre-Cambrian time. A part of this oxygen may have come to be used by the same plants. Then facultatively aerobic bacteria developed, using this oxygen to respire more efficiently. The protozoans evolved in direct respiratory dependence on host plants, active when oxygen was available, and dormant or fueled by fermentation when it was not. Then metazoans of primitive sorts, possibly including ancestral sponges, turbellarians, and coelenterates, evolved under similar conditions of existence, out of reach of the UV rays. They were small and naked, and most unlikely to leave a fossil record. Limited to the "oxygen oases" of individual plant patches, they formed strongly isolated communities, subject to highly divergent evolution (Fig. 2, stage I). It seems likely to me that hundreds of different animal types-- essentially of phylum rank--were thus developed.

The first break in this pattern of isolation came when some animals took photosynthetic algae into their tissues, in zoöchlorellar or zoöxanthellar symbiosis. With this internal

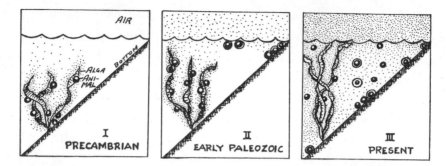

Fig. 2. Stages in animal evolution. *Stage I*, animals living in complete respiratory dependence on host plants; atmosphere essentially devoid of free oxygen. *Stage II*, atmosphere has reached "Pasteur point"; animals may leave plants, but flock toward air-water interface. *Stage III*, atmosphere and water highly oxygenated, animals widely distributed.

oxygen supply, they could cut loose from their external plant hosts to cruise about within the zone of light, but below the UV danger level. Hedgpeth[25] has pointed out that some members of the Ediacara fauna may have lived in this fashion, for Glaessner[8] reports that the jellyfishes lie in an "upsidedown," that is, mouth-up, position, which is the normal one for the living, zoöxanthella-dependent jellyfish Cassiopeia.

The true emancipation of animals came when oxygen over wide areas of the ocean rose to levels capable of supporting oxidizing respiration. The isolation of the many different communities was now broken, the many different phyla were thrown into contact, and many types were eliminated while others survived and evolved.

Presumably, oxygen was the chief limiting factor on the density of animal populations. This placed a premium on living near the sources of supply: plants, and the air-water interface (Fig. 2, stage II). For benthonic animals, the shallowest waters became most attractive. Plants produced greater local concentrations of oxygen, but transitory ones, while the supply from the air remained steady. But the air-water interface and the shallowest bottoms were exposed to dangerous UV radiation.

Some animals may have commuted from deeper-dwelling plants, during the day, to the surface at night. Perhaps the diurnal plankton migrations in our seas hark back to that time. Some benthonic animals in the shallows may have lain buried in the sediment during the day, to become active at night--a pattern followed by various living invertebrates. But many organisms may have developed radiation shielding, in the form of an exoskeleton, to permit them to remain near the surface.

CONCLUSIONS

The main points of the Berkner-Marshall theory are (1) that the widespread appearance of sea animals in

Cambrian time reflects the passage of a significant oxygen threshold in the atmosphere, and (2) that the widespread appearance of forests in the Devonian reflects the reduction of ultraviolet radiation to tolerable levels. The theory thus offers a plausible and internally linked explanation for two striking events in earth history, which have heretofore been riddles.

Berkner and Marshall's suggestions that the appearance of marine animals coincides closely with their origin, and that the appearance of land fossils coincides with the colonization of lands, take recourse to Schindewolfian[9] evolutionary explosions, for which I see no need.

An alternative model proposes that marine animals developed in "oxygen oases"--marine plant communities-- during the latter half of Pre-Cambrian time as small and naked forms, essentially incapable of leaving a fossil record. Attainment of the "Pasteur point" in the atmosphere emancipated them from their host plants, and allowed them to spread widely over the seas. Oxygen dependence forced many of them to the surface and into the shallowest water, where they developed exoskeletons as radiation protection.

An alternative model for the colonization of lands proposes colonization of the soil in Pre-Cambrian time, by soil algae, and the development of a surficial land flora in shaded places during the Cambrian (?), Ordovician, and Silurian. The spread of this flora was triggered by the decrease in radiation, in early Devonian time.

In both cases many independently developed, and therefore highly diverse, biotas were brought together. This presumably led to much extinction, and much evolution among the survivors. If this is correct, then we may expect to find, in the early Cambrian fauna and in the early Devonian flora, various rare or localized organisms of problematic affinities, which have no survivors. The Archaeocyathida, Hyolithida, and Helicoplacoidea may be such types, and more of them may be expected.

In conclusion, the fossil record does not prove the theory of atmospheric evolution; but this theory offers plausible and coherent explanations for two major events which have hitherto been baffling. It encounters some additional support and no obstacles, at the present level of knowledge. The theory holds such important implications to the field of life history that it will surely stimulate the search for pertinent facts and for critical interpretations.

[1]Macgregor, A. M., Trans. Proc. Geol. Soc. S. Africa, 43, 9-16 (1941).

[2]Ginsburg, R. N., Intern. Geol. Congr., 21st, Copenhagen, 1960, Rept. Session, Norden, 26-35 (1960).

[3]Barghoorn, E. S., and S. A. Tyler, Science, 147, 563-577 (1965).

[4]Cloud, P. E., Jr., Science, 148, 27-35 (1965). I am indebted to Dr. Cloud for a preview of his manuscript, for photographs of Pre-Cambrian algae, fungi, and bacteria, and for discussions.

[5]Schindewolf, O. H., in Geotektonisches Symposium zu Ehren von Hans Stille, ed. F. Lotze (Stuttgart: Ferd. Enke, 1956), pp. 455-480.

[6]Seilacher, A., Neues Jahrb. Geol. Palaeontol., Abhandl., 103, 155-180 (1956).

[7]Glaessner, M. F., and B. Daily, "The geology and Late Precambrian fauna of the Ediacara Fossil Reserve," Records S. Australian Museum, 13, 363-401 (1959).

[8]Glaessner, M. F., Biol. Rev. Cambridge Phil. Soc., 37, 467-494 (1962).

[9]Schindewolf, O. H., Z. Deut. Geol. Ges., 105, 153-182 (1954).

[10]Simpson, G. G., in Evolution after Darwin, ed. S. Tax (Chicago: University of Chicago Press, 1960), vol. 1, pp. 117-180.

[11]Macgregor, A. M., S. African J. Sci., 24, 155-172 (1927).

[12]Holland, H. D., "Model for the evolution of the earth's atmosphere," in "Petrologic Studies: A Volume to Honor A. F. Buddington," ed. Engel, James, and Leonard (New York: Geological Society of America, 1962), pp. 447-477. Dr. Holland aroused my interest in atmospheric evolution, and has been a constant source of advice and discussion.

[13]Rutten, M. G., The Geological Aspects of the Origin of Life on Earth (Brussels: Elsevier Pub. Co., 1962).

[14]Monty, C., Ann. Soc. Geol. Belg., Bull., 88, 19-26 (1965).

[15]Blum, H. F., Photodynamic Action and Diseases Caused by Light (New York: Reinholt, 1941); also, Blum, H. F., Carcinogenesis by Ultraviolet Light (Princeton University Press, 1959).

[16]Berkner, L. V., and L. C. Marshall, in The Origin and Evolution of Atmospheres and Oceans, ed. Brancazio and Cameron (New York: John Wiley, 1964), pp. 102-126.

[17]Pérès, J.-M., Océanographie Biologique et Biologie Marine (Presses Universitaires de France, 1961), vol. 1.

[18]For information on cyanophytes in the past and present, I am particularly indebted to Mr. Claude Monty of the University of Liège, who is presently completing a doctoral dissertation at Princeton University on Bahamian algal mats and stromatolites, and their paleontological and geological implications.

[19]Fenton, C. L., and Fenton, M. A., Bull. Geol. Soc. Am., _48_, 1873-1970 (1937).

[20]Pia, J., in _Handbuch der Pala-obotanik_, ed. M. Hirmer (Oldenbourg, 1927), pp. 31-136.

[21]Kesling, R. V., and A. Graham, _J. Paleontol._, _36_, 943-952 (1962).

[22]For advice on the subterranean biota and the colonization of the soil I am indebted to Dr. Bernd Hauser of the Zoological Institute, University of Innsbruck.

[23]Janetschek, H., _Osterr. Akad, Wiss. Math-Naturwiss. Kl., Anz._, _9_, 185-191 (1964).

[24]Pettijohn, F. J., _Sedimentary Rocks_ (New York: Harpers, 1957), 2nd ed.

[25]Hedgpeth, J., in _Approaches to Paleoecology_, ed. Imbrie and Newell (New York: John Wiley, 1965), pp. 11-18.

*

# Cope's Rule and Precambrian and early Cambrian invertebrates

## D. Nicol (1966)

Perhaps no one has been more interested than Percy Raymond in the fascinating problem of the sudden appearance of abundant marine invertebrate fossils at the beginning of Cambrian time. In a presidential address (Raymond, 1935) and in his book Prehistoric Life (Raymond, 1939), he discussed the various theories that have been proposed to explain the abrupt appearance of diverse invertebrate life in Cambrian rocks. The theories he listed in both references follow:

1. Chamberlin's theory: Organisms originated on land in the soil and migrated thence through rivers to the oceans, which they did not reach until Cambrian time.

2. Precambrian fossils were destroyed during the changes which took place in the metamorphism of the rocks.

3. Daly's theory: Precambrian marine organisms lacked calcareous skeletons because of insufficient calcium in the oceanic waters.

4. Lane's theory: The Precambrian oceans were acidic; this condition prevented the formation of calcareous skeletons.

5. Walcott's theory: All Precambrian strata that are now accessible were deposited on land in fresh water of low calcium content.

6. Brooks' theory: Precambrian organisms lacked hard parts because they lived in the surface waters of the oceans, where skeletons were detrimental because of their weight.

7. The writer's (Raymond's) modifications of Brooks' theory: Skeletons appeared after the Precambrian as a result of the adoption of a sessile or sluggish mode of existence.

Cloud (1948, p. 346-348) has suggested that the sudden appearance of diverse invertebrate life in Cambrian time resulted primarily from eruptive or explosive evolution.

Dunbar (1960, p. 126) states that

the development of actively predaceous habits may have been the first great stimulus to the development of protective armor.

The above theories are not all that have been proposed to explain the sudden appearance of fossilized invertebrate animals at the beginning of Cambrian time but are those that are most commonly quoted. Several factors, both physical and biological, undoubtedly contributed to the sudden appearance of many invertebrate animals having shells or skeletons. The biological considerations will be discussed here.

Size is an important biological factor that has been almost completely neglected by the zoologists. However, as early as 1894, Brooks (p. 477) made this concise statement: "The animals of the bottom rapidly increased in size and hard parts were quickly acquired." This was not Brooks' main thesis by any means, but he did realize that size played an important role in the development of animals having shells or skeletons. Brooks considered the adoption of a benthonic mode of existence as being of primary importance in the development of hard parts in animals.

Stokes (1960, p. 186) also mentioned the importance of size in the formation of a shell or a skeleton, as shown by the following statement: "We assume that shells evolved in response to an animal's need for protection against neighboring predators. The appearance of skeletons indicates many animals that originally were rather small have grown larger and heavier."

Increase in size is a common phyletic trend in invertebrate as well as in vertebrate animals and is known as Cope's Rule. Newell (1949) has given many examples of Cope's Rule among fossil invertebrates. Bonner (1965, p. 176-177) has shown this distinct increase in size in plants and animals since the origin of life. Many writers have noted that Cambrian invertebrate fossils are of small size, and not until Middle Ordovician time did some nautiloids attain a length of more than 15 feet (approximately 460 cm.) in length. Dunbar (1960, p. 118) stated that a trilobite, Paradoxides harlani Green, of Middle Cambrian age attained a length of 1½ feet (about 46 cm.). Let us assume that the length of time between Middle Cambrian and Middle Ordovician was 90 million years. Thus, a 10-fold increase of the maximum size amongst animals took 90 million years. Considering the span of time, this is not an exceptionally large increase in size. (For example, the endoceratid nautiloids of Early and Middle Ordovician age increased in size at a more rapid rate.) Let us now go back 90 million years before the Middle

Cambrian to the very Late Precambrian time and let us assume that the rate of increase in size remained constant. This would make the largest Late Precambrian animals 4.6 cm. or somewhat more than 1½ inches in length. At the beginning of Cambrian time, 45 million years before the Middle Cambrian, the maximum size of animals would be 23 cm. or about 9 inches in length.

Like any other organ or organ system, skeletons or shells are generally larger or better developed in large animals than they are in small ones. Conversely, some animals have undoubtedly lost organs for respiration, excretion, and supporting tissues (skeletons or shells). For example, the Ectoprocta (bryozoans) have probably lost respiratory and excretory organs because they are not needed in these tiny colonial animals, but the ancestors of the Ectoprocta, which were very likely larger, had these organs. The small size of Precambrian animals was the main factor in their not having well-developed shells or skeletons. Thin cuticles were undoubtedly present and perhaps a few protists had shells or skeletons, but such feeble development of hard parts prevented the animals from being fossilized. The excellent fossil record of the foraminifera does not negate my thesis. Most living protists do not have a skeleton or shell. Protists are poorly represented in Early Paleozoic time, and production of skeletons in some protistan lineages may be a secondary development in these small organisms. However, tiny fossil Protista in thick sequences of Precambrian sedimentary strata could be easily overlooked.

I feel that the effect of predaceous animals was a relatively minor factor in the development of shells or skeletons. A large predator must have reasonably large prey in order for it to exist. Most predaceous animals are certainly larger than their prey, but some of the largest animals that have ever lived were herbivores. If invertebrate animals had not increased in size, their predators could not have evolved as they did, and the invertebrate animals would not have had an increasing need for skeletal protection against predators. When their body size did increase, however, they developed a need for skeletal material to support their tissues, and the skeletal material evolved in such a way as to be protective as well as supportive--the supportive factor being primary because the larger animals could not have lived without it even if predators were lacking.

James R. Beerbower of McMaster University, who kindly reviewed this manuscript, suggested another possible factor in skeletal development of

Cambrian invertebrates (written communication). His thought on this matter follows: "Ectoparasites are important predators on animals of all sizes and skeletons provide significant protection against ectoparasites even in the absence of the megapredators, which come late in the geologic record at least." This statement may be true but nearly impossible to prove from the evidence in the fossil record.

By the end of Precambrian time probably all of the ecological niches were taken by small animals so that only by a rapid increase in size among members of several different invertebrate phyla could new ecological niches in the sea, both pelagic and benthonic, be colonized. As Bonner (1965, p. 190) so ably pointed out:

> "In a much broader sense the ecological nitches (sic) are related to size. It is simply that a large organism is unlikely to compete with a small one for they will live in separate size worlds or nitches. Bacteria and any large mammal (or small one, for that matter) do not eat the same thing or in any way conflict in their respective existence, except if one happens to be a parasite of the other. This, incidentally, may well be a significant reason why there has been a trend during the course of evolution to increase the size of organisms. Since all the smaller nitches will be occupied, the only way to conquer new worlds is to make larger nitches. It is only through catastrophe or some peculiar change of conditions that the smaller nitches will be vacated, and there might be a selection pressure for a reversion to smaller size."

Invertebrate faunas of Early and Middle Cambrian age show a paucity of colonial animals having skeletons. These faunas contain a few sponges and some archaeocyathids; some doubt exists, however, as to whether sponges having several oscula are truly colonial animals, and the same may possibly be said for the archaeocyathids. It was not until Late Cambrian time when the graptolites appeared that colonial animals having skeletons became more abundant, and not until the Early Ordovician, when the tabulate and rugose corals arose and the ectoprocts appeared, did colonial animals become quite common. In other words, solitary animals like trilobites, brachiopods, and mollusks apparently developed skeletons before colonial animals did. If graptolites were eumetazoans, then the first colonial eumetazoans having skeletons did not arise until Late Cambrian time. If graptolites were not eumetazoans, then the first colonial eumetazoans did not appear until Early Ordovician time. Did colonial eumetazoans actually originate later than solitary eumetazoans?

Evidence for considering size to be the main biological factor in well-developed skeletons or shells in marine invertebrate animals follows:

1. Increasing body size and

diversity of invertebrate animals during the Cambrian and into the Ordovician is well known.

2. The soft-bodied animals found in the Middle Cambrian Burgess Shale are no larger than their contemporaries bearing shells or skeletons.

3. The fact that the trilobites and the archaeocyathids are the largest invertebrates in Lower Cambrian strata and also the dominant or among the dominant fossils found in strata of that age is certainly significant.

Small benthonic animals may have existed for many millions of years before these forms became sufficiently large to develop separate complex organ systems for such functions as respiration, excretion, and rigid supporting tissues (that is, a skeleton). Brooks may be correct in his assertion that the first animals were pelagic or planktonic before some of them became benthonic, but not until the maximum size of some of these benthonic animals became about 6 inches (15 cm.) did many of them secrete a skeleton or shell so that they could be easily fossilized. This size was attained by quite a few representatives in several invertebrate phyla about the beginning of Cambrian time.

REFERENCES

BONNER, J. T., 1965, Size and cycle, an essay on the structure of biology: Princeton, New Jersey, Princeton Univ. Press, 219 p.

BROOKS, W. K., 1894, The Origin of the oldest fossils and the discovery of the bottom of the ocean: Jour. Geology, v. 2, p. 455-479.

CLOUD, P. E., JR., 1948, Some problems and patterns of evolution exemplified by fossil invertebrates: Evolution, v. 2, p. 322-350.

DUNBAR, C. O., 1960, Historical geology: New York, New York, John Wiley & Sons, Inc., 2nd. edition, 500 p.

NEWELL, N. D., 1949, Phyletic size increase, an important trend illustrated by fossil invertebrates: Evolution, v. 3, p. 103-124.

RAYMOND, P. E., 1935, Pre-Cambrian life: Geol. Soc. America, Bull., v. 46, p. 375-391.

---- 1939, Prehistoric life: Cambridge, Massachusetts, Harvard Univ. Press, 324 p.

STOKES, W. L., 1960, Essential of earth history, an introduction to historical geology: Englewood Cliffs, New Jersey, Prentice-Hall, Inc., 502 p.

*

# Oxygen-collagen priority and the early metazoan fossil record

## K.M. Towe (1970)

One of the most striking and enigmatic aspects of paleontology has been the sudden appearance of advanced and diversified metazoan organisms in the early Cambrian. This subject has been the object of considerable research and speculation and numerous hypotheses have been proposed to explain the phenomenon. Most of these hypotheses assume that advanced Precambrian Metazoa existed but have not been preserved in the fossil record. This lack of preservation has been attributed to the unavailability of $CaCO_3$, acid oceans, metamorphism, "Lipalian" interval, or the absence of Precambrian coastal sediments. Simpson[1] has evaluated most of these suggestions, and more recently Cloud[2] has summarized and discussed the problem at length. Among Cloud's more important conclusions are the following: (1) "There are as yet no records of unequivocal Metazoa in rocks of un- doubted Precambrian age." (2) "The availability or lack of $CaCO_3$ is not the explanation for the distributions of fossils observed." (3) "There is no good reason why we should not expect to find records of Precambrian Metazoa if they were present."

With these conclusions as a foundation, Cloud pursued earlier suggestions[3,4] and proposed that the "...more or less simultaneous attain- ment of a metazoan grade of organi- zation by different pre-metazoan stock stocks...may have been brought on by increase in atmospheric oxygen to levels consistent with metazoan oxida- tive metabolism near the beginning of Paleozoic time."

While this hypothesis may explain the rapid evolutionary diversifications in the early Paleozoic, as well as some related inorganic Precambrian events, it still does not explain the sudden appearance in the early Cambrian of highly organized metazoans. Regardless of whether the time of diversification was rapid or not it is still pertinent

Reprinted from Proc.Nat.Acad.Sci.USA 65, 781-788 (1970) by permission of the author and publisher.

to ask where the more primitive
metazoan-grade ancestors to these
already complex organisms are or were.

The idea that ancestral metazoan
stocks existed in the Precambrian but
were not recorded as fossils contradicts one of Cloud's most important and
basic conclusions (3, above). All of
the previously published supporting
ideas for this concept have been
examined[1,2] and found wanting. On
close inspection, however, the crux of
the argument really rests on the words
"no good reason." Clearly, an
explanation is needed that could withstand a reasonable level of criticism
and at the same time be consistent with
the bulk of observable data. Any such
explanation, to be acceptable, must
allow for early metazoan evolution
while at the same time denying or
minimizing the fossil record. On the
other hand the simultaneous and sudden
occurrence of diverse preservable
forms, independent of calcification,
must also be satisfactorily explained.
Recent advances in understanding the
biochemistry of connective tissue may
provide just such an explanation.

DISTRIBUTION OF COLLAGEN

Dense fibrous connective tissues
consist principally of collagen which
is a biochemically unique structural
protein, generally of mesodermal
origin. It is one of the most abundant animal fibrous proteins occurring
in representatives of every metazoan
group, being the dominant extracellular fibrous protein in all metazoan stocks from the most primitive
sponges on up to man.[5] It has even
been reported to occur in a protozoan
Foraminifera.[6] Among the vertebrates
it occurs widely in both the calcified
and uncalcified states. Among the
invertebrates (reports to the contrary
notwithstanding) it occurs only in the
uncalcified state serving primarily as
a connective tissue. Collagen
connects, supports, and surrounds other
tissues and can be considered the
"tape" and "glue" of the metazoan
world.

By way of specific examples of the
examples of the occurrence of collagen
the annelid cuticle is constructed of
this protein.[7,8] The anthozoan body
wall is predominantly collagen, and
keratose and other sponge fibers are
collagen. The coelenterate mesoglea is
collagenous and in some the float is
collagen. The echinoderm body wall is
collagen[5] and echinoderm endoskeletal
calcified elements are held together by
collagen.[9] Pelecypod adductor muscles
are bound and attached to the shell
with collagen[10] and byssus fibers are
collagenous.[7] The brachiopod
lophophore is supported by collagen and
the pedicle is collagenous.[11] The

complex musculature and "tendons" of arthropods involve intimate association with collagen.[7,12] The gill and endocranial cartilages of the horseshoe crab Limulus contain collagen; the gastropod odontophore is collagenous,[13] as are the subcuticular tissues of the spiny lobster and blue crab.[14] The examples of collagenous tissues among the invertebrates are legion and serve to illustrate its structural importance and widespread occurrence.

It is the principal thesis of this paper that in the absence of, or with the minimal use of, connective tissue collagen the chances of preservation of ancestral metazoan organisms so poorly endowed would be so low that from a realistic paleontological point of view they would appear nonexistent.

## STRUCTURE OF COLLAGEN

Our knowledge of the detailed structure of collagen is still in a state of transition although there is general agreement among most workers that the unit tropocollagen molecule is a triple helix, coiled-coil structure made up of three peptide chains held together by one or two interchain peptide hydrogen bonds per three amino acid residue repeat.[15] The structure demands that every third amino acid be glycine. A substantial number of the other amino acids are proline or hydroxyproline. With a few exceptions,[5,16-19] hydroxyproline (and hydroxylysine as well) is often considered unique and characteristic for collagen. This molecule is about 2800 $\overset{o}{A}$ long and 14 $\overset{o}{A}$ wide and is constructed of about 1000 amino acid residues which are 2.86 $\overset{o}{A}$ apart. These basic fibers aggregate to form the larger scale native-type collagen fibrils. The fibrils, when large enough, are characterized by a 600-700 $\overset{o}{A}$ periodicity which is observable by both electron microscopy and small angle X-ray diffraction. When aggregated in this macromolecular configuration, the fibril has a high tensile strength which is responsible for its remarkable properties in animal connective tissues. Most workers agree that for any protein to be classified as collagen it must have a wide angle X-ray diffraction maxima at 2.86 $\overset{o}{\Lambda}$, approximately one third of the total amino acid residues as glycine, and a relatively high content of proline and hydroxyproline.

## BIOSYNTHESIS OF COLLAGEN

One of the most important developments in the study of collagen biosynthesis was the discovery by Stetten[20] that animals fed radioactively labeled free hydroxyproline did not incorporate it into collagen, although earlier reports had demonstrated that similarly labeled proline could act as a precursor to the hydroxyproline. This unusual finding indicated that the hydroxylation of proline was a key step in the biosynthesis of collagen. The parallel observation was also documented for hydroxylysine.[21] Since then it has been found that the proline is incorporated into peptides before it is hydroxylated. Furthermore, it was demonstrated that the oxygen of the hydroxyproline hydroxyl was derived from molecular oxygen rather than from water[22,23] and that this conversion is catalyzed by an oxygenase.[24] Still further observations indicated that in the absence of molecular oxygen a hydroxyproline-deficient protein was found with a collagen-like characteristic-- namely, it was degradable with the collagen-specific enzyme collagenase.[25] This material has been termed "protocollagen,"[26] Furthermore, the oxygenase (proline hydroxylase) has a specificity determined in part, by a peptide sequence. Udenfriend[27] has reviewed much of this work and presented a scheme for hydroxylation.

An interesting corollary is the mechanism acting in the biosynthesis of resilin, a rubberlike structural protein occurring in arthropods. Here the strength of this fibrous material is derived from cross links involving dityrosine and trityrosine.[28] These amino acids appear to be formed through enzymic oxidation of tyrosine involving a peroxidase.[29] Furthermore the tyrosine which is first incorporated into polypeptide chains comes either directly from free tyrosine or from the hydroxylation of phenylalanine.[30] The latter requires an oxygenase, in this case phenylalanine hydroxylase. The important distinction from collagen is that tyrosine can be incorporated by the organism directly whereas hydroxyproline cannot.

An important question is whether or not a complete but unhydroxylated true collagen could be formed under certain circumstances and if so whether it could be aggregated into crosslinked macromolecular fibrils having desirable structural connective tissue properties. As far as the most reasonable structural models for collagen go there is no stereochemically significant reason why an unhydroxylated true collagen could not exist, but autoradiographic evidence has been presented to indicate that "protocollagen" accumulates only intracellularly and is then released for extracellular collagen synthesis.[31] On the other hand, there are considerable data to indicate that the shrinkage and denaturation temperature (and hence stability) of various collagens decreases with a decrease in the total imino acid content[32] which suggests that stabilization may be dependent on total pyrrolidine content. Considerations involving water content and hydrogen bonding[33] between side chains of the triple helix further complicate the picture. Stability could also be provided through hydrogen bonding involving hydroxyproline[34] and intermolecular cross links involving hydroxylysine.[35] Regardless of the role of the imino acids in the stability of collagen or the factors involved in cross-linking, the significant point for the present is that synthesis of a complete, true collagen macromolecule can be specifically inhibited by the lack of molecular oxygen.

## EARLY ATMOSPHERE AND COLLAGEN FORMATION

Several studies, culminating in the most recent works of Cloud[2] have indicated the importance of oxygen to the origin of the Metazoa. Most workers visualize development of photosynthetic organisms liberating free $O_2$ which in the earlier Precambrian was immediately demanded for inorganic reactions but which eventually reached a level "consistent with metazoan oxidative metabolism near the beginning of Paleozoic time." The critical concentration of oxygen is placed somewhere near 3 per cent of the present

atmospheric level--less than 1 per cent oxygen. At this level the ozone absorption of lethal wavelength ultraviolet light becomes significant.[4] This would have allowed a stepwise increase in phytoplankton in surface waters and therefore a concomitant stepwise increase in photosynthetic oxygen. For purposes of comparison with collagen requirements the apparent $K_m$ (Michaelis constant) for $O_2$ in the proline hydroxylase system of the chick embryo is $3 \times 10^{-5}$ $M$ or equivalent to 2.6 per cent oxygen.[36]

The correlation between this scheme of atmospheric evolution and the biosynthesis of collagen is tempting. Inasmuch as the early metazoan stocks would be operating at minimal oxygen level, biochemical efficiency and economy would dictate physiological priorities. In referring to oxygenases in general, Hayiashi[37] has stated that they compete with the conventional electron pathways for oxygen and from a thermodynamic point of view "oxygenation reactions appear to be a waste of energy." Kaufman[38] has similarly concluded that such aerobic hydroxylation reactions are "energetically expensive for the cell." I suggest, therefore, that in the early evolving organisms the much more efficient and physiologically necessary energy-generating reactions would have higher priority for the limited available molecular oxygen than would the formation of dense fibrous collagen through the expensive enzymatic hydroxylation of proline and lysine. This would imply that primitive metazoans might have evolved with dense fibrous connective tissue restricted to high priority use. Such preconditioned and necessarily small organisms could then begin to synthesize the high strength structural collagen with its many adaptive advantages as soon as the oxygen level increased stepwise after the proliferation of uninhibited phytoplankton. Since collagen is evolutionarily primitive, occurring in every metazoan group, all then-existing metazoan stocks would be acted on and evolutionarily pressured by increased oxygen simultaneously and on a world-wide basis. This could result in the sudden appearance in the rock record of the Lower Cambrian of highly evolved and diversified organisms previously present in essentially unpreservable forms. Numerous polyphyletic origins in different groups would not be necessary.

COLLAGEN PRIORITY AND THE FOSSIL RECORD

When careful consideration is given to the early Cambrian faunas it is possible to conclude that functioning soft-bodied representatives of all groups are conceivable.[1,2,39] The presence of a shell, cuticle, or carapace is not a mandatory prerequisite for the metazoan level of organization. In fact, it is a low priority physiological luxury, its functional significance notwithstanding.

The minimizing of connective tissue collagen as a major structural component would tend to inhibit development of all but the most incipient shell, cuticle, carapace, or the like. These protective devices are either made predominantly of collagen or are held together by collagen which acts as basement lamallae as well as serving to bind and connect muscle tissues. Furthermore, since normal muscle contraction requires oxygen to remove the products of anaerobic glycolysis (lactic acid accumulation), it would be uneconomical for an organism evolving under low oxygen levels to develop a significant musculature for metabolically expensive low priority physiological needs such as a shell.

Since a well-developed or enlarged musculature demands the fibrogenesis of collagen the early body wall muscles must have been weak and capable only of limited effects. This condition would effectively limit the organism's size, inhibit extensive locomotion, and would make burrowing impossible. This, in turn, would eliminate the major source of potential Precambrian trace fossils--tracks, trails, or burrows. Clark [40] has discussed such locomotion and burrowing in terms of the necessity of development of an enlarged body-wall musculature in conjunction with a hydrostatic fluid skeleton. Both of these features are independent of a shell or carapace but both require significant formation of dense fibrous connective tissue.

The evolutionary utilization of collagen by primitive organisms in a low oxygen environment would be in accordance with their most immediate priorities. Very broadly, such needs would include the development of the coelom and enlarged musculature followed by segmentation. These requirements are necessary before the development of a shell or carapace could take place.[40] It seems clear, by way of example, that the ancestral trilobites must have constructed a segmented coelom and musculature for a hydrostatic skeleton prior to the development of a rigid exoskeleton with its necessarily complex skeletal muscles.[40] While the latter has a very clear functional and adaptive significance it is a physiologically lower priority and energetically more expensive structure for an evolving organism faced with a limited oxygen budget. This is also true for thick cuticles and, above all, for shells.

Since these external structures would reduce the available surface area for respiration they would be poor investments for primitive organisms lacking advanced circulatory-respiratory mechanisms or those living under near anaerobic conditions.

It is reasonable that primitive organisms in the Precambrian had evolved the preadaptive ability to secrete a necessarily thin shell or carapace, limited muscles, and the mechanism to synthesize the binding connective tissue. However, if the molecular oxygen necessary to catalyze this potentially eruptive adaptive breakthrough is competed for and channelled by physiological priority to more important uses, then the shell-carapace-muscle system would not be able to accumulate and develop to any but the most primitive degree for lack of the ingredients to hold it together, much less operate it. Nevertheless any development at all along these lines, however weak, would be of functional importance both for external protection and as a radiation shield.[41] The sudden environmental change in atmospheric oxygen level would result in a dramatic shift from a clearly preadaptive condition, to a highly eruptive one. The idea of a pre-adaptive condition for collagen synthesis is reinforced by the observation that although specialized cells are usually associated with collagen formation there is increasing evidence that other cells can be stimulated to produce this protein.[42,43] Further support can be derived from the efforts of Millard and Rudall.[44] They were able to show that in earthworm cuticle the organization of collagen depends on vertical cell processes which may be related to "cilia" type structures. This, they interpret, may represent a primitive condition in the development of collagen as an organized extracellular structure. Rudall states: "If Metazoa derived from ciliated protozoan types these could have suffered a change to a collagen pellicle supported by cilia. It is not a great step to the modification of the cilia and their disappearance as motile organelles thus giving rise to separate cells linked together by a collagenous matrix, or to epithelia linked by a collagenous cuticle."

In this way the apparent contradictions presented by Cloud[45] might be resolved. He could not visualize a trilobite and its complicated musculature without its associated carapace nor a brachiopod and its musculature without a shell. Since one is useless without the other, these organisms could have evolved with neither or with both in a much reduced state which was environmentally and physiologically regulated.

Among the diversified organisms those capable of first utilizing oxygen for extensive collagen fibrogenesis would likely have been among the most primitive. These would be those with the fewest organs, least demanding muscles or other energy requiring structures and more importantly, those lacking an advanced circulatory-respiratory system using instead oxygen exchanged through the epidermis by diffusion. Such oxygen might be more readily available for collagen synthesis than would be oxygen partitioned and competed for by a more advanced circulatory-respiratory system. This speculation is supported by the fact that the earliest known metazoan fossils (the Ediacaran fauna) are not those with "hard parts" but those with the hardest or most resistant "soft parts," that is, those with major quantities of collagenous connective tissue. Included in this group are the coelenterates, the various types of worms, and the sponges.

In conclusion, I agree with those who believe that an increased atmospheric oxygen level "triggered" the proliferation of the Metazoa near the beginning of the Paleozoic but I disagree with the conclusion that there is no good reason why we have not found the records of Precambrian ancestors.

If the records of Precambrian Metazoa (the stratigraphic arguments aside) are to be extended with finds similar to the famous Ediacaran assemblage the following words from Cloud[2] are worth serious consideration by all paleontologists interested in this problem: "...so far as a substantial record of Precambrian life is concerned, we have mostly been looking for the wrong things, in the wrong rocks, with the wrong techniques..." It seems to me worth considering what various metazoan ancestors lacking protective coverings might have looked like, how large they might have been, and then, with an open mind and the appropriate tools, examining undoubted Late Precambrian sedimentary sequences for their remains.

The many preservational biases involved in this problem have been emphasized by Simpson.[1] The statistics of preservation is a function of the organisms' structure and composition and of the enclosing sediment type. Most importantly it is a function of the sedimentary processes involved. Catastrophic one-cycle, rapid burial in fine-grained sediments followed by isolation from further sedimentary disturbance for all subsequent time is mandatory for extensive assemblage preservation. At this point organism structure and composition become important. Nonarticulate and colonial sessile calcified organisms together with disarticulated hard parts of other organisms can survive more than one sedimentary cycle (pre- and postdepositional reworking). On the other hand, soft-bodied organisms and trace fossils cannot. The sedimentary bias is clear, as is the preservational hierarchy. A whole gastropod stands a better chance to become a fossil than a complete articulated crinoid, than a cuticled annelid, than a cellular, collagen-deficient animal of any type.

NOTE ADDED IN PROOF

Schopf and Barghoorn (J. Paleontology, 43, 111 (1969)) have argued that the Late Precambrian was more oxygenic than previously assumed (>1% present level). While, as they point out, this would place severe restrictions on the idea of the origin of the Metazoa being coincident with the beginning of the Paleozoic, it is still compatible with the developmental priorities expressed above.

This article was improved through discussion with P. H. Abelson, M. A. Buzas, R. Cifelli, P. E. Hare, P. Malone, S. Udenfriend, H. B. Whittington, and E. Yochelson. Special acknowledgment is due A. J. Bailey, J. Corrigan, and R. F. Fudali for their critical reading of the manuscript.

[1]Simpson, G. G., in *Evolution after Darwin, The evolution of Life*, ed. S. Tax (Chicago: University of Chicago Press, 1960), vol. 1, pp. 117-180.

[2]Cloud, P. E., Jr., in *Evolution and Environment*, ed. E. T. Drake (New Haven: Yale University Press, 1968), pp. 1-72.

[3]Nursall, J. R., *Nature*, *183*, 1170 (1959).

[4]Berkner, L. V., and L. C. Marshall, these PROCEEDINGS, *53*, 1215 (1965).

[5]Gross, J., and K. A. Piez, in *Calcification in Biological Systems*, ed. R. F. Sognnaes (Washington, D.C.: AAAS Publ. 64, 1960), pp. 395-409.

[6]Hedley, R. H., and J. Wakefield, *J. Roy. Microscop. Soc.*, *87*, 475 (1967).

[7]Rudall, K. M., *Symp. Soc. Exptl. Biol.*, *9*, 49 (1955).

[8]Watson, M. R., *Biochem. J.*, *68*, 416 (1958).

[9]Moss, M. L., and M. M. Meehan, *Acta Anat.*, *66*, 279 (1967).

[10]Galtsoff, P. S., *Bur. Comm. Fisheries, Fishery Bull.*, *64*, 1 (1964).

[11]Rudall, K. M., in *Treatise on Collagen, Biology of Collagen*, ed. B. S. Gould (New York: Academic Press, 1968), vol. 2A, pp. 83-138.

[12]Osborne, M. P., *J. Insect Physiol.*, *9*, 237 (1963); Whitear, M., *Phil. Trans. Roy. Soc.*, London, B248, 437 (1965).

[13]Person, P., and D. E. Philpott, *Biol. Rev.*, *44*, 1 (1969).

[14]Kimura, S., Y. Nagaoka, and M. Kubota, *Bull. Japan. Soc. Sci. Fisheries*, *35*, 743 (1969).

[15]Harrington, W. F., and P. H. von Hippel, *Advanc. Protein Chem.*, *16*, 1 (1961).

[16]Lenhoff, H. M., E. S. Kline, and R. Hurley, *Biochim. Biophys. Acta*, *26*, 204 (1957).

[17]Jope, M., *Comp. Biochem. Physiol.*, *30*, 593 (1967).

[18]Hedley, R. H., and J. Wakefield, *Brit. Mus. (Nat. Hist.) Bull.*, *18*, 5 (1969).

[19]Steward, F. C., and J. K. Pollard, *Nature*, *182*, 828 (1958).

[20]Stetten, M. R., *J. Biol. Chem.*, *181*, 31 (1949).

[21]Sinex, F. M., and D. D. Van Slyke, *J. Biol. Chem.*, *216*, 245 (1955).

[22]Fujimoto, D., and N. Tamiya, *Biochem. J.*, *84*, 333 (1962); Prockop, D., A. Kaplan, and S. Udenfriend, *Biochem. Biophys. Res. Commun.*, *9*, 162 (1962); *Arch. Biochem. Biophys.*, *101*, 499 (1963).

[23]Molecular oxygen is also required for the formation of hydroxyproline in the noncollagenous protein "extensin" which occurs in the cell walls of higher plants; Lamport, D. T. A., *J. Biol.*

Chem., 238, 1438 (1963).

[24] Fujita, Y., A. Gottlieb, B. Peterkofsky, S. Udenfriend, and B. Witkop, J. Amer. Chem. Soc., 86, 4709 (1964).

[25] Peterkovsky, B., and S. Udenfriend, J. Biol. Chem., 238, 3966 (1963); Prockop, D. J., and K. Juva, these PROCEEDINGS, 53, 661 (1965).

[26] Juva, K., and D. J. Prockop, in Biochemie et Physiologic du Tissu Conjonctiv, ed. P. Compte (Lyon: Soc. Ormeco et Imprimerie du Sud-Est, 1966), p. 417.

[27] Udenfriend, S., Science, 152, 1335 (1966).

[28] Andersen, S. O., Biochim. Biophys. Acta, 93, 213 (1964).

[29] Coles, G. C., J. Insect Physiol., 12, 679 (1969); Andersen, S. O., Acta Physiol. Scand., 66, 1 (1966).

[30] Andersen, S. O., and B. Kristensen, Acta Physiol. Scand., 59, 15 (1963).

[31] Juva, K., D. J. Prockop, G. W. Cooper, and J. S. Lash, Science, 152, 92 (1966).

[32] von Hippel, P. H., in Treatise on Collagen, Chemistry of Collagen, ed. G. N. Ramachandran (New York: Academic Press, 1968), vol. 1, pp.253-338.

[33] Ramachandran, G. N., and R. Chandrasekharan, Biopolymers, 6, 1649 (1968).

[34] Gustavson, K. H., Svensk Kem. Tidskr., 65, 70 (1953).

[35] Bailey, A. J., and C. M. Peach, Biochem. Biophys. Res. Commun., 33, 812 (1968); Bailey, A. J., L. J. Fowler, and C. M. Peach, Biochem. Biophys. Res. Commun., 35, 663 (1969).

[36] Hutton, J. J., A. L. Tappel, and S. Udenfriend, Arch. Biochem. Biophys., 118, 231 (1967).

[37] Hayiashi, O., in Oxygenases, ed. O. Hayiashi (New York: Academic Press, 1962), pp. 1-31.

[38] Kaufman, S., in Oxygenases, ed. O. Hayiashi (New York: Academic Press, 1962), pp. 129-181.

[39] Glaessner, M. F., Biol. Rev., 37, 467 (1962).

[40] Clark, R. B., Dynamics in Metazoan Evolution (Oxford: Clarendon Press, 1964).

[41] Fischer, A. G., these PROCEEDINGS, 53, 1205 (1965).

[42] Green, H., and B. Goldberg, these PROCEEDINGS, 53, 1360 (1965).

[43] Ross, R., in Treatise on Collagen, Biology of Collagen, ed. B. S. Gould (New York: Academic Press, 1968), vol. 2A, pp. 2-82.

[44] Rudall, K. M., in Structure and Function of Connective and Skeletal Tissue, ed. S. Fitto-Jackson (London: Butterworths, 1965), p. 191.

[45] Cloud, P. E., Jr., Evolution, 2, 322 (1948).

# On the development of metaphytes and metazoans

## J.W. Schopf, B.N. Haugh, R.E. Molnar and D.F. Satterthwait (1973)

When one surveys the panorama of innovations in morphology and bio-chemistry that has appeared during the long course of biological evolution, one is struck by the realization that a surprisingly small number of these developments--probably fewer than a dozen--stand out as having had a profound and lasting impact on sub-sequent evolutionary history. These signal innovations are of the type that Stebbins (1969) has characterized as leading to the emergence of "new levels of organizational complexity," innovations resulting in such major events as the development of auto-trophy, eukaryotic organization, sexuality, multicellularity, homoiothermy, angiospermy, and so forth. Interestingly, almost any such list of major evolutionary events could also serve as an inventory of

"major unsolved problems in paleobiology," for the fossil record has provided only limited insight into the nature of these important evolutionary transitions.

In paleozoology, probably foremost among these unsolved problems is the classic question of "the origin of the Metazoa"; together with its putative stratigraphic corollary, "the Pre-cambrian-Cambrian boundary problem," this topic has long proved a fertile area for interesting, imaginative, but inconclusive speculation. In contrast, the analogous problem in botanical evolution, "the origin of the Metaphyta" (i.e., the development of megascopic, multicellular thallophy-tes), has been largely neglected, chiefly as a result of the emphasis on tracheophytes and tracheophytic evolution traditional to paleobotany. In recent years, an impressive but as yet limited array of fossil evidence bearing on these problems has been reported from sediments of the late

Reprinted from Journal of Paleontology 47, 1-9 by permission of the senior authors and publisher. Copyright 1973 SEPM.

Precambrian. Based on these new data, it now appears likely that a megascopic level of organization was attained more or less concurrently by both plants and animals and that the development and early diversification of metaphytes and metazoans (i.e., the Eumetazoa) may have been closely intertwined. Thus, the origin of these two groups poses a labyrinth of interconnected problems, problems to which satisfactory solutions will probably not be formulated until the late Precambrian fossil record becomes known in substantially more detail. The following scenario therefore represents at most only a plausible "first approximation" of events that appear to have played a role in the evolutionary sequence. Nevertheless, it seems apparent that ultimate understanding of the transition from microscopic to megascopic organization will probably require consideration of four aspects of the problem that have heretofore been generally overlooked --(i) the relevance of the pre-Ediacaran fossil record as a guide to the probable timing and nature of evolutionary events leading to the development of megascopic multicellularity; (ii) the significance of eukaryotic sexuality as a possible "evolutionary trigger" resulting in a marked increase in diversity and evolutionary rate among late Pre-

cambrian eukaryotes; (iii) the evolutionary trend from a "haploid-dominant" to a "diploid-dominant" life cycle in thallophytes, a trend which apparently played a major role in the emergence of megascopic plants; and (iv) the probable interrelatedness of metaphytic and metazoan evolution, the two groups linked both by metabolic characteristics and by their common derivation during the late Precambrian from microscopic, eukaryotic progenitors.

## DEVELOPMENT OF EUKARYOTIC SEXUALITY

All higher organisms are composed of nucleated, eukaryotic cells. The evolutionary progression leading to the development of metaphytes and metazoans therefore had its beginning in the Precambrian derivation of primitive eukaryotes from prokaryotic (anucleate, nonmitotic) ancestors. Although the evolutionary mechanisms involved in this transition have been a subject of controversy in recent years (Allsopp, 1969; Raven, 1970; Taylor, 1970; Margulis, 1970), there is general agreement that the earliest eukaryotes were microscopic unicells and that photosynthetic forms (e.g., rhodophytes and chlorophytes) evolved quite early. Despite the inherent difficulties in determining with certainty the affinities (whether pro- or eukaryotic) of simple, unicellular microfossils, available fossil evidence seems to indicate that unicellular eukaryotes were extant at least as early as 900 million years ago, the approximate time of preservation of the central Australian Bitter Springs microflora (Schopf, 1968; Schopf and Blacic, 1971). Moreover, recent studies suggest that the origin of this cell-type substantially predates deposition of the Bitter Springs cherts. Older presumptive eukaryotes are known from siliceous sediments of the Belt Supergroup of Montana (e.g., Fibularix porulosa and F. funicula; Pflug, 1965; 1966), about 1,100 million years in age, age, and from cherts of the Beck Spring Dolomite of California (Cloud et al., 1969; Licari, 1971 Unpublished Ph.D. thesis, Univ. of Calif., Los Angeles), perhaps 1,300 million years in age. Further, microorganisms at least suggestive of eukaryotic organization have been reported from cherts about 1,700 million years in age from the Belcher Islands of Hudson Bay (e.g., Eomycetopsis sp. and "Type 2" and "Type 4" microstructures described by Hofmann and Jackson, 1969) and from the approximately 1,900 million year-old Gunflint cherts of Ontario, Canada (e.g., Archaeorestis schreiberensis and Eosphaera tyleri described by Barghoorn and Tyler, 1965; and forms shown on p. 39, figs. k, l, in Barghoorn, 1971).

Thus, the origin of the eukaryotic cell apparently occurred earlier than 1,300 million years ago and possibly earlier than 1,700 million years ago (Text-fig. 1).

Judging from the range of life cycles exhibited by extant algae and, in particular, the life cycles of morphologically simple, unicellular varieties (e.g., Chlorella, Chlamydomonas, Tetraspora, Chlorococcum, Porphyridium), it appears probable that the earliest eukaryotes were haploid and asexual, reproducing exclusively by mitotic cell division. Meiosis was apparently a secondary innovation, a derivative of mitosis that presumably evolved as a process for restoration of haploidy in single-celled zygotes produced by gametic fusion. Establishment of the "typical eukaryotic life cycle," involving sexuality and the associated alternation of haploid and diploid generations generations, appears to have required a sequence of at least four distinct steps--(i) the development and refinement of the mitotic apparatus; (ii) the occurrence of syngamy, initially an abnormal and presumably quite uncommon event; (iii) the development of the complex mechanisms of meiosis within diploid cells produced by such gametic fusion; and (iv) selection of mutant strains in which alternation of haploid and diploid phases had become a

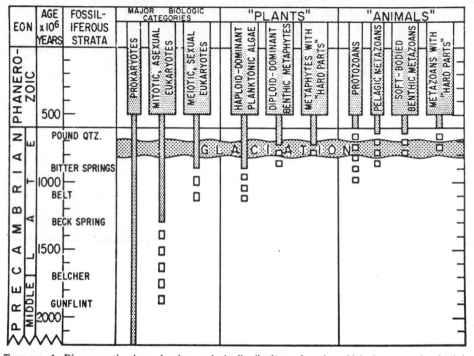

TEXT-FIG. *1*—Diagrammatic chart showing geologic distributions of various biologic groups involved in the late Precambrian transition from microscopic to megascopic organization; dashed bars show (possible or probable) distributions inferred in the absence of definitive fossil evidence.

regularized process. The complexities of this life cycle and the sequential nature of the steps required for its establishment suggest that a considerable segment of geologic time may have elapsed between the origin of the eukaryotic cell-type and the derivation of sexual, meiotic algae.

Once sexual eukaryotes had become established, however, they would have had substantial selective advantage over their asexual contemporaries. The nature of this advantage, resulting primarily from the gene exchange and genetic recombination that occur during the sexual cycle, can be illustrated by the following example (cf. Wilson and Loomis, 1962, p. 314). The occurrence of 10 mutations in an asexual population can result in as many as 11 genotypes, the original type and those of the 10 mutants. In contrast, in a primitive, diploid sexual population (assumed to be homozygous except for normal-mutant gene pairs), 10 mutations could be combined to produce $3^{10}$ (=59,049) distinct genotypes. Obviously, the occurrence of a greater number of mutations in such populations would produce an even more striking disparity (e.g., 100 mutations would yield a potential of 101 genotypes in the asexual population as compared with $3^{100}$ = 5.1 x $10^{47}$ in the sexual population). Thus, the evolution of eukaryotic sexuality must have resulted in a vast increase in genotypic, and hence phenotypic, variability. This increased variability would have been reflected in a proportional increase in the production of advantageous genetic combinations and recombinations. It therefore follows that the establishment of the sexual process should be marked in the fossil record by a sharp increase in both biologic diversity and rate of evolution among eukaryotes.

It is apparent that by the beginning of the Phanerozoic sexual systems were both well established and highly advanced (as evidenced by the diversity and rapid evolution of the Cambrian biota as well as by many other features). The oldest direct indication now known of life cycles apparently involving sexuality is the occurrence of tetrahedral tetrads (Eotetrahedrion princeps) and dispersed sporelike unicells, interpreted as being of meiotic (algal) origin, in the approximately 900 million year-old Bitter Springs cherts (Schopf and Blacic, 1971). This evidence appears reasonably convincing. It is therefore surprising, and seems significant, that the Bitter Springs eukaryotes are of only limited diversity and seem decidedly primitive in organization. Of the nine species of probable eukaryotes recognized in

the assemblage, all are microscopic, all but two are simple spheroidal unicells, and none is known to exhibit features typical of even moderately advanced thallophytes (e.g., the occurrence of tissues, differentiated sporangia, true branching, or polarity of organization). It is possible, of course, that the Bitter Springs cherts do not contain a representative sample of the contemporary biota and that eukaryotes in other environments were diverse, abundant, and relatively complex. This, however, appears rather unlikely. All known biologic assemblages of approximately this age (e.g., Pflug, 1965, 1966; Cloud et al., 1969; Schopf and Barghoorn, 1969; Schopf, 1970; Licari, 1971) are similarly impoverished and none contains evidence of megascopic organisms or of an advanced level of eukaryotic organization. These considerations suggest that eukaryotic sexuality may have been a relatively novel "experiment" in Bitter Springs time, a process not yet widespread in the biota. If so, its evolutionary impact--presumably resulting in an essentially wholesale replacement of asexual members of the biota--must first have become manifest between about 900 million years ago and the beginning of the Phanerozoic. Thus, it is possible, and perhaps likely, that the origin of eukaryotic sexuality served as an "evolutionary trigger" producing a sharp increase in both diversity and evolutionary rate among late Precambrian eukaryotes and that it was the timing of this event (rather than the late appearance of eukaryotic organization per se or a consequence of various postulated environmental factors) that resulted in the relatively late development in evolutionary history of advanced levels of eukaryotic organization.

## DEVELOPMENT OF METAPHYTES

The known fossil record contains little direct evidence of the evolutionary events occurring during the 250 to 300 million year-long interval between deposition of the Bitter Springs cherts and preservation of the oldest known metazoan fauna (the Ediacaran assemblage of South Australia; Glaessner and Wade, 1966; Glaessner, 1971). Nevertheless, the data that are available suggest that this segment of earth history featured a series of highly significant evolutionary innovations, including the diversification of eukaryotic thallophytes; the emergence of protistan heterotrophs; the transition from microscopic to megascopic organization in both plants and animals; and, as a result of the foregoing, a marked adjustment of ecologic relationships.

Comparative studies of extant thallophytes suggest that the earliest sexual eukaryotic algae were, like those those of the Bitter Springs cherts, unicellular in form and planktonic in habit. As in many extant unicellular species, their alternating haploid (gametophyte) and diploid (sporophyte) generations would be expected to have been virtually identical in appearance but of unequal duration, the diploid phase being comparatively short-lived (an isomorphic, "haploid-dominant" life cycle). Although the two phases of such life cycles are genetically similar, they need not be phenotypically identical since only a fraction of the genetic material need be active ("turned on") at any given stage of the developmental sequence. Thus, with the development of molecular mechanisms to regulate genomic activity in life cycles involving alternation of generations, it would have become possible for the two phases to diverge in morphology and physiology, each becoming specialized for performance of a specific role (e.g., vegetative growth or reproduction) in the life cycle. In short, the existence of two distinct generations would have provided potential for an increase in ecologic flexibility, in reproductive efficiency, and, especially in the vegetative phase, in size (resulting in a correlative increase of chlorophyllous surface area, photosynthetic efficacy and nutrient storage capacity as well as improved buffering against changes in the physical environment). Within a moderate segment of geologic time, as this potential began to be realized, algae evolved in which the isomorphic generations were of equal duration. It may be surmised that among the earliest megascopic forms exhibiting such "haploid-diploid coequal" life cycles were sheet-like and encrusting types (cf. modern Ulva, Enteromorpha, Grinnellia). Small "herbaceous" varieties (cf. modern Chondrus, Dictyota, Ectocarpus) presumably evolved somewhat later. As Fritsch (1949) has discussed at length, continued evolution apparently resulted in the appearance of algae in which the diploid generation was large, dominant, and specialized for photosynthetic growth, and in which the haploid generation was much reduced, specialized for reproduction (cf. modern Fucus, Laminaria, Codium). Thus, the evolutionary transition from unicellular to multicellular organization within the algae evidently involved fundamental changes in the relative importance and degree of specialization of the two phases of the life cycle. This trend from a "haploid-dominant" to a "haploid-diploid coequal" and a "diploid-dominant" condition (and from an isogamous to an oogamous sexual process; Fritsch, 1949) was paralleled

by the development of megascopic size and a benthic habit, resulting in the emergence of the Metaphyta. Since algae are necessarily restricted in distribution to the photic zone, diversification of these benthic metaphytes would be expected to have occurred in a shallow, shelf-like, presumable marine setting.

Given the existence of autotrophs and heterotrophs, eukaryotic sexuality, and a neritic milieu, it seems to us probable that the development of megascopic plants and animals was virtually assured. On the other hand, the rate of the evolutionary process could well have been spurred by physical events producing a significant increase in abundance of habitable, shallow-water environments. And, as several workers have suggested (Hicks, 1880; Termier and Termier, 1956; Harland and Wilson, 1956; Rudwick, 1964; Harland and Rudwick, 1964), the well-known "infra-Cambrian" ice age, apparently occurring about 800 to perhaps 650 million years ago (Rhodes and Morse, 1971; pers. com., O. A. Christopherson, 1972), could have provided just such impetus (Text-fig. 1). As summarized by Harland (1964), tillites of this age are extremely widespread (reported from North America, Greenland, Spitsbergen, Scandinavia, the British Isles, Europe, the Soviet Union, China, Australia,

Africa, and, possibly, India and South America).  It is no doubt true that certain of these alleged tillites deserve reinvestigation (cf. Harland, Herod and Krinsley, 1966) and that even if all are regarded as demonstrably glacial in origin, they undoubtedly represent not one but several glacial episodes.  Nevertheless, the occurrence of a relatively "small number of discrete, simultaneous, severe, pro-longed ice ages in late Precambrian time" (Harland, 1964, p. 122) seems well established.  The apparently widespread distribution of "infra-Cambrian" ice sheets (evidently extend-ing into near equatorial regions with the continents arranged in a "pre-drift" configuration; Harland in Spencer, 1969, p. 183) suggests that the glacial episodes would have brought about a substantial (eustatic) fall in sea level.  The "prolonged" nature of the ice ages would have led to exten-sive erosion, possibly approaching peneplanation, of exposed land surfaces.  Thus, the eustatic rise in sea level following glaciation would be expected to have created a widespread, shallow-marine, shelf-like environment with large areas having depths less than that of the photic zone, areas which would be ripe for colonization by benthic metaphytes.

Although, at present, the time of diversification of the Metaphyta cannot be dated precisely, there seems ample evidence that green, red and brown algae were extant prior to the Phanerozoic (Tappan and Loeblich, 1970; Schopf, 1970; Schopf and Blacic, 1971) and that metaphytes were highly diversified at the beginning of the Cambrian (Johnson, 1966).  Further, megascopic plant remains, apparently attributable to ("diploid-dominant"?) brown algae, have been reported from sediments of Vendian age (ca. 680 to 570 million years old) in the Leningrad region of the U.S.S.R. (Gnilovskaja, 1971); megascopic green algae (Papil-lomembrana, probably a "diploid-dominant" dasycladalean) are known from the late Precambrian of southern Norway (Spjeldnaes, 1963) where they occur in sediments closely associated with a tillitic horizon (Spjeldnaes, 1964) that appear to be on the order of 700 million years in age; and drag marks, evidently produced by floating mega-scopic algae, occur in central Australian deposits having an age of about 760 million years (Milton, 1966). Thus, diversification leading to the establishment of megascopic, benthic, "diploid-dominant" metaphytes apparently occurred prior to 700 million years ago and may well have been accelerated by the opening of new niches as a result of glacial activity (Text-fig. 1).  Concomitant with the attainment of this level of organization was the development of holdfasts, anchoring devices that enabled metaphytes to resist dislodgment by wave action and so remain in a favorable habitat. It seems likely, therefore, that the "infra-Cambrian" seas were populated by a diverse assemblage of relatively advanced metaphytes, well adapted to an agitated, shallow-water environment. With these developments, the stage was set for the emergence and diversification of benthic metazoans.

## DEVELOPMENT OF METAZOANS

Diversification of sexual eukaryotic microorganisms during the late Precambrain led to the appearance of heterotrophic, pelagic protists that were primarily dependent on phytoplankton as a food source. It may be surmised that from such protists evolved small, multicellular heterotrophs which initially exhibited only a limited degree of tissue differentiation and primitive sensory and locomotory systems. Like their progenitors, these earliest pelagic eumetazoans would have been dependent upon a planktonic food source; they are therefore presumed to have been suspension-feeders and, in concert with their pelagic habit, to have exhibited radial or spherical symmetry. Such forms presumably gave rise to medusoid coelenterates, a group that had become

well established prior to the preservation of the Ediacaran fauna of South Australia, perhaps 650 to 600 million years ago (Glaessner, 1971). It is conceivable that other of these small pelagic animals could have developed bilateral symmetry, an organization advantageous for active food gathering; if so, these forms might have been precursors of Spinther-like polychaetes (Glaessner and Wade, 1971; Wade, 1972) similar to Dickinsonia of the late Precambrian Ediacaran assemblage. Thus, as has been postulated by earlier workers, we regard the pelagic habit as a primitive feature of eumetazoan organization. However, since adaptive radiation of the group appears to have occurred mainly in a benthic setting, it remains to be suggested what "caused" these early metazoans to "discover the bottom" (cf. Raymond, 1935).

As is summarized above, available data seem to indicate that a megascopic, multicellular level of organization was attained more or less concurrently, and somewhat earlier than 650 million years ago, by both plants and animals. It is probable that this apparent temporal coincidence is primarily a result of the common derivation of these two biologic groups from microscopic, eukaryotic progenitors during the late Precambrian and the presumably somewhat comparable series of genetic, developmental and organizational innovations required for their evolutionary development.  It is also conceivable, however, that ecologic relationships played a role in this apparent coincidence.  Prior to the development of benthic metaphytes, the sediment-water interface may have constituted a relatively lean environment for multicellular heterotrophs (if such existed); available foodstuffs would have been limited to the slow-growing, thin, blue-green algal mats of stromatolitic communities, augmented by a variable influx of detrital organic matter (dead phytoplankton, etc.). This situation, however, would have been altered markedly by the appearance and diversification of encrusting and erect megascopic algae, resulting in a quantum increase in the benthic photosynthetic biomass (and, following death and deterioration, a comparable increase in available small-sized food material).  Thus, it seems possible that the establishment of a benthic metaphytic flora and the colonization by such metaphytes of shallow-water environments, in part created by post-glacial epeiric seas, were approximately synchronous with and provided impetus for the development of benthic eumetazoans (Text-fig. 1).  In addition to representing a potential food source, this photosynthetic community would have provided a refuge from pelagic predator (if such existed), would have served to dampen effects of wave action that might otherwise have proved inimical to the presumably delicate, soft-bodied animals, and would have contributed to the highly oxygenic bottom conditions necessary for aerobic eumetazoan respiration (Rhoads and Morse, 1971).  In short, the "algal forest" would have provided an extensive, favorable habitat for eumetazoan radiation.

Among the earliest benthic eumetazoans were probably sessile, radially symmetrical, suspension-feeding, coelenterate-like forms.  Eumetazoans of this type probably gave rise to pennatulids (octocorals) and other sessile coelenterates, several types of which were apparently abundant at least as early as Ediacaran deposition (Glaessner and Wade, 1966).  Bilateral, vagrant eumetazoans may have developed from sessile coelenterate-like precursors or they may have evolved directly from bilateral swimmers.  In either case, the reported occurrence of trace fossils in sediments predating deposition of the Ediacaran Pound Quartzite (and, with but few exceptions, known to overlie late Precambrian tillites) suggests that mobile metazoans appeared relatively early (Glaessner, 1969).  Apparently the oldest putative trace fossils now known are lobate, mound-shaped structures (<u>Asterosoma? canyonensis</u>) from the Grand Canyon Series which Glaessner (1969) suggests may be on the order of 1,000 million years in age. Cloud (1968), however, regards these and almost all other presumptive pre-Ediacaran trace fossils as demonstrably nonbiogenic. On balance, and recognizing that interpretation of such possible metazoan remains is a matter of some controversy, it seems reasonable to conclude that sedimentary structures assuredly evidencing the existence of bilateral, vagrant eumetazoans are exceedingly rare (if they occur at all) in sediments predating the late Precambrian glacial episodes.

Once they had become established, it is likely that mobile eumetazoans would have diversified rather rapidly, spurred in large measure by the abundance of available foodstuffs (<u>e.g</u>., encrusting and erect metaphytes, algal stromatolites, other metazoans, and bottom detritus). The earliest mobile forms were presumably omnivores, utilizing whatever small-sized food material they encountered, whether plant or animal, living or dead. This food-seeking mode of life, however, would soon have resulted in selection of bilateral eumetazoans having well defined anterior-posterior organization, relatively efficient sensory apparatus, and a concentration of

sensory organs at the "encounter-end" of the organism. Of the earliest known bilateral eumetazoans, <u>Dickinsonia</u> exhibits limited evidence of cephalization (Glaessner and Wade, 1966; Wade, 1972) and may therefore represent a primitive stage in the evolutionary sequence; other members of the Ediacaran fauna, such as <u>Spriggina</u>, were evidently more advanced. Thus, available data seem to suggest that soft-bodied, benthic eumetazoans evolved from pelagic precursors probably during the interval spanned by the "infra-Cambrian" glacial episodes (Text-fig. 1), that the selective pressures resulting in diversification of the group may have been primarily related to metabolic requirements, and that both sessile and mobile forms had become moderately diversified by the time of preservation of the Ediacaran assemblage.

The oldest eumetazoan body-fossils now known occur in sediments, such as the Pound Quartzite, that are situated above "infra-Cambrian" tillites but below the base of the Cambrian; all of these organisms, reported from several, apparently stratigraphically correlative horizons (Glaessner, 1971), appear to have been soft-bodied. It seems probable, therefore, that eumetazoans with supportive skeletons first appeared during the short time interval between preservation of the Ediacaran

assemblage and the beginning of the
Cambrian. Interestingly, metaphytes
with "hard parts" (calcareous green and
red algae) are also first known from
sediments of about this age (Spjeldnaes, 1963, 1964; Tappan and Loeblich,
1970). At least three principal
factors may have played a role in this
"sudden" acquisition of a rather
diverse variety of hard parts by plants
and animals. First, in mobile, benthic
metazoans, acquisition of rigid
appendages (which could act as levers
magnifying muscle displacements) would
have resulted in increased efficiency
of energy utilization; development of a
rigid "body frame" would similarly have
been advantageous, enabling limb movements to propel the organism, rather
than merely flexing the body (Wells,
1968). Second, in sessile forms,
acquisition of a rigid framework would
have permitted a size increase, providing improved buffering against
changes in the physical environment; in
metaphytes, such "hard parts" would
have provided support for an increase
in plant height, advantageous in the
competition for photosynthetic space.
And third, in mobile metazoans and in
sessile forms, whether plant or animal,
acquisition of hard parts and the
associated size increase would have
been advantageous in resisting
predation (cf. Raymond, 1935). Thus,
as cephalization increased in bilater-

al, food-seeking, benthic eumetazoans,
and as eumetazoans became larger,
supportive skeletons may have "become
necessary" for increasing efficiency of
movement; the increased proficiency in
food gathering exhibited by these
moderately advanced heterotrophs presumably resulted in selection of large,
"predator-resistant" metazoans and
metaphytes, many of which possessed
mineralic or organic "hard parts."

SUMMARY

The foregoing essay outlines, in
broad terms, a sequence of evolutionary
events that appear to have played major
roles in the derivation of megascopic
plants and animals during the late Precambrian. This scenario is by no means
wholly novel; the origin of the Metazoa
is a classic problem in paleozoology
for which almost every conceivable
explicatory hypothesis has, at one time
or another, been proposed. Moreover,
it is evident that much remains to be
learned, especially in Precambrian
paleobiology and developmental neo-
biology, before this or any other
speculative scenario can be completely
accepted. Our approach to this problem
problem, however, has differed in
certain respects from that of previous
workers. In particular, we have relied
heavily on the known pre-Ediacaran
fossil record as an index of the timing

and nature of evolutionary innovations leading to the development of mega-scopic multicellularity. Further, we have suggested that the early evolution of metaphytes and metazoans may have been temporally and ecologically inter-related, a significant possibility that has heretofore been largely overlooked. And, finally, while recognizing the fundamental differences in life-style between autotrophs and heterotrophs, we have attempted to formulate a "holist-ic" (and thus necessarily simplified) view of the transition from microscopic to megascopic organization, a view based on the premise that the major trends and principal selective pressures of Phanerozoic evolution can be realistically extrapolated into the Precambrian. Thus, the evolution of the plant life cycle from a "haploid-dominant" to a "diploid-dominant" condition, a trend ultimately result-ing in the derivation of seeds and angiospermy in Phanerozoic tracheo-phytes, is here regarded as a continuum beginning with the late Precambrian origin of sexual algae. And, by anal-ogy with Phanerozoic botanical evolution, the primary selective pressures resulting in diversification of Precambrian thallophytes are here considered to be related to photo-synthetic and reproductive capability. Among heterotrophs, adaptations for effective food acquisition (indirectly

affecting reproductive potential) appear to have been of primary importance in Phanerozoic evolution; we regard this as equally true during the late Precambrian. From this it follows that autotrophs would be expected to have preceded eumetazoans in coloniza-tion of the benthic environment and that that a variety of innovations in early metazoan diversification (e.g., mobility, cephalization, and acquisi-tion of hard parts) may have been pri-marily related to metabolic character-istics, efficiency of food gathering and predator-prey relationships.

The major events postulated in this this sequence, and their approximate times of occurrence, may be summarized as follows (also see Text-fig. 1):

1. Unicellular, planktonic eukaryotes were derived from prokaryo-tic ancestors probably prior to 1,300 million years ago and possibly earlier than 1,700 million years ago; these earliest eukaryotes were haploid, asexual algae that reproduced exclus-ively by mitotic cell division.

2. Sexual, eukaryotic algae, having an alternation of diploid and (meiotically produced) haploid generations, apparently originated about 1,000 million years ago. The advent of this "haploid-dominant" sexual life cycle resulted in a marked increase in both diversity and evolu-tionary rate among late Precambrian

eukaryotes; the development of eukaryotic sexuality may have served as an
"evolutionary trigger" leading to the
emergence of megascopic organization.

3. Apparently between about 900
and 700 million years ago, algal life
cycles evolved through a series of
stages from "haploid-dominant" to
"haploid-diploid coequal" and "diploid-
dominant"; this sequence was paralleled
by the development of multicellular
organization, megascopic size and a
benthic habit (thus resulting in the
origin of metaphytes) and may have been
spurred by the opening of extensive,
shallow-water environments created by
post-glacial flooding of penepianed
continental margins.

4. The late Precambrian diversification of sexual, eukaryotic microorganisms led to the emergence of
pelagic, heterotrophic protists from
which were derived, perhaps as early as
800 to 700 million years ago, primitive
eumetazoans. The earliest eumetazoans
were small, soft-bodied and pelagic,
primarily relying on phytoplankton as a
food source; subsequently, and prior to
about 650 million years ago, eumetazoans "discovered the bottom," a
shallow, shelf-like, oxygenic environment presumably replete with a diverse
assemblage of encrusting and erect
metaphytes.

5. Near the close of the Precambrian, "hard parts," of diverse

types and compositions, evidently
appeard more or less concurrently in
various groups of benthic metaphytes
and metazoans. The origin of this
skeletal material, which served
supportive and protective functions,
seems interrelated with the development
of relatively large body size and with
the appearance of increasingly efficient food gathering capability in
mobile, bilaterally symmetrical
eumetazoans.

## ACKNOWLEDGMENTS

Portions of this study have been
supported by the Earth Science Section,
National Science Foundation, NSF
Grant GA-23741 and by NASA Grant NGR
05-007-292.

## REVERENCES

Allsopp, A. 1969. Phylogenetic rela-
    tionships of the Procaryota and
    the origin of the eucaryotic cell.
    New Phytol. 68: 591-612.
Barghoorn, E. S. 1971. The oldest
    fossils. Scientific American 224
    (5): 30-42.
Barghoorn, E. S., and S. A. Tyler.
    1965. Microorganisms from the
    Gunflint chert. Science 147:
    563-577.

Cloud, P. E., Jr. 1968. Pre-metazoan evolution and the origins of the Metazoa. p. 1-72. In Drake, E. T., ed., Evolution and Environment. Yale Univ. Press, New Haven Haven.

Cloud, P. E., Jr., G. R. Licari, L. A. Wright, and B. W. Troxel. 1969. Proterozoic eucaryotes from eastern California. Proc. Natl. Acad. Sci. (U. S.) 62: 623-630.

Fritsch, F. E. 1949. The ones of algal advance. Biol. Rev. Cambridge Phil. Soc. 24: 94-124.

Glaessner, M. F. 1969. Trace fossils from the Precambrian and basal Cambrian. Lethaia 2: 369-393.

---- 1971. Geographic distribution and time range of the Ediacara Precambrian fauna. Geol. Soc. America Bull. 82: 509-514.

Glaessner, M. F., and M. Wade. 1966. The late Precambrian fossils from Ediacara, South Australia. Palaeontology 9: 599-628.

Gnilovskaja, M. B. 1971. The oldest aquatic plants of the Wendian of the Russion Platform (late Precambrian). Paleontology Jour. 5 (3): 372-378.

Harland, W. B. 1964. Evidence of late Precambrian glaciation and its significance. p. 119-149. In Nairn, A. E. M., ed., Problems on Palaeoclimatology. Wiley, New York.

Harland, W. B., K. N. Herod, and D. H. Krinsley. 1966. The definition and identification of tills and tillites. Earth-Sci. Rev. 2: 225-256.

Harland, W. B., and M. J. S. Rudwick. 1964. The great infra-Cambrian ice age. Scientific American 211 (2): 28-36.

Harland, W. B., and C. B. Wilson. 1956. The Hecla Hoek succession in Ny Friesland, Spitsbergen. Geol. Mag. 93 (4): 265-286.

Hicks, H. 1880. Pre-Cambrian volcanoes and glaciers. Geol. Mag. 7 (11): 488-491.

Hofmann, H. J., and G. D. Jackson. 1969. Precambrian (Aphebian) microfossils from Belcher Islands, Hudson Bay. Can. Jour. Earth Sci. 6: 1137-1144.

Johnson, J. H. 1966. A review of the Cambrian algae. Colorado School of Mines Quart. 61 (1), 162 pp.

Margulis, L. 1970. Origin of Eukaryotic Cells. Yale Univ. Press, New Haven. 349 pp.

Milton, D. J. 1966. Drifting organisms in the Precambrian sea. Science 153: 293-294.

Pflug, H. D. 1965. Organische Reste aus der Belt-Serie (Algonkium) von Nordamerika. Paläont. Zeit. 39: 10-25.

---- 1966. Einige Reste niederer Pflanzen aus dem Algonkium.

Palaeontographica Abt. b 117: 59-74.

Raven, P. H. 1970. A multiple origin for plastids and mitochondria. Science 169: 641-646.

Raymond, P. E.. 1935. Pre-Cambrian life. Geol. Soc. Amer. Bull. 46: 375-392.

Rhoads, D. C., and J. W. Morse. 1971. Evolutionary and ecologic significance of oxygen-deficient marine basins. Lethaia 4: 413-428.

Rudwick, M. J. S. 1964. The infra-Cambrian glaciation and the origin of the Cambrian fauna. p. 150-155. In Nairn, A. E. M., ed., Problems in Palaeoclimatology. Wiley, New York.

Schopf, J. W. 1968. Microflora of the Bitter Springs Formation, late Precambrian, central Australia. Jour. Paleontology 42: 651-688.

---- 1970. Precambrian micro-organisms and evolutionary events prior to the origin of vascular plants. Biol. Rev. Cambridge Phil. Soc. 45: 319-352.

Schopf, J. W., and E. S. Barghoorn. 1969. Microorganisms from the late Precambrian of South Australia. Jour. Paleontology 43: 111-118.

Schopf, J. W., and J. M. Blacic. 1971. New microorganisms from the Bitter Springs Formation (late Precambrian) of the north-central

Amadeus Basin, Australia. Jour. Paleontology 45: 925-960.

Spencer, A. M. 1969. Late Pre-Cambrian glaciation in Scotland. Geol. Soc. London Proc. 1657: 177-198.

Spjeldnaes, N. 1963. A new fossil (Papillomembrana sp.) from the Upper Pre-Cambrian of Norway. Nature 200: 63-65.

---- 1964. The Eocambrian glaciation in Norway. Geologischen Rundschau 54: 24-45.

Stebbins, G. L. 1969. The Basis of Progressive Evolution. Univ. North Carolina Press, Chapel Hill. 150 pp.

Tappan, H., and A. R. Loeblich, Jr. 1970. Geobiologic implications of fossil phytoplankton evolution and time-space distribution. In R. Kosanke and A. T. Cross, eds., Symposium on palynology of the late Cretaceous and early Tertiary. Geol. Soc. Amer. Special Paper 127: 247-340.

Taylor, D. L. 1970. A multiple origin for plastids and mitochondria. Science 170: 1332.

Termier, H., and G. Termier. 1956. Conditions Ecologiques du Cambrien Inférieur. Congr. Géol. Intern., Compt. Rend., 20$^e$, Mexico, 1956: Cambrian Symposium. Pt. I: 417-425.

Wade, M. 1972. _Dickinsonia_: polycha-
ete worms from the late Precambrian Ediacara fauna, South
Australia. Queensland Mus.
Mem. 16: 171-190.

Wells, M. J. 1968. Lower Animals.
McGraw-Hill, New York. 254 pp.

Wilson, C. L., and W. E. Loomis. 1962.
Botany, 3rd ed. Holt, Rinehart
and Winston, New York. 573 pp.

*

# An ecological theory for the sudden origin of multicellular life in the late Precambrian

**S.M. Stanley (1973)**

Students of the history of life have long been troubled by the fact that, after a long Precambrian history of slow unicellular evolution, multicellular organisms arose and radiated suddenly near the beginning of the time interval we have come to call the Cambrian Period.  This fundamental radiation forms our basis for division of the geologic time scale into two major parts. Its cause has been widely regarded as the foremost unresolved problem of paleontology (1-3).  Two approaches have been emphasized in recent attempts to solve this and other riddles in the history of primitive life.  The first is invocation of external physical and chemical controls, such as levels of ultra- violet radiation and atmospheric oxygen (2, 4-6).  The second,

sometimes intertwined with the first, is recognition of major adaptive breakthrough, such as the appearance of autotrophy, eukaryotic organization, sexuality, multicellularity, collagen secretion, and shell formation (3, 5-8). Neither approach has produced a generally accepted, unified explana- tion of relevant ecological theory.

## Chronology of Evolutionary Events

In a recent authoritative evaluation of existing fossil evidence bearing on the origin of multicellular life, Schopf et al. (3) reached the following conclusions:
Eukaryotic organisms arose earlier than 1300 million years ago, and perhaps even earlier than 1700 million years ago, but were probably haploid and asexual at first.  The oldest convincing evidence of sexuality comes from spore-like unicells in

Reprinted from Proc.Nat.Acad.Sci.USA 70, 1486-1489 (1973) by permission of the author and publisher.

the 900 million year-old Bitter Springs fossil flora. The advent of sexuality should have greatly increased genetic variability and correspondingly increased rates of evolutionary diversification. It is then puzzling that the Bitter Springs flora and various other fossil assemblages of similar age are impoverished relative to modern eukaryote assemblages. They have yielded only simple spheroidal unicells, none of which exhibits features typical of even moderately advanced thallophytes (e.g., tissues, differentiated sporangia, true branching, or polarity of organization). Although the precise time of origin of metaphytes is unknown, the fossil record shows that green, red, and brown algae appeared before the beginning of the Cambrian, and perhaps slightly before 650 million years ago. The oldest known metazoan assemblages, the Ediacara fauna of Australia and its equivalents, are of similar age. Animal-like protists must have arisen before metazoans, although we have no definite record of their earlier presence. Schopf et al. suggested that the proliferation of early metaphytes may have contributed to the initial diversification of metazoans. To explain the gap between the advent of sexuality

and the origin of multicellularity, these authors suggested that eukaryotic sexuality was a novel experiment in Bitter Springs time and that its evolutionary impact was somehow delayed at least 250-300 million years (substantially longer if sexuality appeared well before Bitter Springs time). This delay, they suggested, might have been caused by the gradual nature of the trend from a haploid-dominated life cycle to a diploid-dominated cycle of the type that characterized modern algae.

In fact, the idea that the diploid stage, once present, was somehow suppressed is difficult to justify by the "adaptive innovation" approach to Precambrian evolution. Once the diploid stage arose, this approach would predict its immediate rise to dominance if it was a major innovation. Certainly, however, Schopf et al. were correct in stressing a previously ignored fact: that metaphytes and metazoans originated almost simultaneously. The delayed origin of both groups may, I believe, be explained by application of a newly established principle of ecology, which leads to the conclusion that the origin of metazoans brought about the diversification of metaphytes. Further, this approach offers a

likely explanation for the sudden
radiation of both groups in
comparison to the gradual moderni-
zation of unicellular autotrophs,
which required hundreds of millions
of years of Precambrian time.

## Cropping and Diversity

The relevant ecologic principle,
apparently first formulated for
multispecific situations by Paine
(9), holds that the addition of a
trophic level to a given food web
tends to promote increased diversity
at the next lower trophic level.
Increase in diversity at an existing
level can have the same effect.
This generalization has recently
gained strong support from experimental and theoretical studies on
the effects of cropping.  Field
experiments have shown, for example,
that removal of the chief natural
carnivore of an intertidal community
produces a drastic reduction in
herbivore diversity by permitting
one superior competitor (a mussel) to
monopolize space (9).  Similarly,
introduction of predatory fish to
an artificial pond has led to
increased zooplankton diversity (10).
Lower in the food web the same effect
is seen.  Removal of grazing sea
urchins from multispecific communities
of benthic algae has led to dominance

by a single algal species (11).
There is also evidence that the grazing of diverse herbivores on seeds
and seedlings in tropical forests
has contributed to the high
diversity of trees because many
herbivores tend to feed on high-
density populations, which naturally
occur near parents; increased
dispersion and reduced density of
populations has then permitted a
large variety of tree species to
coexist within habitats (12).
Even in the deep sea, remarkably
intense predation appears to
contribute to the characteristically high diversity of benthic
life (13).

Phillips (14, 15) has provided
a mathematical analysis of the
cropping phenomenon.  His theory,
which applies to simple well-mixed
aquatic systems containing plankton,
is particularly applicable to Precambrian aquatic communities of
unicellular autotrophs.  The theory
assumes a region of finite volume,
with spatially uniform populations
that react to variations in nutrient
supply at rates that are slow
relative to their own rates of
reproduction.  Under these conditions,
it turns out, the number of species
at equilibrium must be equal to or
less than the number of removable
nutrients, and for every species

there will be at least one domain of values of removable nutrients in which only this species will grow. Under most conditions the system will be stable. In a stable system, change in rate of supply of certain nutrients will cause a decrease in the number of species surviving. Introduction of herbivores changes the system drastically. With cropping, the competitive exclusion principle fails, and a much greater diversity of producers is possible. Although the exact predictions of this theoretical model must be expected to differ somewhat from events in natural systems, where the assumptions will be violated to varying degrees, the experimental evidence cited above indicates that the cropping principle holds for systems in general in at least a qualitative way. Furthermore, the single-trophic-level systems of the Precambrian would seem to approximate the un-cropped system of Phillips far better than present-day natural systems, in which cropping is nearly universal. Polluted, eutrophic systems form perhaps the best modern examples.

## The Delay of Heterotrophy

The Precambrian trend from simple producer communities to producer-herbivore and producer-herbivore-carnivore communities can be viewed as a long-term natural cropping experiment resembling those performed by living ecologists. The following scenario can be envisioned:

Since we are considering the original adaptive radiation of eukaryotic life, initial diversity was very low. We have good evidence that planktonic eukaryotic autotrophs were well established by 1300 million years ago, and it seems reasonable that similar benthic forms were also present, although they were clearly subordinate to blue-green algae (3, 17). As both habitat groups of autotrophs developed, they must rapidly have attained quite substantial biomasses, for in the absence of cropping, abundances would have been limited only by environmental resources, including light, space, and nutrient supply. Blue-green algae, having evolved first, apparently tended to exclude early eukaryotes from benthic settings (17). The fact that well-preserved floras contain fewer planktonic than benthic species, although the former are largely eukariotic (17), probably reflects the greater uniformity of the pelagic realm. There has been much debate about the relationship between food web complexity and stability (16), but as Phillips (14) has shown mathematically, a simple aquatic producer system of the type just described should have been relatively stable. Not only would within-habitat diversity have been constrained, with relatively few species dominating, but like modern unicellular aquatic forms, most species must have had broad geographic distributions. Even freshwater and marine forms were probably quite closely related, as they are today (18). In short, worldwide diversity must have been extremely low, especially for the dominantly planktonic eukaryotes. We can envision an all-producer Precambrian world that was generally saturated with producers and biologically monotonous. Morphologic diversity must also have been suppressed, for speciation can occur only in the presence of available niches. Rates of speciation must have been very low relative to those that have characterized more modern multitrophic communities. Speciation events may have been confined largely to replacement of lineages terminated by extinction, but low worldwide diversity must have made even these events very rare. Gradual changes may have occurred within species as adaptations for more efficient exploitation of their

existing niches, but such changes also could only have produced slow differentiation. Even if metaphytes had somehow appeared before the advent of cropping, their diversification too would have been greatly restricted. Heterotrophs had to appear before resource limitation could be relieved and diversification accelerated. The important point is that the changes required to produce unicellular heterotrophs, to say nothing of metazoans, had to arise slowly because diversification was so slow. The self-limiting nature of this system would seem to account for the impoverished nature of eukaryote floras of Bitter Springs age and for the delay in origin of metaphytes, as well as heterotrophs.

When they finally arose in the late Precambrian (we do not know the exact time), cell-eating heterotrophs may have fed on bacteria. There would seem to have been no distinct barrier to feeding on algae, however, as the ability to ingest larger food particles evolved. The first algae-eaters must have relieved resource limitation of algal growth in certain habitats, although modern animal-like protists cannot ingest some kinds of benthic algal cells, and they are also relatively unsuccessful as zooplankters, even in lakes (18). One might suggest that, in the initial absence of carnivores, the diversification of algae-eaters and rise of carnivores might have been suppressed in a way parallel to the earlier suppression of autotroph diversification. The difference is that no marked change was required in certain heterotrophic groups to permit them to develop carnivorous habits. Feeding on unicellular heterotrophs differs little from feeding on unicellular autotrophs, and many living protist groups (especially ciliate taxa) do both [19]. The Flagellata, generally regarded as the ancestral protists, include a very limited variety of free-living heterotrophs today [20]. Presumably the origin of the Ciliata was an important step in diversification.

Thus, the adaptive breakthrough to algal feeding, when it finally came, rapidly led to the addition of successive trophic levels. Not only autotrophs, but also heterotrophs below top carnivore levels were permitted to diversify. A key point is that, in general, diversification at any trophic level promotes diversification, not only of the level below by cropping, but also of the level above by providing more feeding options. Thus, when one trophic level diversifies, a mutual feedback system is set up with both super- and subadjacent trophic levels. In diverse communities this system is damped. When algae-eating protists first arose, however, self-propagating systems of this type, operating in numerous azoic environments, must have produced explosive rates of evolution. Rapid advancement to multicellularity must also have ensued. Whether they arose slightly earlier or slightly later than metaphytes, metazoans must have played a crucial role in permitting metaphytes to differentiate, for protists could not have grazed effectively on metaphytes. Being more efficient algal grazers than animal-like protists, metazoans probably fostered even greater diversification of unicellular algae as well. The earliest metazoans must have been herbivores that fed on unicellular taxa. Like herbivorous protists, however, such creatures intergrade with carnivores. In fact, many extant species of the Coelenterata and Rotifera, which perhaps represent levels of organization not far removed from late-Precambrian metazoans, are difficult to classify unequivocally as suspension feeders or carnivores.

## Independent Evidence from Stromatolites

To test the above arguments we can consider what fossil evidence exists, in addition to that already cited, indicating limitation of Precambrain algal growth by resources rather than by cropping. Dominantly planktonic assemblages like the Bitter Springs flora represent unusual preservation conditions and are too rare to provide useful data for this purpose. The most abundantly fossilized Precambrian producer communities are those forming stromatolites (laminated sedimentary structures produced chiefly by blue-green algal mats that trap and bind sediment). Stromatolites were in fact extremely widespread in the Precambrian, living in a wide range of intertidal and subtidal marine environments. Garrett (21) has convincingly argued that their rapid decline at the start of the Phanerozoic was a result of their destruction by grazing and burrowing animals. Taxa with the potential to form stromatolites have persisted, but in greatly reduced biomass. Today they are able to flourish well enough to form stromatolites only in hostile environments, like hyper-saline lagoons and some supratidal areas that are largely devoid of metazoans. Garrett found experimental-ly that persistent mats formed in more hospitable environments only when metazoan grazers and burrowers were excluded. Also where surficial

mats were forming naturally, introduction of grazing snails resulted in their destruction. Garrett concluded that marine blue-green algae grew rampantly in the Precambrian because of the absence of metazoans (most species that form stromatolites, being filamentous and covered by a mucilaginous sheath, are inedible to unicellular herbivores). After the origin of eukaryotic organization, blue-green algae were relegated to an unimportant role in the diversification of plant life. Nonetheless, their abundant growth throughout the Precambrian provides strong evidence that they and other photosynthetic autotrophs saturated aquatic environments until very late in the Precambrian, being limited only by resources, including nutrient supply. In fact, the period during which only prokaryote life existed may have been prolonged by the total absence of cropping, just as eukaryote diversification was later delayed.

## DISCUSSION

In summary, the explosive radiation of life in the late Precambrian (giving rise to our distinction between the Precambrian and the Phanerozoic) was produced by a kind of self-propagating mutual feedback system of diversification between trophic levels, which was initiated by the advent of heterotrophy. The nearly simultaneous origin and explosive radiation of heterotrophic protists, metaphytes, and metazoans were inevitable. In contrast, the self-limiting, saturated autotrophic system of the earlier Precambrian had strongly inhibited diversification, to delay crossing of the "heterotroph barrier" for hundreds of millions of years.

The seemingly explosive appearance of skeletons, giving rise to the diverse fossil record of the Cambrian, was no more sudden than the rest of the radiation, which apparently produced most currently recognized multicellular phyla during only a few tens of millions of years. Although earliest appearances of fossil taxa are difficult to delimit precisely because of the limitations of preservation and stratigraphic correlation, it seems that most phyla known from the Cambrian appeared during the first 10 million years or so of the period (22). Some of these may have arisen as soft-bodied taxa in the late Precambrian. Many important phyla and classes did not contribute to the known fossil record until later Cambrian or Ordovician time. Skeleton formation was, in fact,

probably in part a response to predation pressure, as Hutchinson and Brooks each suggested (23). Furthermore, the overall rate of diversification, which, after all, was operating during the invasion of previously vacant adaptive zones, was not inordinately high. For comparison, it is worth noting that most major types (i.e., orders) of Cenozoic placental mammals, from bats to whales, arose during a period of only about ten million years (24).

The theory presented above has four attractive features: first, it seems to account for what facts we have about Precambrian life; second, it is simple, rather than complex or contrived, as some earlier explanations have tended to be; third, it is purely biological, avoiding ad hoc invocation of external controls; and fourth, it is largely the product of direct deduction from an established ecological principle.

I thank my colleagues Jeremy B. C. Jackson and Owen M. Phillips for evaluating the original manuscript of this paper.

1. Simpson, G. G. (1960) "The history of life," in Evolution after Darwin, ed. Tax, S. (Univ. of Chicago Press, Chicago), Vol. 1, pp. 117-180.

2. Fischer, A. G. (1965) "Fossils, early life, and atmospheric history," Proc. Nat. Acad. Sci. USA 53, 1205-1215.

3. Schopf, J. W., Haugh, B. N., Molnar, R. E. & Satterthwaite, D. F. (1973) "On the development of metaphytes and metazoans," J. Paleontol. 47, 1-9.

4. Berkner, L. V. & Marshall, L. C. (1964) "The history of oxygenic concentration in the earth's atmospher," Discuss. Faraday Soc. 37, 122-141.

5. Cloud, P. E. (1968) "Pre-metazoan evolution and the origins of the Metazoa," in Evolution and Environment, ed. Drake, E. T. (Yale Univ. Press, New Haven), pp. 1-72.

6. Towe, K. M. (1970) "Oxygen-collagen priority and the early metazoan fossil record," Proc. Nat. Acad. Sci. USA 65, 781-788.

7. Margulis, L. (1970) Origin of Eukaryotic Cells (Yale Univ. Press, New Haven).

8. Schopf, J. W. (1970) "Precambrian organisms and evolutionary events prior to the origin of vascular plants," Biol. Rev. 45, 319-352.

9. Paine, R. T. (1966) "Food web complexity and species diversity," Amer. Natur. 100, 65-75.

10. Hall, D. J., Cooper, W. E. &

Werner, E. E. (1970) "An experimental approach to the production dynamics and structure of freshwater animal communities," Limnol. Oceanogr. 15, 839-928.

11. Paine, R. T. & Vadas, R. L. (1969) "The effects of grazing by sea urchins, Strongylocentrotus spp., on benthic algal popula-tions," Limnol. Oceanogr. 14, 710-719.

12. Janzen, D. H. (1970) "Herbivores and the number of tree species in tropical forests," Amer. Natur. 104, 501-528.

13. Dayton, P. K. & Hessler, R. R. (1972) "Role of biological disturbance in maintaining diversity in the deep sea," Deep Sea Res. 19, 199-208.

14. Phillips, O. M. (1973) "The equilibrium and stability of simple marine biological systems. I. Primary nutrient consumers," Amer. Natur. 107, 73-93.

15. Phillips, O. M. (1973) "The equilibrium and stability of simple marine biological systems. II. Herbivores," Amer. Natur., in press.

16. Hurd, L. E., Mellinger, M. V., Wolf, L. L. & McNaughton, S. J. (1971) "Stability and diversity at three trophic levels in terrestrial succesional

ecosystems," Science 173, 1134-1136.

17. Schopf, J. W. & Blacic, J. M. (1971) "New microorganisms from the Bitter Springs Formation (late Precambrian) of the north-central Amadeus Basin, Australia," J. Paleontol. 45, 925-960.

18. Hutchinson, G. E. (1967) A Treatise on Limnology (John Wiley, New York), Vol. 2.

19. Fenchel, T. (1968) "The ecology of marine microbenthos. II. The food of marine benthic ciliates," Ophelia 5, 73-121.

20. Mackinnon, D. L. & Hawes, R. S. J. (1961) An Introduction to the Study of Protozoa (Oxford Univ. Press, London).

21. Garrett, P. (1970) "Phanerozoic stromatolites: noncompetitive ecologic restriction by grazing and burrowing animals," Science 167, 171-173.

22. Kummel, B. (1970) History of the Earth (W. H. Freeman, San Francisco).

23. Hutchinson, G. E. (1961) "The biologist poses some problems" in Oceanography, ed. Sears, M. (Amer. Ass. Advan. Sci. Publ. 67), pp. 85-94.

24. Romer, A. S. (1966) Vertebrate Paleontology (Univ. Chicago Press, Chicago).

# Permo-Triassic extinctions (extract)

## F.H.T. Rhodes (1967)

7.  CAUSES OF EXTINCTION

Extinction itself is seldom simple.  The extinction of a population involves as many delicate and complex physical and biological interactions as its survival.  Apart from the intricacy of the factors involved, and the apparent uniqueness of each specific case, extinction may reflect either evolutionary change or cessation.  As a taxon undergoes phyletic evolution, it produces a new taxon by its own extinction.  In other cases, extinction involves the disappearance of the lineage which the taxon represents.

It seems unlikely that any single factor is sufficient to account for all cases of extinction or that all larger episodes of extinction must be the results of similar causes.  It has also been shown that Late Permian extinction was not conspicuously more severe than

---

Extract from a paper in The Fossil Record (W. B. Harland et al., eds.), pp. 66-75.  Reprinted by permission of the publisher.  Copyright 1967 by the Geological Society of London.

that of earlier periods of time, and may therefore require no special causal explanation.  But it may be that unusual factors were involved in the lack of rapid replacement, which seems to be the most conspicuous feature of Permo-Triassic faunal changes.

(i)  'Neocatastrophism'

'Neocatastrophist' hypotheses are best represented by the writings of Schindewolf (1962, and earlier references therein).  The term is a little ambiguous, but it is used here in a restricted sense to denote the postulate of instantaneous mass extinction.  Schindewolf has argued that, in at least some areas, there is no evidence of physical unconformity at era or period boundaries characterized by major faunal changes.  Schindewolf suggests that this apparent sedimentary continuity implies that the organic discontinuity is real, rather than apparent, representing a true biological hiatus, rather than non-preservation of a once continuous lineage. Newell (1956; 1962, p. 606) has argued convincingly against this view, stressing the common occurrence of paraconformities in what are apparently continuous rock successions. One need not argue that all such successions must conceal breaks, however, although Newell's arguments are very strong. Even if some successions of Permo-Triassic marine strata were shown to be continuous, this would in no way prevent the possibility of migration of new faunas from other areas. Extinction may well have been very rapid in some late Permian groups, but it was certainly not instantaneous, nor, as Schinde-wolf's own tables show, was replacement instantaneous (Schindewolf 1962, fig. 2). This does not affect the validity of Schindewolf's views on the mechanism of replacement, which are discussed below (p. 68); but it does contradict his interpretation of the nature of replacement, as sudden, widespread and more or less instantaneous.

(ii)  Extra-Terrestrial Factors

A number of authors have, at various times, suggested that Permo-Triassic faunal changes were so extreme that some extra-terrestrial agency must be involved to account for them (e.g. Schindewolf 1954, 1962; Stechow 1954; Wilser 1931). There are good reasons for rejecting all such explanations, quite apart from the Ockhamian economy to which we should remain committed. Certainly such cosmic variations may, and probably have, occurred, but the length of time over which the extinct-ions took place makes it seem wholly unlikely that any such factor could have been the chief agent. If the effects were so lethal as the propon-ents of these hypotheses claim, it seems unlikely that certain families, genera and species could have withstood their influence for millions of years longer than others. Furthermore, if radiation was the agent, its effects should be far more conspicuous upon terrestrial organisms than upon aquatic ones. Clarke (1939) and others have shown that water of only a few metres depth is an effective filter of such radiation. But it was marine organisms which were most affected: terrestrial organisms were little influenced at this period of time.

Schindewolf (1954, p. 182) has suggested another possible effect of radiation would be to increase mutation rates and thus produce the 'rapid differentiation' of Triassic times; but the big problem is the lack of early Triassic differentiation, not the excess of it. Even then, it seems to be stretching the argument beyond its limits to suggest that radiation, lethal to so many taxa, may affect others only as a mutational stimulant. Nor can we accept the premise that increased mutation rate, if it ever occurred, would necessarily produce an increase in **evol**utionary rate. Newell (1956) and George (1958) have also dealt with these problems.

### (iii) Plankton Reduction

Bramlette's hypothesis of a sudden reduction in plankton supply (Bramlette 1965A, B) has much to make it attractive. A major break in the oceanic food chain would have far-reaching consequences for other groups, but whatever its merits for other periods, there is no evidence available for or against such a relationship in Permo-Triassic times.

### (iv) Climatic Changes

One common suggestion is that changes in climate may have produced late Palaeozoic extinction; but **again** terrestrial organisms (especially plants), which should indicate the existence of extremely rapid or intense climatic changes in late Palaeozoic times, provide little or no support for this view. Stokes (1960, pp. 438-40) has suggested that diastrophic changes may have influenced oceanic circulation, which might in turn have produced a lowering of sea-water temperature.

Such changes would, however, inevitably control land climates, and, if they were sudden, should be reflected by land plants.

In a more general sense, late Palaeozoic times were indeed marked by climatic extremes. The major glaciation of Gondwanaland, and the widespread development of continental red beds, are two reflections of this. How far these factors, as such, influenced marine faunas, is difficult to determine; the only valid comparison we can make is with the Pleistocene glaciation, which involved changes in sea level of the order of 300 ft. This had marked effects on the distribution and survival of many marine organisms, but its net effects seem of a different order of magnitude from those of late Palaeozoic times.

Some have suggested the Permo-Triassic climatic changes may have reduced the salinity of the oceans (e.g. Beurlen 1965). This might have resulted from melting of the late Palaeozoic glaciers, from a reduction in terrigenous salt supplies from arid lands, or perhaps even from the locking-up of salt in late Palaeozoic evaporites. The greatest effects of these changes would be upon stenohaline forms; yet some of the most stenohaline groups, such as cephalopods, were least influenced by Permo-Triassic changes (Ruzhentsev & Sarytcheva 1965).

Fischer (in Nairn, 1964) argued for reduced oceanic salinity, as a result of evaporite formation producing 'pockets' of dense brine on the ocean floors. He argued that those aquatic forms least affected by the late Permian extinctions were euryhaline, whereas those most affected were stenohaline. This is broadly true, but it certainly cannot be the whole explanation. Nautiloids, which were almost totally unaffected, can scarcely be regarded as very different in their tolerance from ammonoids; ostracods are generally amongst the more euryhaline forms, and yet were 'decimated' (Fischer's phrase).

Fischer has argued that the cosmopolitan and meagre Lower Triassic faunas reflect uniform and peculiar conditions. The great rarity or virtual absence from the Lower Triassic of many groups well represented in Upper Permian and Middle Triassic faunas, tends to support this. Most Lower Triassic marine faunas are virtually restricted to bivalve molluscs, ammonoids and inarticulate brachiopods. Fischer regards the early Triassic 'marine faunas' as derived largely from brackish or lagoonal Permian ancestors. Here again, ammonoids and pelecypods make strange bed-fellows in terms of salinity tolerances. Teichert (in Nairn 1964,

p. 576) has stressed that the presence of a rich echinoid fauna in the lowest Triassic of the Salt Range scarcely indicates reduced salinity, and Lowenstam (in Nairn 1964, p. 576) has shown that the scanty oxygen isotope ratios available for the period imply no salinity changes.

(v)  Orogeny

The view that world-wide, instantaneous diastrophism or orogeny brought about faunal change (both in extinction and rapid evolution, e.g. Strakhov (1948)) was once popular; but there is now overwhelming evidence that such physical changes (e.g. volcanic poisoning of Pavlov (1924) and the isotopic effects of Ivanova (1955)) are neither universal nor instantaneous (e.g. Henbest 1952; Spieker 1956; Westoll 1954; George 1958). For example, to correlate the Variscan orogeny, which lasted from early Carboniferous to late Permian times, with late Permian extinction, implies a plasticity of interpretation of both time and the geographical effects of orogeny that are unacceptable. Indeed, the most severe deformational phases of the Variscan were completed in most areas by early Permian times; while in the western United States mid-Triassic tectonism was far more severe than that of Permo-Triassic transitional time.

## (vi)  Racial Senescence

The concept of racial senescence is more of a straw man than a serious hypothesis, yet it seems to have influenced much thinking (see Hawkins 1950, p. 6).  Certainly some Permian groups were small in numbers and of limited distribution, some had had a long earlier history (the trilobites are perhaps the best examples), but, even if this is what is meant by 'senility', the phrase is wholly inappropriate for many of the groups which suffered extinction.  Permian crinoids, blastoids, foraminifera, dalmaneloids, productoids and other groups display an evolutionary 'vigour' which is far more remarkable than the questionable 'senility' of a few restricted groups.

## (vii)  Biological Factors

Nicol (1961) has stressed the need to assess ecologic relationships between organisms involved in periods of extinction.  This is never easy, although some food chains have been tentatively reconstructed.  On balance, Permo-Triassic extinctions seem to have had a more profound effect upon benthic than upon planktonic or nektonic organisms.  This may be a reflection of the fact that at that period, as in later periods, the numbers of taxa of benthic forms far exceed those of pelagic forms, although the latter involve enormous numbers of individuals.  Certainly ammonites, which were largely pelagic, shared a similar fate to benthic forms.  Whether sessile benthic forms were more affected by extinction than vagrant forms is a particularly difficult question.  There is some suggestion that they were (gastropods and pelecypods, for example) being much less influenced than were corals, brachiopods, crinoids and bryozoa; but whether this is a case of cause and effect is more difficult to judge.

## (viii)  Eustatic Changes in Sea Level

The effects of eustatic changes in sea level were conspicuous in late Palaeozoic and early Triassic times. Various authors have pointed out that the restriction of shallow seas, and the high emergent continents of those times, combined to produce a quantitatively unique environment (Newell 1956, 1962; Moore 1954, 1955B; Kummel 1961).  Moore has argued that "regression of a shallow sea inevitably produces crowding together of populations so that weaker, less well adapted, marine invertebrates should be weeded out. It seems reasonable to construe times of marine regression as more significant in terms of accelerated evolution than times of marine transgression." (Moore 1954, p. 259; Moore 1955B, p. 483). Moore's 'accelerated evolution' in this sense implies more rapid extinction.

Rutten (1955) has argued against this thesis, partly on the grounds that transgression of shelf seas does not necessarily supply the genetically isolated niches supposed (by Rutten) to be essential for evolution. Moore (1955) has challenged the view that speciation necessarily requires either a very long period of time or particular conditions of geographical isolation. Newell (1956) has argued that if the shallow shelf seas were destroyed by a eustatic drop in sea level of, say, 100 fathoms, it seems unlikely that neritic benthic faunas would re-establish themselves on the continental slope; even though the shallow depths would still be available, the steeply sloping bottoms would produce a quite different environment. Reduction of optimum population size has particularly severe effects in those aquatic groups in which the chance union of gametes is involved, which will vary logarithmically with population density. This may produce an ultimate biological cause of extinction in a population

which has been reduced in numbers of physical causes (see also Kummel (1961, p. 214) and Ruzhentsev & Sarytcheva (1965)).

It seems to the writer that the piecemeal pattern of extinction of, say, the fusulinids is far too extensive in time and uneven in its effects to attribute to a single period of eustatic change in later Permian times.

Perhaps the only relatively recent comparison that we can make is that of the changes in sea level involved during the Pleistocene, which were of some 300 ft. These did not bring about any substantial mass extinctions of either terrestrial or marine organisms.

This neither proves, nor even implies, that there is no connexion between diastrophism and evolution. Local fluctuations in sea level are clearly of major importance in creating and destroying a great variety of different environments. Hallam (1961) and others have shown the influence of such movements on ammonite evolution. But the effects are neither simple nor uniform. The origin of some groups, the extinction of others and the acme of still others have all been correlated with similar eustatic changes. This variety of effects, as well as the piecemeal character of extinction, seem to make the dependence of mass extinction on extreme eustatic fall altogether too simple, though there may be a more general--and much more complex--connexion between them.

## 8. TOWARDS A SYNTHESIS

What then has to be explained? Firstly, the degree and extent of late Permian and early Triassic extinction. This was certainly not instantaneous, it was taxonomically uneven in its effects, and it involved some groups that were already in decline or limited in their geographical distribution, or sometimes both. But it was unusually widespread in its effects. Permo-Triassic extinctions are, like C. S. Lewis' view of miracles, not so much unique, but magnified, and perhaps accelerated, examples of events that are commonplace in the everyday history of life. Perhaps we need no 'special' explanation over and above the explanation of normal extinction (imponderable as that is).

Secondly, the Triassic was also relatively unremarkable. Triassic first appearances are less difficult to interpret, given Permian extinction. Indeed, to the writer it is not the appearance of new taxa, but the time-lag between the disappearance and replacement of the old that is unusual. But such time lags are known in other parts of the column. The ecological replacement of ocean-going reptiles by marine mammals is a familiar example. In Triassic times, even in forms which 'replaced themselves' by the rise of new groups within the same phylum, a conspicuous time lag appears to have existed (corals, bryozoa, brachiopods and crinoids are examples; ammonites are a major exception). To what extent limitation in outcrops of early Triassic rocks are responsible for this is not clear, but it is regarded as an unlikely 'total' explanation. The discovery of Permian elements in the Transcaucasian and possibly the Himalayan Lower Triassic faunas greatly reduces both the size and the regularity of the gap, but does not altogether remove it.

It seems unlikely that there is a simple answer to the difficult questions of extinction. The restriction of late Permian sea-ways, and extreme geographic and oceanographic conditions were unusual, if not unique, terrestrial conditions. Ruzhentsev & Sarytcheva (1965) have shown that Dzhulfian sediments lack rudaceous deposits and are often very thinly developed (measuring from a few score to a few hundred metres in thickness) in contrast to those of earlier stages of the Permian, which are measured in thousands of metres. Certainly these, and probably other, factors interacted to produce the piecemeal extinction of Permo-Triassic

times, but there is no evidence that
any single factor was of decisive
importance.

The homogeneity and lack of
diversity of early Triassic faunas
seems to be a real feature, and not
only an effect of limited outcrops or
collecting, although the Transcaucasian
faunas have changed our understanding
of the exact nature of this.  This may
be the more direct result of palaeo-
geographical conditions, though clearly
in intricate association with biological and perhaps other physical
components.  Early Triassic faunas may
reflect partial adaptation to the
limited deeper waters and perhaps
reduced salinity.  Examples are the
changes in molluscs, the virtual
absence of corals in most areas and the
survival of some of the more tolerant
ostracod species (Ruzhentsev &
Sarytcheva 1965).

The homogeneity and lack of
variety of Lower Triassic faunas may
also be a reflection of the limited
extent of shelf seas.

## 9.   CONCLUSION

This analysis has shown that
Permo-Triassic times were not marked
by any 'instantaneous' extinction of
large numbers of organisms.  Certainly
several major taxa of Palaeozoic
invertebrates became extinct, but the

pattern of extinction shows little
uniformity.  Geographical differences,
piecemeal survival of different genera
and species, the marked absence of
major changes in terrestrial biotas,
the recognition of paraconformities in
areas once thought to display continuous sedimentation, the suggested
extended ranges of many 'Permian'
groups into the 'Lower Triassic' of
Transcaucasia and the provisional and
imprecise nature of Permo-Triassic
zonation all argue against any
catastrophic view of Permian-Triassic
extinction.  They also seem to make any
single cause of extinction unlikely.
Late Palaeozoic extinction rates were
by no means unique, and the character
of early Triassic faunas is best
explained by the multiple interactions
of a wide variety of physical and
biological factors; but to select any
one of these as decisive seems to
neglect the complexity of adjustment
and response involved in both
extinction and survival.

Furthermore, it is particularly
difficult in the Permo-Trias to
distinguish the 'real' biological
extinction of organisms from the
variety of effects of geological non-
preservation, which have led to
oversimplification in the interpretation of the character of extinction.
The increasing recognition that late
Palaeozoic extinction was neither

uniform in its effects, nor instantaneous in its action, the presence of
numbers of typical Palaeozoic faunal
elements as early Triassic survivors,
and the virtual absence of any
undisputed Permo-Triassic transitional
marine sequences, all suggest that the
sharpness of the faunal break has been
exaggerated by non-preservation.  In
this sense, but in this sense only,
can restricted late Palaeozoic seas as
yet be accepted as the 'cause' of this
rapid extinction.

This is not to deny that there was
a major decline of Palaeozoic taxa in
Permo-Triassic times.  This seems to me
to be an inevitable conclusion from our
present data, however inadequate their
detail may be.  Furthermore, it was a
decline whose total magnitude was
probably greater than any that had
preceded it, and also perhaps greater
than any that followed it (though that
of the late Cretaceous was broadly
comparable).  The decline was followed
by the rise of new taxa, whose bulk
diversity (measured in generic and
family totals) was less than that of
its immediate predecessors.  None of
this is in dispute.  But there is no
evidence that Palaeozoic extinction was
qualitatively, rather than quantitatively, different from that of any
other period.  Indeed, most periods
seem to end with just such an
episode of extinction and replacement,

though its cause is far from clear and
it is probably, at least in part,
exaggerated by both our stratigraphical
practice and by non-preservation.  The
pattern, as opposed to the extent, of
Permo-Triassic extinction is not
different from any of the vast number
of other episodes of extinction that
make up the fossil record, and probably
demands no additional explanation.
Even in the case of recently extinct
species there is little agreement as to
the particular cause of extinction, and
it is scarcely surprising that we
remain ignorant of the causes of this
'routine, every-day' extinction of
fossil species.  Our difficulty in
interpreting Permo-Triassic faunal
changes appears to stem directly from
this basic ignorance, rather than from
the operation of any unique physical
or biological influences in Permo-
Triassic times.

ACKNOWLEDGEMENTS

I am deeply grateful to Dr. Martin
Rudwick, of the University of
Cambridge, who read this paper for me
at the symposium for which it was
written; to Dr. Raymond C. Moore, of
the University of Kansas, for this
advice on Permo-Triassic crinoid
faunas; and to Dr. Bernhard Kummel, of
Harvard University, for his advice on
the Permo-Triassic succession of West

Pakistan. The paper was written during the early part of the tenure of a National Science Foundation Senior Visiting Scientist Research Fellowship of Ohio State University, which I acknowledge with gratitude.

## 10. REFERENCES

Beurlen, K. 1965. Der Faunenschnitt an der Perm-Trias Grenze. Z. dt. geol. Ges. 108, 88-9.

Bramlette, M. N. 1965A. Mass extinctions of Mesozoic biota. Science, N.Y. 150, 1240.

---- 1965B. Massive extinctions in biota at the end of Mesozoic time. Science, N.Y. 148, 1696-9.

Clarke, G. L. 1939. The utilization of solar energy by aquatic organisms. In Problems in Lake Biology. Publs. Am. Ass. Advmt. Sci. 10, 27-38.

George, T. N. 1958. Rates of change in evolution. Sci. Prog., Lond. 46, 409-428.

Hallam, A. 1961. Cyclothems, transgressions and faunal change in the Lias of north-west Europe. Trans. Edinb. geol. Soc. 18, 124-174.

Hawkins, H. L. 1950. Earth movements and organic evolution: Introduction. Int. Geol. Congr. 18(12), 5-6.

Henbest, L. G. 1952. Significance of evolutionary explosions in geologic time. J. Paleont. 26, 298-318.

Ivanova, E. A. 1955. Concerning the problem of the relations of evolutionary stages of the organic world with evolutionary stages of the earth's crust. Dokl. Akad. Nauk SSSR 105, 154-157.

Kummel, B. 1961. History of the Earth. San Francisco (Freeman).

Moore, R. C. 1954. Evolution of late Paleozoic invertebrates in response to major oscillations of shallow seas. Bull. Mus. comp. Zool. Harv. 112, 259-286.

---- 1955A. Invertebrates and geologic time scale. In Poldervaart, A. (Ed.): Crust of the earth. Spec. Pap. geol. Soc. Am. no. 62, 547-74.

---- 1955B. Expansion and contraction of shallow seas as a causal factor in evolution. Evolution, Lancaster, Pa. 9, 482-3.

Nairn, A. E. M. (Ed.) 1964. Problems in Palaeoclimatology. New York Interscience, Wiley).

Newell, N. D. 1952. Periodicity in invertebrate evolution. J. Paleont. 26, 371-85.

---- 1956. Catastrophism and the fossil record. Evolution, Lancaster, Pa. 10, 97-101.

---- 1962. Paleontological gaps and geochronology. J. Paleont. 36, 592-610.

---- 1965. Mass extinctions at the end of the Cretaceous period. Science, N.Y. 149, 922-4.

Nicol, D. 1961. Biotic associations and extinction. Syst. Zool. 10, 35-41.

Pavlov, A. P. 1924. About some still little studied factors of extinction. In Pavlov, M. B. Causes of animal extinction in past geologic epochs. Moscow (State Publishing House), pp. 89-130.

Rutten, M. G. 1955. Evolution and oscillations of shallow shelf seas. Evolution, Lancaster, Pa. 9, 481-2.

Ruzhentsev, B. E. & Sarytcheva, T. T. 1965. Development and change of marine organisms at the Paleozoic and Mesozoic boundary. Trudy paleont. Inst. 108, 1-431.

Schindewolf, O. H. 1954. Uber die moglichen Ursachen der grossen erdgeschichtlichen Faunenschnitte. Neues Jb. Geol. Palaont Abh. 10, 457-65.

---- 1962. Neokatastrophismus? Z. dt. geol. Ges. 114, 430-45.

Simpson, G. G. 1952. Periodicity in vertebrate evolution. J. Paleont. 26, 358-70.

Spieker, E. M. 1956. Mountain-building chronology and nature of geologic time scale. Bull. Am. Ass. Petrol. Geol. 40, 1769-815.

Stechow, E. 1954. Zur Frage nach Ursache des grosen Sterkens am Ende der Kreidezeit. Neues Jb. Geol. Palaont. Mh. 183-6.

Stokes, W. L. 1960. Essentials of Earth History. Englewood Cliffs, N.J. (Prentice-Hall).

Strakhov, H. M. 1948. Principles of Historical Geology, pt. II. Moscow (State Geological Publishing House).

Westoll, T. S. 1954. Mountain revolutions and organic evolution. In Huxley, J., Hardy, A. C. & Ford, E. (Eds.): Evolution as a process. London (Allen & Unwin).

Williams. A. 1957. Evolutionary rates in brachiopods. Geol. Mag. 94, 201-11.

Wilser, J. L. 1931. Lichtreaktionen in der Fossilen Tierwelt. Berlin (Verlag von Gebruder Borntraeger).

*

# Extinction of perched faunas

## J.G. Johnson (1974)

INTRODUCTION

Since the time of Cuvier, the
causes of faunal extinctions have been
investigated by many scientists.
Biological and physical (environmental)
factors influence the extinction of as
small a biologic unit as a single
species. When many species of
diverse systematic position become
extinct at about the same time, it is
reasonable to seek an explanation in
the evolution of the environment and
thus to ask why instead of how. An
event of environmental change may or
may not leave evidence of its
existence and timing in the sedimentary
record. When it does and the timing
of particular types of physical and
biological events can be demonstrated
to recur concomitantly, a direct causal
relationship may be supposed.

Large-scale regressions of
epicontinental seas are obvious

possible causes of the extinction of
benthic marine animals, simply
because of the loss of those habitats.
However, examination of the Phanerozoic
record of North America does not
support a one-for-one relationship
between major times of regression and
times of widespread extinction. In
North America and on the Russian
Platform, the Paleozoic-Mesozoic record
is partitioned by regressive episodes
that define cratonic sequences
(Sloss, 1963, 1972), but recognized
times of major extinction do not
coincide with regressions that end
sequences. Instead, the extinctions
coincide with midsequence times.

EXTINCTIONS

For benthic marine faunas,
extinctions came abruptly for large
numbers of the inhabitants of epicontinental seas during at least three
times: (1) the end of Ordovician
time, (2) the end of Frasnian (end of
early Late Devonian) time, and (3) the

end of Permian time. Wholesale extinctions at the end of Permian time are well known on a world-wide scale; the earlier two are less well known but are convincingly evident to specialists who deal with Ordovician and Devonian benthic marine fossils.

These times, viewed in terms of sequences (Fig. 1), occurred near or at sequence maxima. These were times when relatively "sudden" regressions, probably due to eustatic control, altered paleogeography and thus paleoenvironments on a large scale. The regressions occurred when benthic marine faunas had migrated cratonward and had adapted to the environment of epicontinental seas where they were "perched," and therefore vulnerable

to elimination of the habitats.

Regression as a cause for extinctions has long had its adherents, and some of their arguments have been persuasive (see, for example, Newell, 1967). To my knowledge it has never been specifically stated that regression must come when epi-continental seas are large to be an effective cause of extinction, but that requirement may be self-evident.

Why then are there not major extinctions at each regression that is a sequence minimum? Each sequence does end with a craton-wide regression. Empirical evidence indicates that the length of time involved in regression is critical. Sequence-terminating regressions are much slower, judging

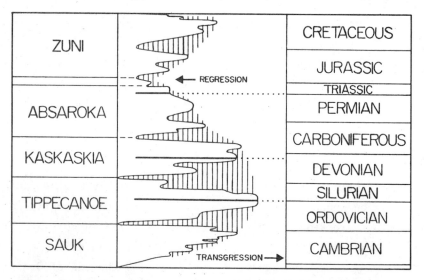

Fig. 1. Transgression-regression curve for western North America, showing major cratonic cycles (sequences) that are inferred to be eustatic. Heavy horizontal lines within Tippecanoe, Kaskaskia, and Absaroka Sequences are times of important faunal extinction, inferred to be times of relatively sudden regression followed by transgression, involving time interval too short to record on vertical scale of this figure. Vertical lines indicate demonstrable hiatus of larger magnitude. No hiatus that widens toward a continental margin is included.

from the time value of the marine hiatus produced, than are the regressions to which the three times of extinction mentioned seem to correspond.

## END-OF-ORDOVICIAN EXTINCTIONS

A sizable amount of data suggests that an African glacial epoch occurred at the end of Ordovician time; it would have caused a glacio-eustatic retreat of marine waters from cratonic platforms, resulting in extinction of benthic faunas of the Late Ordovician epicontinental seas (Boucot and Johnson, 1973; Sheehan, 1973; Berry and Boucot, 1973).

The Richmondian brachiopod fauna of the epicontinental seas of North America had evolved into a zoogeographic province unto itself. Richmondian seas were widespread--the height, or maximum, of the Tippecanoe Sequence--and this situation was punctuated by a glacio-eustatic regression at the end of Ordovician time. When Llandovery seas returned, they were repopulated from the relatively cosmopolitan faunas of the marginal seas and geosynclines. The resulting Silurian fauna, of geosyncline and platform alike, was largely cosmopolitan, at least until Ludlow time, when land-sea relations were probably responsible for some

significant constraints on faunal migration to and from eastern North America, heralding Devonian provinciality of that region.

The end-of-Ordovician faunal extinctions seem a good example of extinction of a perched fauna as defined above.

## END-OF-FRASNIAN EXTINCTIONS

A major extinction at the end of early Late Devonian time seems to have long been known, but little recorded. Modern references come from western Canada (Crickmay, 1957; McLaren, 1959, p. 747, 748), and additional evidence of the magnitude of this extinction has been presented by Johnson and Boucot (1973, p. 94, 95). Only 10 of 71 Frasnian "taxonomic units" (a single genus, or several genera of related morphology) of brachiopods survived this extinction, which caused the demise of the Atrypoidea, Pentameroidea, and the orthid and stropheodontid brachiopod groups.

These extinctions also occurred near a sequence maximum, that is, after Taghanic onlap had spread Devonian seas far onto the craton (Johnson, 1970, 1971a). Regarding western Canada, Bassett and Stout (1968, p. 748) pointed out that the upper Frasnian Woodbend shale and limestone represent "the maximum transgressive phase in the Upper Devonian."

But does the end of Frasnian time coincide with abrupt and widespread regression?  Again, for western Canada, Bassett and Stout (1968, p. 748, 749, and Fig. 4) documented a through-going unconformity between the Frasnian and the Famennian.  However, in the western United States, a Frasnian-Famennian unconformity is not obvious.  This is why I did not earlier seriously consider that level when analyzing mid-sequence events and their relation to orogeny (Johnson, 1971b, p. 3276).  A lithologic break occurs at the top of the Frasnian across the northern Rocky Mountains and plains of the western United States (Sandberg and Mapel, 1968, Fig. 2); it consists of evaporite on evaporite (Logan Gulch on Birdbear). Perhaps that is why the record is so difficult to read on the western craton.

An unconformity may be at the top of the Frasnian in the Mississippi Valley and Central Lowlands--the lower Famennian is poorly represented there, but some Famennian is widespread (Saverton, Maple Mill, upper Chattanooga, upper but not uppermost New Albany).  The Upper Devonian of the Appalachian Basin does not provide evidence of a mid-Late Devonian hiatus, but this at least is understandable in terms of "orogenic override":  the Acadian highlands were generating so

great a volume of clastic sediment that downwarp was probably assured in spite of any eustatic lowering of sea level.

End-of-Frasnian extinctions were effective on perched faunas of the cratonic interior, and a short-lived regressive-transgressive cycle caused a lowering of sea level that either exposed vast areas to erosion briefly (as in western Canada) or created a shallow-water hypersaline environment deleterious to most marine animals (western craton of the United States).

END-OF-PERMIAN EXTINCTIONS

That many higher taxonomic groups of marine animals suffered extinction at the end of Permian time may be taken as established.  That a hiatus of small duration (producing a paraconformity) exists at all but a few of even the most complete Permian-Triassic sections can be taken as an assumption of high probability, if the conclusions of knowledgable specialists are accepted (for example, Tozer, 1972; Kummel and Teichert, 1970, p. 77, and articles published with these workers or noted in their lists of references).  Tozer (1972, p. 648) stated the conclusion clearly:  "The boundary may mark a relatively sudden eustatic change in sea level."

It is also true that the end of the Permian is no sequence boundary.

There is no great regression followed by a stage-by-stage transgression such as marks the initiation of a new sequence; the Lower Triassic is tied, according to its depositional-tectonic situation, to the Permian that lies beneath it. In fact, the hiatus between Permian and Triassic time occurred near a sequence maximum, as in other examples cited above. The Permian benthic marine faunas that were to become extinct were "perched faunas" in the same sense as those of the Late Ordovician period.

The phrase "<u>near</u> a sequence maximum" to describe the end of the Permian was chosen advisedly. The Absaroka Sequence began about at the start of Pennsylvanian time and continued through early Triassic time without especially well-marked peaks of cratonic inundation. If there was an Absaroka maximum, it probably was during the Pennsylvanian; thus, the postulated end-of-Permian eustatic event occurred as a sequence cycle was waning. Early Triassic seas of North America had barely re-established their positions when the end-of-Absaroka regression became dominant. Possibly, the marine faunas never had the opportunity to recover. This would explain the long Triassic pause in expansion of animal diversity noted by Newell (1967, p. 78), following end-of-Permian extinctions, which contrasts

with a situation of increasing diversity eventually following major extinctions as new niches again become available.

BIOMERES

Having considered three major events of extinction in the Phanerozoic record of brachiopods and having noted the occurrence of these events near sequence maxima, it is appropriate to examine some lesser, yet significant, extinctions of a similar nature. These are extinctions of epicontinental sea faunas followed by repopulation of the same environmental sites by stocks that migrated cratonward from deeper water, marginal seas, or the geosynclines.

In the Upper Cambrian of North America, the boundaries of trilobite biomeres (Palmer, 1965a) are good examples. Palmer defined a biomere as a "regional biostratigraphic unit bounded by abrupt non-evolutionary changes in the dominant elements of a single phylum." Palmer characterized biomere boundaries as times of "annihilation," and he pointed out that the initial fauna of a biomere was more like the initial fauna of the preceding biomere than the youngest fauna of the preceding biomere that was temporally juxtaposed. This was accounted for logically by assuming that annihilated cratonic faunas were replenished from a slow-evolving basic stock that occupied "oceanic" regions.

Palmer (1965a) believed that the youngest fauna of a biomere, toward craton center, was contemporaneous with the initial fauna of the succeeding biomere, away from craton center. This was not demonstrated and is unlikely, except that a fauna very similar to an initial biomere fauna must have existed earlier off the craton--but at a place where the youngest fauna of the previous biomere did not exist. Thus, it should always be assumed that where zones of two biomeres occur in succession, the higher zone is younger than any part of the zone of the subjacent biomere, at any locality.

Palmer (1965a) did not hypothesize the cause of the major extinctions that became his biomere boundaries, although evidence available to him must have suggested consideration of environmental shifts. Elsewhere, Palmer (1965b, p. 6) suggested that lowering of the temperature of marine waters over the shelves could have caused the Late Cambrian extinctions. In that paper Palmer pointed out that the faunal change at the base of his Pterocephaliid biomere "is not directly related to a consistent physical change or interruption in sedimentation" (my italics). Use of "consistent" amounts to a trick with words, because during transgression or regression of an epicontinental sea, the physical changes in sedimentation ensuing from a single event must differ at different localities. Knowing Palmer's conclusion, it is surprising to find, upon examination of data published earlier (Palmer, 1960, Fig. 1), that the faunal change from the Crepicephalid biomere to the Pterocephaliid biomere coincides with a lithologic change at every locality depicted in an array of ten columnar sections in eastern Nevada and western Utah in which both faunas were recognized.

Lochman-Balk (1970) analyzed Upper Cambrian faunal patterns of the craton and noted (p. 3212) that biomere boundaries coincide "with the onset of regressive conditions." Her Figure 8, reproduced here with modifications as Figure 2, illustrates this suggestion well, with the exception of the top of the Conaspid biomere. Lochman-Balk went on to combine regression and cool water in a glacio-eustatic model. Whether or not the hard-to-prove, coolwater part of this model is correct, the striking fact is that the great times of extinction of cratonic trilobite faunas took place when Late Cambrian epicontinental seas were at or near their maximum extent--not at regressive minima.

Lochman-Balk did not suggest that the Late Cambrian regressions emptied

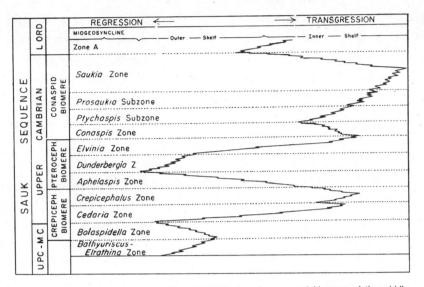

Fig. 2.  Lateral and temporal relations of trilobite faunal zones and biomeres of the middle part of the Sauk Sequence, with transgressive and regressive phases of sedimentation.  Modified from Lochman-Balk (1970, Fig. 8).

epicontinental seas; Great Basin sections studied by Palmer are probably relatively complete, but it has been noted (McKee, 1945; Robison, 1960, p. 43, 44) that only regression produced certain sedimentary rock types.  Although obvious, it should be noted that a universally wide spread regression event may not produce distinct changes in sedimentation at all localities, but still can be devastating to faunas. Thus, to discover a major faunal change without a corresponding lithologic change at some locality is no argument against regression as the cause of faunal change.  It is therefore reasonable to believe that regression either removed or altered nearly all environments to which Late Cambrian trilobite faunas had become accustomed.  Thus,

at least three Late Cambrian extinctions are additional examples of the extinction of perched faunas such as occurred near sequence maxima several other times during and at the end of the Paleozoic Era.

The end-of-biomere extinctions have some inescapable parallels with evolutionary trends demonstrated for Cretaceous molluscs by Kauffman (1972). He showed that the initial event of a regression seems to have a "shock effect" on the biotas, triggering speciation.  In other words, environmental stress results as soon as environments begin to decrease in size. The degree to which any group of organisms will be affected by such an event depends on the degree to which it is stenotopic.  For the Cambrian, it seems reasonable to suggest that as trilobite faunas migrated into epicontinental seas from oceanic regions, they became stenotopic through stabilizing selection. It is also reasonable to assume that trilobites of a biomere attained unusually high stenotopic levels, because they spread into very large and moderately varied environments in which they encountered little competition for living space from other animal groups--a situation unique to Cambrian-Early Ordovician time, when the Sauk Sequence was being formed. Adaptation to life in an epicontinental sea left biomere trilobites "perched" in the sense discussed above.

## CONCLUSIONS

It appears that several factors, which can follow one another in a natural order, must occur before a significant extinction event can take place. First, the sustained enlargement of epicontinental seas provides environments marine animals can migrate into and adapt to so that an equilibrium is established. Those animals are then stenotopic to variable degrees and are perched subject to the continued existence of their environment. Second, extinction occurs proportionate in magnitude to the speed at which regression occurs, to the stenotopic level attained, or to both.

Relatively rapid regression can occur most easily when epicontinental seas are most widespread, because a combination of gentle slope and small eustatic fluctuation can affect large areas. This situation existed during mid-sequence times.

Regression at the end of sequences was much slower, judging from the stratigraphic record, and this must have provided time for organisms that inhabited epicontinental seas to radiate successfully at variable rates (depending on the organism), thus lowering the number of actual extinctions and scattering numerous occurrences of speciation and extinction over a long time interval-- too long to identify as a single extinction event.

## REFERENCES CITED

Bassett, H. B., and Stout, J. G., 1968, Devonian of western Canada, in International Symposium on the Devonian System, Calgary 1967 (Proc.), Vol. 1: Calgary, Alberta Soc. Petroleum Geologists, p. 717-752 (imprint 1967).

Berry, W. B. N., and Boucot, A. J., 1973, Glacio-eustatic control of Late Ordovician-Early Silurian platform sedimentation and faunal changes: Geol. Soc. America Bull., v. 84, p. 275-283.

Boucot, A. J., and Johnson, J. G.,
   1973, Silurian brachiopods, in
   Hallam, A., ed., Atlas of palaeo-
   biogeography: Amsterdam, Elsevier,
   p. 59-65.

Crickmay, C. H., 1957, Elucidation of
   some western Canada Devonian formations: Calgary, published by
   author, 15 p., 1 pl.

Johnson, J. G., 1970, Taghanic onlap and
   the end of North American Devonian
   provinciality: Geol. Soc. America
   Bull., v. 81, p. 2077-2105, 4 pls.

---- 1971a, A quantitative approach to
   faunal province analysis: Am. Jour.
   Sci., v. 270, p. 257-280.

---- 1971b, Timing and coordination of
   orogenic, epeirogenic, and
   eustatic events: Geol. Soc. America
   Bull., v. 82, p. 3263-3298.

Johnson, J. G., and Boucot, A. J.,
   1973, Devonian brachiopods, in
   Hallam, A., ed., Atlas of palaeo-
   biogeography: Amsterdam, Elsevier,
   p. 89-96

Kauffman, E. G., 1972, Evolutionary
   rates and patterns of North American Cretaceous Mollusca: Internat.
   Geol. Cong., 24th, Montreal 1972;
   sec. 7, p. 174-189.

Kummel, Bernhard, and Teichert, Curt,
   1970, Stratigraphy and paleontology
   of the Permian-Triassic boundary
   beds, Salt Range and Trans-Indus
   Ranges, West Pakistan: Kansas Univ.
   Dept. Geology Spec. Pub. 4, p. 1-
   110.

Lochman-Balk, Christina, 1970, Upper
   Cambrian faunal patterns of the
   craton: Geol. Soc. America Bull.,
   v. 81, p. 3197-3224.

McKee, E. D., 1945, Stratigraphy and
   ecology of the Grand Canyon Cambrian: Carnegie Inst. Washington
   Pub. 563, p. 5-168.

McLaren, D. J., 1959, The role of fossils in defining rock units with
   examples from the Devonian of western and Arctic Canada: Am. Jour.
   Sci., v. 257, p. 734-751.

Newell, N. D., 1967, Revolutions in the
   history of life: Geol. Soc. America
   Spec. Paper 89, p. 63-91.

Palmer, A. R., 1960, Some aspects of the

early Upper Cambrian stratigraphy of White Pine County, Nevada and vicinity, in Intermtn. Assoc. Petroleum Geologists, Eastern Nevada Geol. Soc. Guidebook, 11th Ann. Field Conf.: p. 53-58.

---- 1965a, Biomere--A new kind of biostratigraphic unit: Jour. Paleontology, v. 39, p. 149-153.

---- 1965b, Trilobites of the Late Cambrian Pterocephaliid biomere in the Great Basin, United States: U.S. Geol. Survey Prof. Paper 493, 105, p. 20 pls.

Robison, R. A., 1960, Lower and Middle Cambrian stratigraphy of the eastern Great Basin, in Intermtn. Assoc. Petroleum Geologists, Eastern Nevada Geol. Soc. Guidebook, 11th Ann. Field Conf.: p. 43-52.

Sandberg, C. A., and Mapel, W. J., 1968, Devonian of the Northern Rocky Mountains and plains, in International Symposium on the Devonian System, Calgary 1967 (Proc.), Vol. 1: Calgary, Alberta Soc. Petroleum Geologists, p. 843-877 (imprint 1967).

Sheehan, P. M., 1973, The relation of the Late Ordovician glaciation to the Ordovician-Silurian changeover in North American brachiopod faunas: Lethaia, v. 6, no. 2, p. 147-154.

Sloss, L. L., 1963, Sequences in the cratonic interior of North America: Geol. Soc. America Bull., v. 74, p. 93-113.

---- 1972, Synchrony of Phanerozoic sedimentary-tectonic events of the North American craton and the Russian Platform: Internat. Geol. Cong., 24th, Montreal 1972, sec.6, p.24-32.

Tozer, E. T., 1972, The earliest marine Triassic rocks: Their definition, ammonoid fauna, distribution and relationship to underlying formations: Bull. Canadian Petroleum Geology, v. 20, no. 4, p. 643-650.

ACKNOWLEDGMENTS

Reviewed by P. Bretsky and G. Klapper. Supported by the Earth Sciences Section, National Science Foundation Grant GA-41332.

# Man
## Section 6.

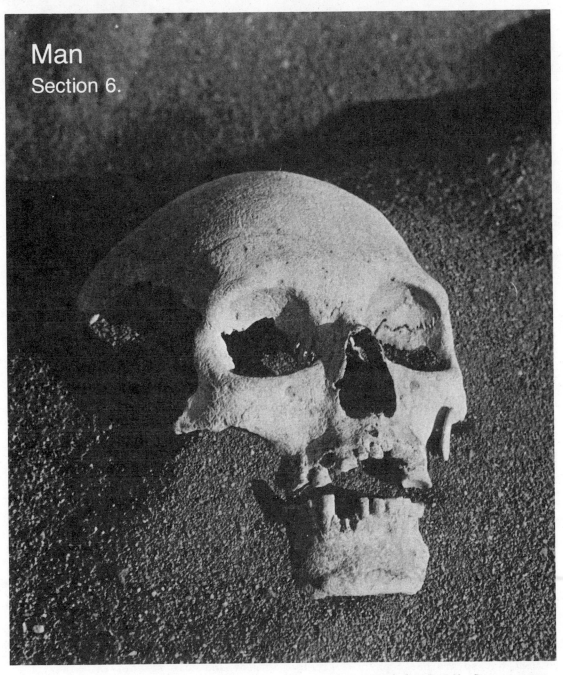

Photograph by Jon W. Branstrator.

We end our readings with a Section on mankind, particularly his future, because that future is intimately linked with the biological and geological resources of the earth. Man has come a long way in his development,

especially in the last few centuries.
He now seems threatened by this very
development for it has produced an
expanding society critically dependent on earth's limited resources.
Mankind must come to terms with these
problems somehow.

We have selected four papers that
indicate how this might come about.
They reflect aspects of two basic ass-
ertions: one, that man is heading rapidly towards ultimate calamity unless
he does something quickly; and two,
that since he probably won't do anything at all, the natural course of
evolution will proceed. Large segments
of the world's population that are ill-
prepared to withstand disaster will be
wiped out, leaving a few who were
better prepared. Civilization as we
know it may survive, but perhaps at a
reduced level.

Our first reading, a report on
studies of the !Kung people of Africa,
gives us an idea of how early human
populations lived in balance with their
environment and of the magnitude of
changes that can affect a society as
its life style changes. Not only are
matters of comfort and resources involved, but so are basic physiological
responses like reproduction.

The reading by STEBBINS reviews
the history of mankind. He predicts
that changes in life style necessitated
by population increase and resource

depletion will not happen, because
human behavior patterns are more or
less fixed: catastrophes are sure to
come. But here Stebbins in one sense
is optimistic -- he doesn't see doom
for all of the human race, but he does
foresee catastrophes for smaller, less
powerful and less affluent segments of
it.

Of particular concern to us is
whether our own society is likely to
undergo one of these large-scale catastrophes. McKELVEY examines our future
from the critical point of view of the
non-renewable resources found in the
earth. He is optimistic that we can
find enough raw materials for the developed societies to continue their present life styles, and probably enough
to help the underdeveloped nations too.
His main point is that we can only
plan properly if we have an accurate
idea of the resources available to us
now and in the future.

These opinions should not make us
complacent, because there is another
point of view based on scientific
reasoning that is just as valid. This
view is expressed by the so-called
"prophets of doom", and they include
eminent and able scientists. They
believe that man is heading directly
and rapidly into a world-wide crisis
from which he may not recover. Books
have been written on this subject and
all are agreed that doom lies ahead:

the only question is when. We reproduce a summary of one aspect of this
argument, co-authored by one of the
best-known "prophets", Paul Ehrlich.
He and HOLDREN paint a gloomy picture
because there are many inconspicuous
and insidious side effects of man's
various activities that might already
have sealed our fate. Yet even these
authors hold out a faint ray of hope
for some of us, provided we take vigorous action now.

Of all the controversies we have
presented, this one is the most far-
reaching. We hope that all human
beings will take cognizance of these
problems and that the controversy is
circumvented and never resolved. For
one of the possible resolutions would
mean that we would not be around to
discuss it.

FURTHER READINGS

EHRLICH, P.R. 1968. The population
    bomb. Ballantine, New York.
EHRLICH, P.R. and EHRLICH, A.H. 1970.
    Population, resources, environment
    -- Issues in human ecology. W.H.
    Freeman, San Francisco.
National Research Council/National Acad-
    emy of Sciences 1969. Resources
    and Man. W.H.Freeman, San Franc-
    isco.
WATT, K.E.F. 1974. The Titanic Effect.
    Sinauer Assoc.Inc., Stamford, Conn.

*

# !Kung hunter-gatherers: feminism, diet and birth control

## G.B. Kolata (1974)

If results from recent studies of the !Kung* people apply to other societies, anthropologists may now have some new clues as to the social, dietary and demographic changes that took place during the Neolithic Revolution when people forsook lives of hunting and gathering and began to farm and to keep herds of domestic animals. The !Kung have lived as hunters and gatherers in the Kalahari Desert of South Africa for at least 1,000 years; but recently they have begun to live in agrarian villages near those of Bantus. Investigators who are documenting this change find that, among other things, the settled !Kung women are losing their egalitarian status, the children are no longer brought up to be nonaggressive, and the size of the !Kung population is rapidly

*The exclamation point refers to an alveolar palatal click. The tongue tip is pressed against the roof of the mouth and drawn sharply away, producing a hollow popping sound.

increasing rather than remaining stable.

The !Kung's very existence is anomalous since they have lived by hunting and gathering since the Pleistocene. In his archeological studies, John Yellen of the Smithsonian Institution in Washington, D.C., find artifacts from Late Stone Age hunter-gatherers, of about 11,000 years ago, at the same water holes where modern !Kung set up camp. According to Yellen, these prehistoric hunter-gatherers even hunted the same animals as the contemporary !Kung, including the nocturnal springhare which must be hunted by a special technique because it spends its days in a long deep burrow.

As recently as 10 years ago many of the !Kung still lived by hunting and gathering. Now, however, less than 5 percent of the 30,000 !Kung live in this way; the remainder live in agricultural villages. This period of rapid social change coincided with extensive study of these people by numerous investigators throughout the world and from many disciplines.

It is difficult to distinguish between changes due to settling down and changes due to acculturation to Bantu society. Investigators have drawn on extensive long-term studies of the nomadic !Kung in their documentation of the effects of the !Kung's adoption of an agrarian life, but cannot conclusively state the causes of these effects.

One aspect of the settled !Kung society that has aroused considerable interest among social scientists is the role of women. Patricia Draper of the University of New Mexico reports that !Kung women who belong to the nomadic bands enjoy higher status, more autonomy, and greater ability to directly influence group decisions than do sedentary !Kung women. This loss of equality for the agrarian women, Draper believes, may be explained in terms of the social structure of nomadic, as compared to sedentary, groups.

Draper postulates that one reason for the higher status of !Kung hunter-gatherer women is that the women contribute, by gathering, at least 50 percent of the food consumed by a band. Since food gathered by women is so important to the group, the women, of necessity, are as mobile as the men (who hunt), and women and men leave the camp equally often to obtain food. Both the women and men who do not seek food on a given day remain in the camp and share in taking care of the children.

The women in sedentary !Kung societies have far less mobility than the men and contribute less to the food supply. The men leave the village to clear fields and raise crops and to care for the cattle of their Bantu neighbors. The women remain in the village where they prepare food and take care of the shelters. Since the men work for the Bantus, they learn the Bantu language. Thus when the Bantus deal with the !Kung, they deal exclusively with the men. This practice, together with the !Kung's emulation of the male dominated Bantu society, contributes to increasingly subservient roles for !Kung women.

Also contributing to a loss of female egalitarianism is the different way that agrarian, as compared to nomadic, !Kung bring up their children. Draper points out that the nomads live in bands consisting of very few people so that a child generally has no companions of the same age. Thus play groups contain children of both sexes and a wide variety of ages. This discourages the development of distinct games and roles for boys and girls.

Unlike the nomadic children, the sedentary children play in groups consisting of children of the same sex and similar ages.  The boys are expected to help herd cattle, so they leave the village where they are away from adults and on their own.  The girls, according to Draper, have no comparable experience but remain in the village and help the adult women with chores.

In addition to promoting sexual egalitarianism by their child rearing practices, the nomadic !Kung also discourage aggression among their children.  This is no longer the case when the !Kung become sedentary.  The nomadic children observed by Draper do not play competitive games.  She attributes this to the wide range of ages of children in a group which would make competitiveness difficult.  Moreover, since these children are constantly watched by adults, the adults can and do quickly stop aggressive behavior among children.  The children rarely observe aggressive behavior among adults because the nomadic !Kung have no way to deal with physical aggression and consciously avoid it.  For example, according to Richard Lee of the University of Toronto, when conflict within a group of adults begins, families leave for other bands.  Lee observed that the sedentary !Kung, who cannot easily pick up and leave, rely on their Bantu neighbors to mediate disputes.

In addition to studying social changes taking place when the !Kung settle down, investigators are studying dietary and demographic changes.  The !Kung diet is of interest because the nomadic !Kung are exceedingly healthy and are free from many diseases thought to be associated with the diets of people in more complex societies.  The sedentary !Kung have substantially altered their diets, thus providing investigators with a unique opportunity to document the effects of diet on the health of these people.  The demographic changes taking place among the !Kung are of interest because the settled !Kung seem to have lost a natural check on their fertility rates.

The diet of the completely nomadic !Kung, which has been analyzed by geneticists, biochemists, and nutritionists, consists of nuts, vegetables, and meat and lacks milk and grains.  All the investigators agree that the diet is nutritionally well balanced and provides an adequate number of calories.  They found very few people with iron deficiency anemia, even when they included pregnant and lactating women in their sample.  They also discovered that the nomadic !Kung have a very low incidence of deficiency of the vitamin folic acid and that the concentrations of vitamin $B_{12}$ are higher in their serums as compared to concentrations considered normal for other populations.  These findings led Henry Harpending of the University of

New Mexico and his associates to suggest that Stone Age people probably had no deficiencies of these vitamins and that deficiencies first appeared when people settled down into agrarian societies.

In addition to being well nourished, the nomadic !Kung are free from many common diseases of old age. For example, Lee and others have found little degenerative disease among elderly !Kung, although it is common-place for these people to live for at least 60 years and some live for as long as 80 years. A. Stewart Truswell of the University of London also finds that the nomadic !Kung are one of only about a dozen groups of people in the world whose blood pressure does not increase as they grow older.

The medical effects of the altered diet and way of life of the sedentary !Kung are not yet well established. In contrast to the hunter-gatherers, these people consume a great deal of cow's milk and grain. In his studies of a generation of !Kung brought up on such a diet, Lee finds that they are, on the average, taller, fatter, and heavier than the nomadic !Kung. Nancy Howell of the University of Toronto finds that the agrarian women have their first menstrual periods (menarches) earlier than the nomadic women.

The average age of menarche among nomadic !Kung is late--at least age 15.5 according to Howell. Although these women marry at puberty, they have their first children when they are, on the average, 19.5 years of age. This late start to reproductive life helps limit the growth of the population. However, a more significant curb on the size of nomadic populations is the low fertility of the women. Howell finds that the average length of time between giving birth for a nomadic !Kung woman is 4 years. These women have fewer children than any other women in societies that do not practice contra-ception or abortion. The low fertility of nomadic !Kung contradicts previously held theories that the sizes of hunter-gatherer populations were limited solely by high mortality rates. The !Kung population size remains stable because there are so few children born. Combining her studies of the fertility and mortality rates of !Kung hunter-gatherers, Howell concludes that the long-term growth rate for such a population is only 0.5 percent per year. This is in sharp contrast to the sedentary !Kung whose population is growing rapidly.

The population growth among the sedentary !Kung results from both a decrease in the age of menarche and a decrease in the average time between births. Lee has found that the birth intervals drop 30 percent when !Kung women become sedentary. The causes of these reproductive changes are unknown, but some investigators suspect that these decreased birth intervals may result from changes in nursing or dietary habits.

Nomadic !Kung women have no soft food to give their babies, and so they nurse them for 3 or 4 years, and during this time the women rarely conceive. Sedentary !Kung women, on the other hand, wean their babies much sooner by giving them grain meal and cow's milk. Irven DeVore of Harvard University believes that a contraceptive effect of the long lactation period is not unexpected, since investigators have observed the same phenomenon in many animals, including monkeys and the great apes. A woman who begins to supplement her infant's diet while the child is very young would not experience this effect because her child would require less and less milk.

Howell and Rose Frisch of the Harvard Center for Population Studies believe that an explanation of the decrease in the age of menarche and in the birth intervals of sedentary !Kung women may involve the diet of the sedentary !Kung. They base this idea on a study by Frisch and Janet McArthur of the Massachusetts General Hospital in Boston. These investigators showed that the amount of body fat must be above a certain minimum for the onset of menstruation and for its maintenance after menarche. Howell points out that

the !Kung hunter-gatherers are thin, although well nourished. When women from these bands lactate, they need about 1000 extra calories a day. Thus, during the 3 or 4 years that a woman nurses a baby, she may have too little body fat for ovulation to take place. The shorter birth intervals for sedentary !Kung women would follow from their shorter periods of lactation and larger amounts of body fat. Howell notes that this explanation of the low fertility of nomadic !Kung women cannot be verified until more extensive medical studies are performed with these people.

Although no one claims that the changes taking place in the !Kung society necessarily reflect those that took place when other hunter-gatherer societies became agrarian, studies of the !Kung are providing anthropologists with clues relative to the origins of some features of modern societies. Many findings, such as the social egalitarianism, lack of aggression, and low fertility of nomadic !Kung are leading to new perspectives on the hunting and gathering way of life which was, until 10,000 years ago, the way all humans lived.

ADDITIONAL READING

1. R. B. Lee and I. DeVore, Eds.,
   Kalahari Hunter-Gatherers
   (Harvard Univ. Press, Cambridge,
   Mass., in press).

# The natural history and evolutionary future of mankind

## G.L. Stebbins (1970)

A naturalist who is going to some distant country to study a little-known species of animal or plant begins by seeking answers to the following four basic questions:

1. <u>Where</u> does the species occur? How does it fit into an ecosystem?

2. <u>Why</u> does it have a particular distribution pattern? Why does it exist in some places and not in others?

3. <u>When</u> did this species first occupy its present niche?

4. <u>How</u> did it evolve the ability to occupy this niche?

I should like to begin by answering these four questions as well as possible, about <u>Homo sapiens</u>, our own species. The answer to the first question is, of course, "nearly everywhere." Temporarily, man has occupied all parts of the earth, from the depths of its oceans to its highest summits, as well as outer space and the surface of the moon. His permanent habitations encompass nearly as wide a range as those of all land animals put together. In all of these habitats, he is actually or potentially in complete control of the biotic community in which he exists. If we should devote to the elimination of any other species of animal the same amount of time, money, and energy that we devote to preparations for destroying each other in the name of defense and security, that species could be eliminated in a relatively short time. Man's ecological niche is, therefore, complete dominance over the biotic community. Furthermore, as Wallace (1914) and others pointed out many years ago, man is the only being that is aware of his position or niche in the biotic community. From the ecological point of view, modern naturalists can still reaffirm the statement by Wallace that "man is as much above, and as different from, the beasts that perish as they are above and beyond the inanimate masses of meteoric matter which, as we now know, occupy the apparently vacant spaces of our solar system."

The basic answer to the question-- "Why does man occupy this worldwide and universally dominant niche?"--also given by Wallace, is that by the use of his greatly superior mind, man has continually modified the environment to meet his needs, so that "he would cease to be influenced by natural selection in his physical form and structure." As Dobzhansky (1962, 1967) has pointed out, this statement is an exaggeration. Nevertheless, the general conclusion of Wallace, that in early man the action of natural selection was largely transferred from the bodily structure to the mind, is still valid.

The implication of this answer is that the structural and biochemical differences between man and his nearest relatives among the animals must be regarded in an entirely different context from differences with respect to his ecological niche and his way of life. Recent comparisons between the hemoglobins and other proteins of man and the great apes have emphasized more and more our biochemical similarity to these animals. At least one authority, Allan Wilson (Wilson and Sarich 1969) maintains that from the biochemical point of view, man and chimpanzees are as much alike as are donkeys and horses. As evidence of our biological and biochemical similarity to other animals is accumulating, the ecological gap between man and his biological relatives is widening. The last vestiges of human societies which depend for their existence upon their relationships to wild animals and plants persist only in a few remote corners of the earth, such as the desert wastes of Australia and the jungle fastnesses of the Amazon and Congo basins. Even there, human societies that form part of more or less natural biotic communities persist largely because more "advanced" forms of society are sympathetic to their continued existence. Man has, therefore, broken contact almost entirely with the ecological universe that existed before his culture developed. He no longer occupies ecological niches; he makes them. If modern men wish to become associated again with the world of nature, they must do so as visitors from outside.

The answer to the question, "When did this happen?" must have two parts. Man's widespread geographical and ecological distribution came relatively early in his evolutionary history. Africa and Eurasia appear to have been entirely occupied by the species of hominid from which man descended, H.

erectus, at least half a million years ago. The occupations of Australia and America were much later, but were probably completed by 15,000 years ago (Haynes 1969). At this time, however, human life was still completely interwoven with the natural biotic communities of which his societies formed a part. Mankind still carved out his ecological niche, as do other animals, by accommodating himself to his natural surroundings. He still preyed upon wild animals, gathered wild plants, and had to defend himself against natural predators.

The true "human revolution" began with the dawn of agriculture several thousands of years later, and can be regarded as complete only in our own time, with the appearance of the industrial and atomic revolutions; the mechanization of agriculture; the conscious improvement by controlled breeding of domestic animals and cultivated plants; the harnessing of power to replace human and animal labor; and the virtual elimination of the more common diseases and plagues.

On a comparative time scale, beginning with the first tool using hominids, the time required for man to reach his maximum geographic and ecological distribution represents approximately the first 98% of the period of his existence. The total time span of the human revolution is only the final 1%

Cowen & Lipps—Earth Sciences—25

of this period, and it has affected the majority of human populations only during the last 0.2%-0.3% of the total period. In relation to the evolutionary time scale, within which even a million years is a relatively short interval, the human revolution has taken place almost instantaneously.

Obviously, therefore, the question, "How did this come about?" must have two answers. We must first answer the question, "What factors are responsible for man's widespread geographical and ecological distribution?" We must give a different answer to the question, "What factors induced man to transform his environment into a special niche for himself?"

The key qualities that are responsible for both man's wide distribution and his radical transformation of the earth's environment are his intelligence and foresight, his ability to use tools, and his social organization. These three qualities evolved together as an adaptive syndrome. Modern studies of a wide variety of other species of primates, as reviewed by Crook (1968, 1970) have shown that cohesive societies exist in most of them, and that in all species that are adapted to savannas or other types of open country, rather than dense forests, some kind of social organization is the rule. Their patterns of organization can be highly complex, and can vary adaptively in relation to the environment. Furthermore, although the tendency to form societies may be innate and genetically conditioned, their structure and the role played in them by each individual animal are largely transmitted from parents to offspring by teaching and learning. Washburn, Jay, and Lancaster (1965) note that "the same species develops quite different learned habits in different environments."

These facts suggest that man's ancestors, even when they were no more intelligent than other primates and did not use tools at all, already had a well-developed social organization, based upon learning as well as instinct. Moreover, recent observations of chimpanzees in nature have shown that they often use sticks for opening termite nests and for other purposes (Goodall 1964). Hence the use of simple tools could well have been characteristic of man's ancestors at a very early stage, even while they were in the process of leaving the forests and colonizing open territory. We must conclude, therefore, that man did not start to use tools and form organized societies because he had first acquired a superior intelligence. On the contrary, the adaptive trigger that first placed a high selective value upon genes for increased

intelligence was probably the fact that his apelike ancestors had already become adapted to life in open savannas by virtue of their social structure and their ability to use unfashioned tools. In the gelada baboon of Ethiopia, learned patterns of behavior play an important role not only in the reproductive process, but also in defense against predators. Their basic social structure is a group of "harems," each one of which consists of a dominant male and several females. The "bachelor males," who have not been able to acquire females, form separate bands which remain near the periphery of the colony, and often defend it against potential predators, such as hyenas and village dogs. Under these circumstances, predators which could easily overcome a single baboon are defeated by collective action.

One can easily imagine how a comparable social structure would have been of benefit to primates that were migrating from forests into open country. The harem social structure would mean that in times of scarcity of food, males would be expendable without reducing the reproductive capacity of the population. Furthermore, those family groups in which young males were taught to band together for defending the colony before they could acquire females would be more successful in defending and perpetuating themselves than would those families which had not acquired this trait. Since throwing stones for defending is an almost automatic reflex action on the part of many modern primates (Hall 1968), the use of such tools for defense could have begun very early. If, now, these defenders should occasionally kill one of their attackers with a well-aimed missile, the use of these same tools for hunting and attacking prey would come naturally. Once this way of life had been instituted, those family bands in which, by mutation and genetic recombination, a genetic capacity for greater intelligence and learning ability had been acquired, would be able to make better tools, would be better coordinated with respect to both defense against predators and hunting for food, and so would give rise to more efficient and reproductively prolific families among their off-spring.

Based upon this reasoning, I believe that the syndrome of character-istics--closely knit social structure, tool using, and intelligence--was already evolving as a synchronous unit before man's ancestors acquired their wide distribution. The dispersal of mankind to many different habitats, therefore, came about not as an inevitable consequence of the establishment of this syndrome, but by virtue of certain special properties

that resulted from it. When they migr-ated into cold climates, man's ancest-ors must have protected their naked bodies by covering them with the skins of animals that were as large as or larger than themselves. Killing enough large animals for this purpose would certainly be done much more easily by bands of hunters than by single indiv-iduals. Moreover, building shelters in the form of huts, or acquiring shelter by ridding caves of the bears and other large carnivores that inhabited them, would also have been much more easily done by coordinated bands than by single individuals or pairs. Consequ-ently, we cannot ascribe man's wide distribution to his intelligence alone. Our social structure, as well as the ability to teach and learn, are not recent acquisitions of H. sapiens in his present advanced state. They already existed when our ancestors were nothing more than another species of primate. Furthermore, they were probably the prior conditions that made possible man's increase in intell-igence and his acquisition of dominance over the world of life.

The answer to the second part of the fourth question--"How did man transform his environment in order to make his own environmental niche"--must be based on the now well-recognized fact that long before this transformation began, while all of mankind was still in the first Stone Age, the average intelligence of the human species may already have been as high as it is today. This fact is evident both from the exquisite crafts-manship required for making the more advanced Paleolithic tools, and from the excellence of the well-known paintings found in the caves of western Europe. The presence of graves containing evidences that an elaborate ceremony of burial had been performed indicates that these men who existed 15,000 or more years before the dawn of agriculture possessed a highly organized and elaborate social structure, and were capable of abstract thinking and symbolism. Consequently, the beginning of the "human revolution" in global ecology was not the inevitable result of man's advance to a high state of social organization, tool using, and intelligence, but of certain special circumstances that appeared long after this state had been acquir-ed. Anthropologists all agree that the most important of these circumstances was the dawn of agriculture and live-stock raising.

However, man's transformation of his habitat by these practices quickly became a boomerang. Valleys that are cultivated no longer support the wild plants that were originally there. Hills that are systematically grazed by livestock can no longer support an abundance of wild game. Consequently, soon after the former hunters and gatherers had learned to raise domestic animals and cultivated plants, they began to depend upon such products for their existence. In this way, they severed their previously essential connection with the natural environ-ment, and created a new biotic environ-ment based upon human artifacts and upon man-directed transformations of living beings.

I would like to suggest that the destructive effects of man's earliest exploration of nature produced disasters even in prehistoric times. Legends of great floods are prevalent in the folklore of many tribes and nations; the Hebrew version, with Noah as its hero, is the one familiar to us. Archeologists who have studied the ruins of the ancient cities of Mesopotamia (Wooley 1954) have produced evidence that, in fact, a great flood, or perhaps several floods, did sweep down this valley about 3,500 years before Christ, and destroyed all the cities that had been built there. Isn't it possible that these floods were the result not only of unusually rainy periods, but also of the previous denudation of the vegetation on the surrounding hills by cultivation and overgrazing? I believe that Noah's flood was, as the biblical author says, a retribution for mankind's sins.

These were not, however, the immoralities of city people. They were, rather, the sins of overexploitation that farmers and herdsmen had been committing for centuries against their life-giving environment.

Obviously, these first disasters interrupted only briefly man's progress toward increasing control over and transformation of his environment. With the advent of writing, permanent templates for the design of his social organization became possible, and city-states came into being. From then on, the spread of the human revolution over the entire earth was only a matter of time.

We now find ourselves at exactly that instant of time in the evolution of the earth when our own species has transformed virtually the entire global ecosystem into habitats that are designed to favor his continued reproductive capacity and increase in numbers. At the same time, the evils that have accompanied this transformation have by no means diminished. The exploitation of one individual by another, that may have begun in our apelike ancestors when dominant males began to deny their brothers or cousins any access to females, has continued through the universal slavery that existed in the early city-states, the feudalism of the Middle Ages, the vicious exploitation of labor that accompanied the beginning of the Industrial Revolution, the African slave trade of the eighteenth and nineteenth centuries, up to the present exploitation of the people in the "developing nations," either directly or indirectly, by the citizens of nations that consider themselves more advanced.

Moral hypocrisy, exemplified by the formulation of laws, constitutions, and doctrines of rights, which are violated whenever a group of men is powerful or clever enough to do so with impunity, has probably existed also ever since the beginning of writing and the dawn of civilization. Finally, natural biotic communities all over the earth bear scars of wounds inflicted upon them by men for centuries or millennia, and these scars are continually becoming larger and more ugly. What can we expect for the future? We are now potentially able, on the basis of our technology, to keep the human population down to a level that the earth can tolerate, to do this without the tragedies of infant mortality, disease, and war, and to feed populations of a reasonable size as well as previous civilizations fed their exploiting rulers. Will we be able to reorganize our society to use this technology for human betterment, or will the spiral of continuing exploitation and degradation continue

forever, until we leave as our legacy
a lifeless planet?

One can hardly overemphasize the
probability that, as the modern
Jeremiahs have told us with inceasing
emphasis, the critical events that will
decide between these two alternatives
are likely to take place during the
next twenty-five or fifty years.  John
Platt (1969) has appropriately placed
mankind in a rocket that has just taken
off to an unknown destination, at
breakneck speed, which is engendering
previously unimaginable strains and
crises.  Our ability to withstand these
stresses will determine whether we will
progress or degenerate.  What are the
positive and negative aspects of our
present situation?

The first item on the negative
side is our capacity for thoughtless
destructiveness.  With thoroughness and
efficiency, we destroy our environment
by slashing down our forests, scarring
our land with mine strips, quarries,
and suburban bulldozing, and polluting
water and air alike with the effluents
of our industrial "progress."  With
equal enthusiasm and disregard for the
future, we believe that the ultimate
solution of the world's political and
social problems is the killing of other
men--whom we regard as evil--and to use
for this purpose the most powerful
engines of destruction that we can
create.  For the most part, moreover,

none of this destructiveness is overtly
evil or malicious.  The despoilers and
polluters of our natural environment
regard themselves as "developers," who,
at considerable financial risk to
themselves, are making the world a
better place for their fellowmen, and
who, therefore, deserve large profits
for their efforts.  The national
leaders who have ordered mass
destruction of a supposedly dangerous
"enemy," from Napoleon, Bismarck, the
German Kaiser Wilhelm II, the World War
I generals on both sides, to Mussolini,
Hitler, Tojo, Truman, and the Vietnam
generals; all these men regarded and
regard themselves as heroes, who, by
doing the right thing at the right
time, were helping to make the world a
better place for decent people to live
in.  The calculating, power-mad leader,
whose only thoughts are for his own
aggrandizement, are a fiction of
mystery stories and television dramas,
or when present in real life can be
handled with relative ease.  The world
is threatened not by small groups of
willful egotists, but by would-be
heroes who have misguided views about
how they can save their country, and
how they can lead the world to a better
future.

Hardly less ominous for our future
is mankind's incredible ability for
self-deception.  The main thread of
United States history during the

twentieth century is a succession of self-deceptions on the part not only of our leaders but also of the majority of the citizens who followed them. Before 1914, the peace and prosperity of our country seemed to be assured forever, as long as we minded our own business and kept out of entangling foreign alliances. Then came the first shock, World War I, which we made palatable to ourselves by converting it, in our imaginations, to the "war to end wars, and "to make the world safe for democracy." Our victory led our president to believe that these goals could be achieved by the simple device of projecting the rationalist-legalist approach of the United States Constitution to encompass and overcome international rivalries and in-equalities that had existed for millennia. Meanwhile, by means of prohibition, our moralists hoped to legislate morals into people. As the folly of both of these attempts became evident, we reverted to the "normalcy" of private gain via inflated values on the stockmarket. Once this bubble was pricked, a new president deceived himself and many others by his belief that he could eliminate greed and profit seeking by a simple plan, NIRA, and by castigating the "malefactors of great wealth." On top of this failure and disillusionment came Hitler, and a war in which Democracy had to fight

hard for its very existence. Even then, we deceived ourselves into thinking that this was the ultimate battle, and listened to radio crooners giving out with "there'll be love and laughter and peace forever after, tomorrow when the world is free."

After the victory came another series of deceptions. The world, in our imagination, was now divided into two opposite camps, one of them con-taining the "bad guys" who belonged to the "totalitarian Communist conspiracy" aimed at destroying everything of value, and the other consisting of the "good guys" who belonged to the "free world" (including, of course, Franco, Chiang Kai-shek, and a dozen or so of minor dictators who had to be tolerated as "bulwarks of defense"). The world's problems were resolved into making sure that the "good guys" would eventually win over the "bad guys." Our present national dilemma appears to be the difficulty of extricating ourselves, without losing face, from the imposs-ible position which resulted from our behaving as if this myth were the ultimate truth.

Self-deception, however, has by no means been the monopoly of twentieth-century Americans. During the same period, the major nations of western Europe paid the price of their belief, during the nineteenth century, that by assuming the "white man's burden," and civilizing "barbarians" and "savages" throughout the world by means of economic and colonial exploitation, they could establish worldwide peace and prosperity. The deception involved in Marxist dreams of social democracy and economic equality has been made so obvious to westerners that it hardly needs mentioning.

This capacity for self-deception is, moreover, nothing new. In the earliest days of civilization, the pharaohs of Egypt believed that they could gain immortality by erecting gigantic tombs, which would attract the attention of people throughout the ages, and in which their own remains would forever be safe from vandalism and depredation. Judging from the folkways and customs of "primitive" people, self-deception about the causes of natural phenomena was an accepted way of life over tens of thousands of years, from the dawn of man's desire to understand the world around him up to the advent of modern science. If phenomena like the phases of the moon, the passage of the seasons, and the basis of thunder and lightening could not be explained, the invention of a god who was responsible for each phenomenon was an easy and acceptable deception that could replace the impossible solution. In fact, if we can see any hope that this sorry

history of self-deception will finally end, it is that during the past two centuries the area of deception has been restricted, so that it no longer encompasses natural phenomena, but only relationships between men.

On the positive side, the first item to note is man's astonishing inventiveness. This quality has been responsible for our dominant position on the earth. Moreover, new inventions and discoveries have repeatedly extricated civilized man from the dilemmas into which his folly had led him. As each of the empires of antiquity fell apart or was conquered because of internal defects and inconsistencies, a new empire, based on the invention of new machinery and ideas, took its place. In modern times, we have repeatedly faced the prospect that vital natural resources would become exhausted in a few years, only to find that at the end of the predicted period new discoveries have postponed the critical date. We cannot by any means be sure that such inventions will always save us from disaster, but they have been successful so far.

The second item on the positive side is man's ability to devise templates by means of which his inventions and the elements of his organization can be perpetuated indefinitely in a very precise form. This ability has particular meaning for the naturalist and biologist. It permits us to make meaningful analogies between organic and cultural evolution. Biologists have demonstrated that living organisms all possess the same genetic code and the same mechanism for translating this code into the working molecules of the cell. Once DNA molecules, as well as their colinear relationships to RNA and proteins, had become self-perpetuating properties of living matter, their value as templates was so great that their basic organizational pattern has persisted without change for billions of years. By analogy, we can assume that writing, photography, and tape recording have now provided us with templates for civilization that are equally permanent. Even though individual societies or cultures may be partly or completely destroyed in the future, as long as there are men on the earth who can read, as well as depositories of books, photographs, and recordings, new cultures will be able to arise quickly from the ruins of older ones.

A most encouraging item on the positive side is a phenomenon which has arisen only during the last decade. This is the new awareness on the part of many people, particularly those of the younger generation, of the potential disasters that lie ahead. Linked to this new awareness is a refusal to be deceived by the old panaceas. The younger generation in America and many other parts of the world is like a man who, as a result of psychoanalysis, has suddenly been able to see himself as others see him. We of the older generation may be impatient with some of the excesses which this new awareness has brought about, but we must accept it as the most promising sign on mankind's troubled horizon. Just as the new man, after successful psychoanalysis, becomes better adjusted to his fellow-men and makes more of his life than he ever has previously, so this no-nonsense, undeceivable new generation may provide us with the key to adjustment between different classes, different races, and different nations having divergent cultural backgrounds and conflicting designs for the future.

To a naturalist, a heartwarming feature of this new awareness is its gravitation toward the previously neglected discipline of ecology. We may bemoan the fact that much of this new devotion is highly unscientific, and in some instances appears to pervert or even degrade the discipline which is espoused. We must, nevertheless, recognize a genuine motive; an earnest, devoted, and persistent attempt to set mankind's oikos, or house, in order. It is our task to show the younger generation that ecology is much more than picking up empty cans from beaches and campsites, or preventing factories from spewing their effluents into rivers and lakes. Through combining the enthusiasm of youth with the experience of age, we may discover scientific methods of organizing and interpreting the enormous mass of interrelated facts which ecologists have already gathered about the web of life and man's position in it. We may also develop more effective ways of impressing upon all men the need for preserving our environment, to obtain more effective action in this direction.

An equally and perhaps even more encouraging positive item is the degree to which the seriousness of the situation is recognized by important segments of the establishment. A highly important volume which the National Academy of Sciences is now preparing on the life sciences in the United States contains a long chapter entitled "Biology and the Future of Man," which I have had the opportunity to read, in advance of its publication, through the courtesy of Academy President Handler. This chapter strongly emphasizes the critical seriousness of the present situation and the need for taking drastic steps immediately in order to improve it.

## THE THINGS THAT CAN AND MUST BE DONE

The leaders of this new awareness are telling us in no uncertain terms about the various things that we need to do now in order to save mankind. We hear the slogans from every direction: "Defuse the population time bomb," "End nationalism before a nuclear holocaust ends us," "Eliminate ghettos and ghetto nations," "Eliminate waste and pollution," "Stop using fossil fuels before they are all gone," "Use more fertilizer and better varieties of plants so as to increase our food supply."

To a citizen who is listening to all of these cries, and who has not adopted one or two of these slogans as all encompassing panaceas, the confusion becomes more confounded as the cries grow louder and more strident. In many instances, the Jeremiahs tell us that proposed solutions other than their own should be suppressed, since they will only interfere with the genuine way out. Thus we are told that it is wrong to supply more food to people who are living on the edge of starvation, because this will only increase the number of mouths to feed. We can then ask in good conscience, "Is it right or even practical to hold the club of starvation over a nation of people in the hope that this will make them practice birth control?"

The naturalist who has followed the scientific study of ecology, and more particularly evolution during the past few decades, sees in these conflicting cries a remarkable resemblance to the babel of separate and often conflicting voices that have clouded his scientific horizon during this period. We have had ecologists who maintained that the key to the understanding of biotic communities is a careful analysis of the various factors of their environment: climate, soil, availability of food, etc. Others have favored making estimates of biomass, or the analysis of food chains, or the temporal studies of plant succession. In the discipline of evolution, we have had mutationists, Lamarckian environmentalists, and selectionists, each of whom has insisted that his own avenue was the only way toward a true understanding of the way in which evolution works.

In the scientific disciplines of ecology and evolution, these conflicting voices have begun to subside. They have been largely replaced by the approach which Dobzhansky (1968) has very appropriately designated compositionist. It consists of two separate steps. First, the individual contributing factors are analyzed. In order to understand evolution, we must have precise knowledge about the nature and rates of mutation, the effect of various factors on the extent of gene interaction and genetic recombination, the dynamics of natural selection, and the ways in which reproductive isolation can interrupt gene flow between populations. After this analysis comes the process of synthesis or composition. The synthesis never appears automatically from the results of analyzing separate factors. It requires a distinct kind of thinking in comparative terms. We must ask and find answers to questions such as these: "Does selection usually act on individual genes or on combinations of them?" "Is reduction in population size more likely to cause change via random alterations in gene frequency or by altering the adaptive properties of certain genes and causing changes in the intensity and direction of natural selection?" "To what extent can natural selection affect directly the degree of reproductive isolation between populations?"

Is it not possible that the two-step compositionist approach is the best way to attack the problems connected with restoring mankind's deteriorating environment and rebuilding his degenerating social order? The essential tasks of the first step, analyses of the various contributing factors, are now well under way. Many of them have been carried far enough so that most intelligent people who know the facts can agree on the kinds of action that must be taken. This is particularly true of those actions to which technology makes a large contribution. The technological means of reducing the birthrate are now easily available, if legal restrictions against them have not been imposed and would-be parents are motivated to use them. Moreover, nearly everyone agrees that their much wider use should be promoted by every possible means. In other words, the problem of population control is no longer a biological and technological problem, but a sociological one. For its solution, mankind needs an extensive removal of age-old taboos that were established in times when increase in numbers was a source of strength as well as an increase in the motivation and desire of couples to have fewer offspring.

Means of reducing pollution are becoming increasingly available and practical. Their increased use depends largely upon persuading society in general, and particularly those groups that are most active in causing polution, to pay the high price that control will cost or to forego the comforts and conveniences of better transportation or more goods and services that are supplied by manufacturing processes responsible for pollution. The spectacular progress made by agricultural experts during the past few years in countries such as Mexico and India, where food is in critically short supply, have shown us that these countries can stave off starvation caused by excess population for much longer than many prophets have predicted. We can conserve our supply of fossil fuels, provided that we are ready to pay the extra cost involved in using other kinds of energy, particularly the limitless energy provided by the sun, as well as the fuels that are made from it by contemporary plants growing in fields and waste places. On the other hand, the problems involved in eliminating rivalries and inequalities between nations and classes appear to be as formidable and insoluble as they have always been. Nevertheless, they may be easier to solve once we have achieved a better balance between the number of people on the earth and the means available for providing them with a decent living.

In order to take the second steps, those of synthesis or composition, we need first to ask specific questions about the relationships between the factors involved. Will increasing the food supply intensify or reduce the desire of people to have more offspring? Will population control increase or reduce the rivalry between nations or the capacity of individual nations to make war? If we divert to peaceful uses the money, manpower, and technological skill that are now spent on defense, can we then devise a practical and perpetual source of energy for keeping our civilization going? Will the problem of inequality between nations and classes be solved more easily if we first increase greatly the total supply of goods and services, or does this increase depend upon first reducing social inequalities, so that a larger number of skilled and well-motivated workers will be available?

Obviously, the first requirement for obtaining answers to these questions is to gather together and compare knowledge that has been derived from several different disciplines. Moreover, armchair thinking and speculation will never be sufficient for this purpose. In order to obtain adequate solutions, scholars must join forces with men of action, who understand the devious and often irrational aspects of human nature. Opportunities must constantly be sought for trying out on a small scale prospective improvements in individual factors, and then analyzing in depth and as

impartially as possible the entire effect of such changes on various aspects of society.

## CAN WE CHANGE HUMAN NATURE?

An essential question that we must ask before we raise our hopes too high is, "Do men have innate, genetically determined tendencies for aggression and acquisitiveness that will forever thwart our attempts at social progress unless we can breed a genetically different race of mankind?" A number of authorities on animal behavior, particularly Konrad Lorenz (1966) and his popularizer Robert Ardrey (1967), have maintained that this is the case. Nikko Tinbergen (1968) has also reached this conclusion, although with less certainty. Their evidence is based largely upon their profound knowledge of behavior in a great variety of animals. In various species of birds and mammals, Lorenz appears to have recognized a fountain of aggression that is constantly building up pressure in our nervous systems, and that must have an outlet. He believes that this innate aggressive pressure is genetically conditioned in most higher animals, including mankind. Antagonisms or frustrations imposed by other animals or people may increase its intensity, but absence of such stimuli will not cause the aggressive tendency to disappear.

How well is this hypothesis supported by modern studies of behavior in human beings and in our closest relatives, the primates? With respect to the latter, a comprehensive answer has recently been given by John Crook (1968, 1970). His reviews of the extensive observations on primate behavior that have been made by many workers during the past ten years emphasize above all the great variability that exists in the amount of aggression, not only between species of primates, but also between races of the same species. Furthermore, careful observations in the field of the ways in which adult monkeys and baboons teach patterns of behavior to their young, fortified by experiments in which the behavior of individual animals has been radically altered by raising them separately from their accustomed societies, have rendered highly plausible his conclusion that behavioral differences between races and species of primates are determined to a large extent by learning rather than heredity. Moreover, in the closest genetic relative of mankind, the chimpanzee, both aggression and defense of territory are weakly, or not at all, developed.

With respect to human behavior, positive evidence for innate, genetically determined tendencies to aggression is very indirect. It is based only upon analogies to animal behavior and on anecdotes. Lorenz tells his readers about the aggressive behavior of his aunt toward her servants in old Vienna; Ardrey emphasizes the war games of American youth and the eagerness with which Germans marched to wars in 1914 and 1939. Each of these anecdotal accounts can, of course be countered by anecdotes of an opposite nature: wealthy widows who have been very kind to their servants; youths who played war games at the age of ten and became pacifist draft resisters at the age of eighteen; and the reversal of behavior in the majority of the youth of Germany and Japan between 1939 and 1959. This does not mean, of course that genetic tendencies for aggression are totally absent from the human species. It does, however, indicate that their presence has not been clearly demonstrated. Furthermore, even if present, they can probably be overcome by proper education and conditioning.

Nevertheless, we must not delude ourselves into believing that if human beings can be made less aggressive and more cooperative without altering their genes, such transformations of human nature become an easy task. Much of the learning process that determines behavioral patterns undoubtedly takes place in the home, and passes from parents, chiefly mothers, to their children at a very early age. In most societies, the home and family are among the most conservative of all institutions. Attempts to produce radical and wholesale alternations in them, particularly those made in recent years by Communist countries, have resulted more in disruption than in effective change.

In a recent book (Stebbins 1969), I have suggested that one of the most important features of all living organisms, from bacteria to mankind, is the conservatism of complex structures and patterns of organization that have acquired a high adaptive value. In most organisms, this conservatism slows down progressive evolution. Nevertheless, by preventing or greatly reducing the probability that genetic changes of a degenerative nature will become established, it actually increases the probability that progress to more complex levels of biological organization will eventually be achieved. By analogy, one can argue that complex patterns of social organization, once they have acquired a high adaptive value, will usually promote conservatism but can, in times of crises, be the foundation upon which new and more successful patterns can be erected.

This line of reasoning is in accord with the conclusion of Kingsley Davis (1965) that patterns of social structure are as resistant to change as are organized phenotypic patterns that are determined by our genes. It leads one to conclude, nevertheless, that radical improvements in human society do not depend upon breeding a genetically better race of mankind, but largely upon adjusting to the radically new conditions of modern life the age-old practices upon which the stability of the home and family have been built. The survival of our society depends upon radical and successful adjustments not only in the larger home that is shared by all mankind and is the object of study by human ecologists, but also in the more intimate and actual home of each family. Interactions between alterations of these two kinds of "homes" must, like all other interactions, be a prime target of study for all of us who wish to overcome the maladjustments of modern human society.

Radical changes in home life and the early training of children, though admittedly difficult to achieve, must not be regarded as impossible. The child-rearing experiment now being carried out in the kibbutzim of Israel, and reported by Bruno Bettelheim, is an outstanding example.

THE IMMINENCE OF CATASTROPHE

I believe that anyone who has followed my argument this far will agree with me that I have not been able to suggest an easy or quick solution to the problems of mankind. Even if the awesome prophesies of the "modern Jeremiahs" are only half true, the prospects of avoiding a major catastrophe that will decimate mankind in many parts of the earth, and will seriously disrupt our maladjusted social structure, seem dim indeed. Furthermore, the "solutions" proposed by these "prophets of doom," such as Garrett Hardin (1968) and Paul Ehrlich (1968), are idealistic but impractical. They propose such drastic alterations in the way of life practiced by any modern society, that they could not be enforced. Their enforcement would require intrusions into people's private lives of such severity that they would not be tolerated anywhere. That major catastrophes will come appears to be almost certain. Only the dates of their arrival and the parts of the earth that will be most affected are in doubt. Nevertheless, the prediction that these catastrophes will completely obliterate the human species appears to me to be equally unrealistic and defeatist. More likely they will resemble the catastrophes that have happened in the past, and that mankind has survived, as Thornton Wilder has so aptly stated, "by the skin of our teeth." This is the most constructive wager for us to make. We must bet also that under the stress of recovery from these catastrophes, men will be more ready to adjust institutions and social attitudes to meet actual conditions than they are now. If, under these conditions, some of the survivors of the catastrophes can present new methods of technological and social adjustment that have been the product of careful multidisciplined research, they may provide guidelines for rebuilding society into a healthier state.

In conclusion, I cannot see any conditions, short of the complete destruction of life on the earth by atomic fallout, that will cause the human species to become extinct at any time in the future. By developing technological and sociological templates that are as widespread and permanent as are the biological templates that appeared with the origin of life, man has assured the permanent survival on this planet either of his own species or of species directly descended from mankind that are even more adept at achieving dominance through the use of these new templates. The critical question for us to decide, however, is whether mankind can build a new society that is more in harmony with the new physical world that he has created, or whether he will stumble forever from one morass of maladjustment, misery, and partial destruction to another similar quagmire. The technological means for resolving this problem in favor of our descendants are now available. The sociological problems, however, appear to be as far from solution as ever.

I can, however, see a ray of hope. This is partly because of the new awareness; the refusal of many young people to accept the hypocrisy and self-deception of the past. In addition, the catastrophes that will come may teach us lessons that mankind can never learn from the prophets. We owe it to future generations of mankind to build upon this hope, by working together in every way that we can.

The ideas for this paper were largely developed while I was a fellow at the Institute for Advanced Studies in the Behavioral Sciences, Stanford. I wish to acknowledge with thanks the helpful suggestions which were made by various ones of the fellows for 1968-1969, particularly Ray Birdwhistell, John H. Crook, Pierre Noyes, and Robert Sears. An additional review of the manuscript was made by Dr. Herbert Bauer of Davis, with many helpful suggestions that are gratefully acknowledged. Nevertheless, I assume full responsibility for all of the ideas that have been expressed.

Ardrey, R. 1967. The territorial imperative. Collins, London.

Crook, J. H. 1968. The nature and function of territorial aggression, p. 141-178. In F. M. Ashley-Montagu (ed.), Man and aggression. Oxford Univ. Press, New York.

---- 1970. The socio-ecology of primates. In J. H. Crook (ed.), Social behaviour in birds and mammals. Academic, London (in press).

Davis, Kingsley. 1965. Sociological aspects of genetic control, p. 173-204. In J. D. Roslansky (ed.), Genetics and the future of man. Appleton-Century-Crofts, New York.

Dobzhansky, T. 1962. Mankind evolving. Yale Univ. Press, New Haven, Conn.

---- 1967. Changing man. Science 155: 409-415.

---- 1968. Are they compatible? On Cartesian and Darwinian aspects of biology. Grad. J. 8: 99-117.

Ehrlich, P. R. 1968. The population bombs. Ballantine, New York.

Goodall, J. 1964. Tool using and aimed throwing in a community of free-living chimpanzees. Nature 201: 1264-1266.

Hall, K. R. L. 1968. Tool-using performances as indicators of behavioral ability, p. 131-148. In P. C. Jay (ed.), Primates, studies in adaptation and variability. Holt, Rinehart & Winston, New York.

Hardin, G. 1968. The tragedy of the commons. Science 162: 1243-1248.

Haynes, C. V., Jr. 1969. The earliest Americans. Science 166: 709-715.

Lorenz, K. 1966. On aggression. Methuen, London.

Platt, J. 1969. What we must do. Science 166: 1115-1121.

Stebbins, G. L. 1969. The basis of progressive evolution. Univ. North Carolina Press, Chapel Hill.

Tinbergen, N. 1968. On war and peace in animals and man. Science 160: 1411-1418.

Wallace, A. R. 1914. Social environment and moral progress. Funk & Wagnalls, New York and London.

Washburn, S. L., Phyllis C. Jay, and Jane B. Lancaster. 1965. Field studies of old world monkeys and apes. Science 150: 1541-1547.

Wilson, A. C., and V. M. Sarich. 1969. A molecular time-scale for human evolution. Nat. Acad. Sci., Proc., 63: 1088-1093.

Woolley, Sir Leonard. 1954. Excavations at Ur. Crowell, New York.

# Mineral resource estimates and public policy

## V.E. McKelvey (1973)

Not many people, I have found, realize the extent of our dependence on minerals. It was both a surprise and a pleasure, therefore, to come across the observations of George Orwell in his book The Road to Wigan Pier. When describing the working conditions of English miners in the 1930s he evidently was led to reflect on the significance of coal:

> Our civilization...is founded on coal, more completely than one realizes until one stops to think about it. The machines that keep us alive, and the machines that make the machines are all directly or indirectly dependent upon coal ... Practically everything we do, from eating an ice to crossing the Atlantic, and from baking a loaf to writing a novel, involves the use of coal, directly or indirectly. For all the arts of peace coal is needed; if war breaks out it is needed all the more. In time of revolution the miner must

go on working or the revolution must stop, for revolution as much as reaction needs coal... In order that Hitler may march the goosestep, that the Pope may denounce Bolshevism, that the cricket crowds may assemble at Lords, that the Nancy poets may scratch one another's backs, coal has got to be forthcoming.

To make Orwell's statement entirely accurate--and ruin its force with complications--we should speak of mineral fuels, instead of coal, and of other minerals also, for it is true that minerals and mineral fuels are the resources that make the industrial society possible. The essential role of minerals and mineral fuels in human life may be illustrated by a simple equation,

$$L = \frac{R \ x \ E \ x \ I}{P}$$

in which the society's average level of living ($L$), measured in its useful consumption of goods and services, is seen to be a function of its useful consumption of all kinds of raw materials ($R$),

Reprinted from U.S.Geol.Survey Prof. Paper 820, 9-19 by permission of the author and the Director of the U.S. Geological Survey, who happens to be the same person.

including metals, nonmetals, water, soil minerals, biologic produce, and so on; times its useful consumption of all forms of energy (E); times its useful consumption of all forms of ingenuity (I), including political and socio-economic as well as technologic ingenuity; divided by the number of people (P) who share in the total product.

This is a restatement of the classical economists' equation in which national output is considered to be a function of its use of capital and labor, but it shows what capital and labor really are. Far from being mere money, which is what it is popularly thought to mean, capital represents accumulated usable raw materials and things made from them, usable energy, and especially accumulated knowledge. And the muscle power expended in mere physical toil, which is what labor is often thought to mean, is a trivial contribution to national output compared to that supplied by people in the form of skills and ingenuity.

This is only a conceptual equation, of course, for numerical values cannot be assigned to some of its components, and no doubt some of them--ingenuity in particular--should receive far more weight than others. Moreover, its components are highly interrelated and interdependent. It is the development and use of a high degree of in-

genuity that makes possible the high consumption of minerals and fuels, and the use of minerals and fuels are each essential to the availability and use of the other. Nevertheless, the expression serves to emphasize that level of living is a function of our intelligent use of natural resources, and it brings out the importance of the use of energy and minerals in the industrial society. As shown in Figure 2, per capita Gross National Product among the countries of the world is, in fact, closely related to their per capita consumption of energy. Steel consumption also shows a close relation to per capita GNP (Fig. 3), as does the consumption of many other minerals.

Because of the key role that minerals and fuels play in economic growth and in economic and military security, the extent of their resources is a matter of great importance to government, and questions concerning the magnitude of resources arise in conjunction with many public problems. To cite some recent examples, the magnitude of low cost coal and uranium reserves has been at the heart of the question as to when to press the development of the breeder reactor--which requires an R & D program involving such an enormous outlay of public capital that it would be unwise to make the investment until absolutely necessary. Similarly, estimates of

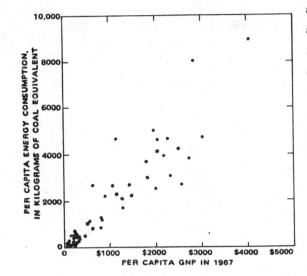

Fig. 2. Per capita energy consumption compared to per capita Gross National Product (GNP) in countries for which statistics are available in the United Nations "Statistical Yearbook" for 1967.

and parks, the construction of dams, and other matters related to land use involve appraisal of the distribution and amount of the resources in the area. The questions of the need for an international regime governing the development of seabed resources, the character such arrangement should have, and the definition of the area to which it should apply also involve, among other considerations, analysis of the probable character, distribution, and magnitude of subsea mineral resources.

And coming to the forefront is the potential oil and gas resources are needed for policy decisions related to the development of oil shale and coal as commercial sources of hydrocarbons, and estimates are needed also as the basis for decisions concerning prices and import controls.

Faced with a developing shortage of natural gas, the Federal Power Commission is presently much interested in knowing whether or not reserves reported by industry are an accurate indication of the amount of natural gas actually on hand; it also wants to know the extent of potential resources and the effect of price on their exploration and development. At the regional or local level, decisions with respect to the designation of wilderness areas

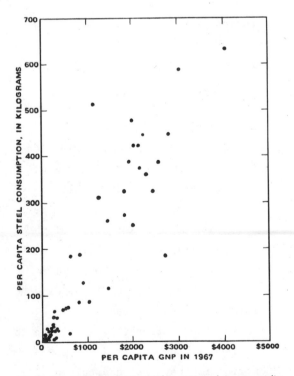

Fig. 3. Per capita steel consumption compared to per capita Gross National Product (GNP) in countries for which statistics are available in the United Nations "Statistical Yearbook" for 1967.

most serious question of all--namely, whether or not resources are adequate to support the continued existence of the world's population and indeed our own. The possibility to consider here goes much beyond Malthus' gloomy observations concerning the propensity of a population to grow to the limit of its food supply, for both population and level of living have grown as the result of the consumption of nonrenewable resources, and both are already far too high to maintain without industrialized, high-energy, and high mineral-consuming agriculture, transportation, and manufacturing. I will say more about this question later, but to indicate something of the magnitude of the problem let me point out that, in attaining our high level of living in the United States, we have used more minerals and mineral fuels during the last 30 years than all the people of the world used previously. This enormous consumption will have to be doubled just to meet the needs of the people now living in the United States through the remainder of their lifetimes, to say nothing about the needs of succeeding generations, or the increased consumption that will have to take place in the lesser developed countries if they are to attain a similar level of living.

## CONCEPTS OF RESERVES AND RESOURCES

The focus of most of industry's concern over the extent of mineral resources is on the magnitude of the supplies that exist now or that can be developed in the near term, and this is of public interest also. Many other policy decisions, however, relate to the much more difficult question of potential supplies, a question that to be answered properly must take account both of the extent of undiscovered deposits as well as deposits that cannot be produced profitably now but may become workable in the future. Unfortunately, the need to take account of such deposits is often overlooked, and there is a widespread tendency to think of potential resources as consisting merely of materials in known deposits producible under present economic and technologic conditions.

In connection with my own involvement in resource appraisal, I have been developing over the last several years a system of resource classification and terminology that brings out the classes of resources that need to be taken into account in appraising future supplies, which I believe helps to put the supply problem into a useful perspective. Before describing it, however, I want to emphasize that the problem of estimating potential resources has several built-in uncertainties that make an accurate and complete resource inventory impossible, no matter how comprehensive its scope.

One such uncertainty results from the nature of the occurrence of mineral deposits, for most of them lie hidden beneath the earth's surface and are difficult to locate and to examine in a way that yields accurate knowledge of their extent and quality. Another source of uncertainty is that the specifications of recoverable materials are constantly changing as the advance of technology permits us to mine or process minerals that were once too low in grade, too inaccessible, or too refractory to recover profitably. Still another results from advances that make it possible to utilize materials not previously visualized as usable at all.

For these reasons the quantity of usable resources is not fixed but changes with progress in science, technology, and exploration and with shifts in economic conditions. We must expect to revise our estimates periodically to take account of new developments. Even incomplete and provisional estimates are better than none at all, and if they differentiate known, undiscovered, and presently uneconomic resources they will help to define the supply problem and provide a basis for policy decisions relating to it.

The need to differentiate the known and the recoverable from the undiscovered and the uneconomic requires that a resource classification system convey two prime elements of information: the degree of certainty about the existence of the materials and the economic feasibility of recovering them. These two elements have been recognized in existing terminology, but only incompletely. Thus as used by both the mining and the petroleum industries, the term reserves generally refers to economically recoverable material in identified deposits, and the term resources includes in addition deposits not yet discovered as well as identified deposits that cannot be recovered now (e.g. Blondel and Lasky 1956).

The degree of certainty about the existence of the materials is described by terms such as proved, probable, and possible, the terms traditionally used by industry, and measured, indicated, and inferred, the terms devised during World War II by the Geological Survey and the Bureau of Mines to serve better the broader purpose of national resource appraisal. Usage of these degree-of-certainty terms is by no means standard, but all of their

definitions show that they refer only
to deposits or structures known to
exist.

Thus one of the generally accepted
definitions of <u>possible</u> ore states that
it is to apply to deposits whose existence is known from at least one exposure, and another definition refers to
an ore body sampled only on one side.
The definition of <u>inferred</u> reserves
agreed to by the Survey and the Bureau
of Mines permits inclusion of completely concealed deposits for which there
is specific geologic evidence and for
which the specific location can be
described, but it makes no allowance
for ore in unknown structures of
undiscovered districts.  The previous
definitions of both sets of terms also
link them to deposits minable at a
profit; the classification system these
terms comprise has thus neglected deposits that might become minable as the
result of technologic or economic developments.

To remedy these defects, I have
suggested that existing terminology be
expanded into the broader framework
shown in Figure 4, in which degree of
certainty decreases from left to right
and feasibility of recovery decreases
from top to bottom.  Either of the
series of terms already used to describe degree of certainty may be used
with reference to identified deposits
and applied not only to presently

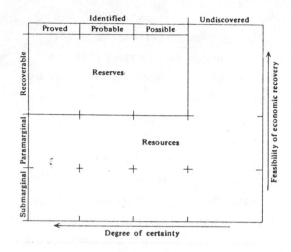

FIGURE 4. — Classification of mineral reserves and resources.
Degree of certainty increases from right to left, and feasibility of
economic recovery increases from bottom to top.

minable deposits but to others that
have been identified with the same
degree of certainty.  Feasibility-of-
recovery categories are designated by
the terms <u>recoverable</u>, <u>paramarginal</u>,
and <u>submarginal</u>.

Paramarginal resources are defined
here as those that are recoverable at
prices as much as 1.5 times those prevailing now.  (I am indebted to Stanley
P. Schweinfurth for suggesting the pre-
fix <u>para</u> to indicate that the materials
described are not only those just on
the margin of economic recoverability,
the common economic meaning of the term
<u>marginal</u>.)  At first thought this price
factor may seem to be unrealistic.  The
fact is, however, that prices of many
mineral commodities vary within such a
range from place to place at any given

time, and a price elasticity of this order of magnitude is not uncommon for many commodities over a space of a few years or even months, as shown by recent variations in prices of copper, mercury, silver, sulphur, and coal. Deposits in this category thus become commercially available at price increases that can be borne without serious economic effects, and chances are that improvements in existing technology will make them available at prices little or no higher than those prevailing now.

Over the longer period, we can expect that technologic advances will make it profitable to mine resources that would be much too costly to produce now, and, of course, that is the reason for trying to take account of submarginal resources. Again, it might seem ridiculous to consider resources that cost two or three times more than those produced now as having any future value at all. But keep in mind, as one of many examples, that the cutoff grade for copper has been reduced progressively not just by a factor of two or three but by a factor of ten since the turn of the century and by a factor of about 250 over the history of mining. Many of the fuels and minerals being produced today would once have been classed as submarginal under this definition, and it is reasonable to believe that continued

technologic progress will create recoverable reserves from this category.

EXAMPLES OF ESTIMATES OF POTENTIAL RESOURCES

For most minerals, the chief value of this classification at present is to call attention to the information needed for a comprehensive appraisal of their potential, for we haven't developed the knowledge and the methods necessary to make meaningful estimates of the magnitude of undiscovered deposits, and we don't know enough about the cost of producing most presently noncommercial deposits to separate paramarginal from submarginal resources. Enough information is available for the mineral fuels, however, to see their potential in such a framework.

The fuel for which the most complete information is available is the newest one--uranium. As a result of extensive research sponsored by the Atomic Energy Commission, uranium reserves and resources are reported in several cost-of-recovery categories, from less than $8 to more than $100 per pound of $U_3O_8$. For the lower-cost ores, the AEC makes periodic estimates in two degree-of-certainty categories, one that it calls reasonably assured reserves and the other it calls additional resources, defined as uranium surmised to occur in unexplored extensions of known deposits or in undiscovered deposits in known uranium districts. Both the AEC and the Geological Survey have made estimates from time to time of resources in other degree-of-certainty and cost-of-recovery categories.

Ore in the less-than-$8-per-pound class is minable now, and the AEC estimates reasonably assured reserves to be 143,000 tons and additional resources to be 167,000 tons of $U_3O_8$-- just about enough to supply the lifetime needs of reactors in use or ordered in 1968 and only half that required for reactors expected to be in use by 1980. The Geological Survey, however, estimates that undiscovered resources of presently minable quality may amount to 750,000 tons, or about two and a half times that in identified deposits and districts. Resources in the $8-$30-a-pound category in identified and undiscovered deposits add only about 600,000 tons of $U_3O_8$ and thus do not significantly increase potential reserves.

But tens of millions of tons come into prospect in the price range of $30-$100 per pound. Uranium at such prices would be usable in the breeder reactor. The breeder, of course, would utilize not only $U^{235}$ but also $U^{238}$, which is 140 times more abundant than $U^{235}$. Plainly the significance of uranium as a commercial fuel lies in its use in a breeder reactor, and one may question, as a number of critics have (for example, Inglis, 1971), the advisability of enlarging nuclear generating capacity until the breeder is ready for commercial use.

Until recently the only information available about petroleum resources consisted of estimates of proved reserves prepared annually by the American Petroleum Institute and the American Gas Association, plus a few estimates of what has been called ultimate production, i.e. the total likely to be eventually recovered. A few years ago, however, the API began to report estimates of total oil in place in proved acreage, and the Potential Gas Committee began to estimate possible and probable reserves of natural gas, defining them as consisting of gas expected to be found in extensions of identified fields and in new discoveries in presently productive strata in producing provinces. It also introduced another category, speculative resources--equivalent to what I have called "undiscovered"-- to represent gas to be found in nonproducing provinces and in presently unproductive strata in producing provinces.

In 1970 the National Petroleum Council released a summary of a report on Future Petroleum Provinces of the United States, prepared at the request of the Department of the Interior, in which it reported estimates of crude oil in the combined probable-possible class and in the speculative category. In addition, NPC estimated the amounts that would be available under two assumptions as to the percent of the oil originally in place that might be recovered in the future (Table 2). NPC did not assess the cost of such recovery, but the average recovery is now about 30 percent of the oil in place, and NPC expects it to increase gradually to about 42 percent in the year 2000 and to 60 percent eventually. The NPC estimates do not cover all potentially favorable areas either on land or offshore but, even so, in the sum of these various categories NPC sees about twelve times as much oil remaining to be discovered and produced as exists in proved reserves alone.

Table 2. Some estimates of U.S. crude-oil reserves and resources, in billion barrels

| | In identified fields or structures | | In undiscovered fields and structures (speculative) |
| | Proved | Probable-Possible | |
|---|---|---|---|
| Recoverable at present rate[1] | 31 (API) | 74 (NPC) | 67 (NPC) |
| Additional at 42 percent recovery | 47 (NPC) | 22 (NPC) | 21 (NPC) |
| Additional at 60 percent recovery | 69 (NPC) | 40 (NPC) | 37 (NPC) |
| Oil originally in place | 388 (API)[2] | 227 (NPC) | 209 (NPC) |
| Total oil originally in place | 2200 (Hendricks and Schweinfurth, 1968). | | |
| Ultimate production[2] | 190 (Hubbert, 1969); 353 (Moore, 1966); 433 (Weeks)[3]; 450 (Elliott and Linden, 1968); 432-495 (NPC); 550 (Hendricks and Schweinfurth, 1968). | | |

[1] Average recovery is 30 percent of oil in place.
[2] Includes past production of 86 billion barrels.
[3] See McKelvey (1968, Table II, p. 18).

The Potential Gas Committee's estimates of potential gas resources similarly do not cover all favorable areas, but they indicate that resources in the probable, possible, and speculative categories are about twice that of proved reserves and past production. Because about 80 percent of the gas originally in place is now recovered, paramarginal and submarginal resources in ordinary gas reservoirs are not as large as for crude oil. Paramarginal and submarginal gas resources may be significant, however, in kinds of rocks from which gas is not now recoverd, namely impermeable strata and coal. In the Rocky Mountain province, for instance, Haun, Barlow, and Hallinger (1970) recently estimated potential gas resources in ordinary reservoirs to be in the range of 100-200 trillion cubic feet, but pointed out that gas in impermeable strata which might be released by nuclear stimulation would be several times that amount. Gas occluded in coal--now only a menace in this country as a cause of explosions-- is already recovered in some European mines and is also a potentially large resource.

The uncertainties concerning potential coal resources center not on their total magnitude, as they do for oil and gas, but on the amounts available at present prices. Because coal beds have great lateral continuity, geologic mapping and stratigraphic studies make it possible to project them long distances from their outcrops and to categorize them in terms of thickness of beds, thickness of overburden, rank of coal, and other features that affect cost. The Geological Survey has prepared such estimates, but the cost of recovering coal in the various categories has yet to be determined. Coal in beds more than 14 inches thick totals at least 3.3 trillion tons in the United States, but estimates of the amounts minable at present prices have ranged from 20 to 220 billion tons (see U.S. Office of Science and Technology, Energy R & D and National Progress). Because 20 billion tons represents nearly a 40-year supply at present rates of consumption, it is easy to see why the studies needed to determine how much would be available at various costs have not been undertaken. The question is by no means only of academic interest, for the nuclear power development program was justified in part in its early years on the assumption that reserves of low-cost coal were extremely limited, and part of the continued growth of the nuclear power industry is said to be the result of the difficulty power companies are having in acquiring low-cost reserves.

QUANTIFYING THE UNDISCOVERD

Considering potential resources in the degree-of-certainty, cost-of-recovery framework brings out the joint role that geologists, engineers, mineral technologists, and economists must play in estimating their magnitude. Having emphasized the importance of the economic and technologic side of the problem, I want now to turn to the geological side and consider the problem of how to appraise the extent of undiscovered reserves and resources.

It is difficult enough to estimate the extent of unexplored resources of the inferred or possible class. In fact, it is even difficult to estimate measured or proved reserves with a high degree of accuracy until they have been largely mined out. Thus estimates of proved reserves prepared in advance of appreciable production commonly have an error of about 25 percent, and the error in estimates of incompletely explored deposits is usually much larger. Generally the combination of the geologist's inherent conservatism and the lack of information on the geology of concealed areas leads to estimates that err in being too low rather than too high.

One eminent mining geologist reported that, having recognized these effects, he once arbitrarily tripled his calculations to arrive at an estimate of the ore remaining in a producing district; twice the amount of his inflated estimate, however, was found and mined over the next 20 years, and more was in prospect. To match many such stories are at least a few prematurely deserted mills and mine installations built on the expectation of finding ore that did not materialize. Both kinds of experiences emphasize the difficulty of appraising the extent of mineral deposits even in partly explored areas. In the light of such experiences one is justified in asking--as many well-informed people have--whether estimates of the magnitude of undiscovered deposits can have enough reliability to make them worthwhile.

The fact that new districts are still being discovered for nearly every commodity and that large areas favorable for the occurrence of minerals of all kinds are covered by alluvium, volcanics, glacial drift, seawater, or other materials that conceal possible mineral-bearing rocks or structures assures us that undiscovered deposits are still to be found. Qualitatively, at least, we know something about the distribution of minerals with respect to other geologic phenomena and, if this is so, we have a chance of developing quantitative relations that will give us at least a start.

Two principal approaches to the problem have been taken thus far. One is to extrapolate observations related to rate of industrial activity, such as annual production of the commodity; the other is to extrapolate observations that relate to the abundance of the mineral in the geologic environment in which it is found.

The first of these methods has been utilized by M. K. Hubbert (1969), C. L. Moore (1966), and M. A. Elliott and H. R. Linden (1968) in estimating ultimate reserves of petroleum. The essential features of this approach are to analyze the growth in production, proved reserves, and discovery per foot of drilling over time and to project these rate phenomena to terminal values in order to predict ultimate production. Hubbert has used the logistic curve for his projections, and Moore has utilized the Gompertz curve, with results more than twice as high as those of Hubbert. As Hubbert has pointed out, these methods utilize the most reliable information collected on the petroleum industry; modern records on production, proved reserves, number of wells drilled, and similar activities are both relatively complete and accurate, at least as compared with quantitative knowledge about geologic features that affect the distribution of petroleum.

The rate methods, however, have an inherent weakness in that the phenomena they analyze reflect human activities that are strongly influenced by economic, political, and other factors that bear no relation to the amount of oil or other material that lies in the ground. Moreover, they make no allowance for major breakthroughs that might transform extensive paramarginal or submarginal resources into recoverable reserves, nor do they provide a means of estimating the potential resources of unexplored regions. Such projections have some value in indicating what will happen over the short term if recent trends continue, but they can have only limited success in appraising potential resources.

Even the goal of such projections, namely the prediction of ultimate production, is not a useful one. Not only is it impossible to predict the quantitative effects of man's future activities but the concept implies that the activities of the past are a part of an inexorable process with only one possible outcome. Far more useful, in my opinion, are estimates of the amounts of various kinds of materials that are in the ground in various environments; such estimates establish targets for both the explorer and the technologist, and they give us a basis for choosing amoung alternative ways of meeting our needs for mineral supplies.

The second principal approach

taken thus far to the estimation of un-discovered resources involves the extrapolation of data on the abundance of mineral deposits from explored to unexplored ground on the basis of either the area or the volume of broad-ly favorable rocks.  In the field of metalliferous deposits, Nolan (1950) pioneered in extrapolation on the basis of area in his study of the spatial and size distribution of mineral deposits in the Boulder Dam region and in his conclusion that a similar distribution should prevail in adjacent concealed and unexplored areas.  Weeks (1958, 1965) and Pratt (1950) played similar roles with respect to the estimation of petroleum resources--Weeks extrapolat-ing on the basis of oil per unit volume of sediment and Pratt on the basis of oil per unit area.  Many of the esti-mates of crude oil that went into the NPC study were made by the volumetric method, utilizing locally appropriate factors on the amount of oil expected per cubic mile of sediment.  Olson and Overstreet (1964) have since used the area method to estimate the magnitude of world thorium resources as a func-tion of the size of areas of igneous and metamorphic rocks as compared with India and the United States, and A. P. Butler (written commun., 1958) used the magnitude of sandstone uranium ore reserves exposed in outcrop as a basis for estimating the area in back of the

outcrop that is similarly mineralized.

Several years ago, Zapp (1962) and Hendricks (1965) introduced another approach, based on the amount of drill-ing required to explore adequately the ground favorable for exploration and the reserves discovered by the footage already drilled--a procedure usable in combination with either the volumetric or areal approach.  Recently J. B. Zimmerman and F. L. Long (cited in "Oil and Gas Journal, 1969") applied this approach to the estimation of gas re-sources in the Delaware-Val Verde basins of west Texas and southeastern New Mexico; and Haun, Barlow, and Hallinger (1970) used it to estimate potential natural gas resources in the Rocky Mountain region.  In the field of metals, Lowell (1970) has estimated the number of undiscovered porphyry copper deposits in the southwestern United States, Chile and Peru, and British Columbia as a function of the propor-tion of the favorable pre-ore surface adequately explored by drilling, and Armstrong (1970) has similarly esti-mated undiscovered uranium reserves in the Gas Hills area of Wyoming on the basis of the ratios between explored and unexplored favorable areas.

I have suggested another variant of the areal method for estimating reserves of nonfuel minerals which is based on the fact that the tonnage of minable reserves of the well-explored elements in the United States is roughly equal to their crustal abundance in percent times a billion or 10 billion (fig. 5). Obviously this relation is influenced by the extent of exploration, for it is only reserves of the long-sought and well-explored minerals that display the relation to abundance. But it is this feature that gives the method its greatest usefulness, for it makes it possible to estimate potential resources of elements, such as uranium and thorium, that have been prospected for only a short period. Sekine (1963) tested this method for Japan and found it applicable there, which surprised me a little, for I would not have thought Japan to be a large enough sample of

the continental crust to bring out this relationship.

The relation between reserves and abundance, of course, can at best be only an approximate one, useful mainly in order-of-magnitude estimates, for obviously crustal abundance of an element is only one of its properties that lead to its concentration. That it is an important factor, however, may be seen not only in its influence on the magnitude of reserves but also in other expressions of its influence on the concentrations of the elements. For example, of the 18 or so elements with crustal abundances greater than about 200 parts per million, all but fluorine and strontium are rock-forming in the sense that some extensive rocks are composed chiefly of minerals of which each of these elements is a major constituent. Of the less abundant elements, only chromium, nitrogen, and boron have this distinction. Only a few other elements, such as copper, lead, and zinc, even form ore bodies composed mainly of minerals of which the valuable element is a major constituent, and in a general way the grade of minable ores decreases with decreasing crustal abundance. A similar gross correlation exists between abundance of the elements and the number of minerals in which they are a significant constituent.

Members of a committee of the

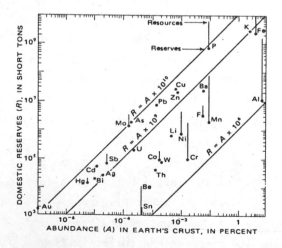

FIGURE 5.—Domestic reserves of elements compared to their abundance in the earth's crust. Tonnage of ore minable now is shown by a dot; tonnage of lowergrade ores whose exploitation depends upon future technological advances or higher prices is shown by a bar.

Geology and the Conservation of Mineral Resources Board of the Soviet Union have have described a somewhat similar method for the quantitative evaluation of what they call "predicted reserves" of oil and gas, based on estimates of the total amount of hydrocarbons in the source rock and of the fraction that has migrated into commercial reservoirs --estimates that would be much more difficult to obtain for petroleum than for the elements. Probably for this reason not much use has been made of this method, but it seems likely that quantitative studies of the effects of the natural fractionation of the elements might be of some value in estimating total resources in various size and grade categories.

Some studies of the grade-frequency distribution of the elements have, in fact, been undertaken by geochemists in the last couple of decades, and, taking off from Nolan's work, several investigators have studied the areal and size-frequency distribution of mineral deposits in conjunction with attempts to apply the methods of operations research to exploration (for example, Allais, 1957; Slichter, 1960; Griffiths, 1964; DeGeoffroy and Wu, 1970, and Harris and Euresty, 1969). None of these studies has been concerned with the estimation of undiscovered reserves, but they have identified two features about the distribution of

mineral deposits that may be applicable to the problem.

One is that the size distribution of both metalliferous deposits, expressed in dollar value of production, and of oil and gas, expressed in volumetric units, has been found to be log normal, which means that of a large population of deposits, a few contain most of the ore (e.g. Slichter 1960, Kaufman 1963). In the Boulder Dam area, for example, 4 percent of the districts produced 80 percent of the total value of recorded production. The petroleum industry in the United States has a rule of thumb that 5 percent of the fields account for 50 percent of the reserves and 50 percent of fields for 95 percent. And in the USSR, about 5 percent of the oil fields contain about 75 percent of the oil, and 10 percent of the gas fields have 85 percent of the gas reserves.

The other feature of interest is that in many deposits the grade-tonnage distribution is also log normal, and the geochemists have found this to be the case also with the frequency distribution of minor elements.

These patterns of size- and grade-frequency distribution will not in themselves provide information on the magnitude of potential resources, for they describe only how minerals are distributed and not how much is present. But if these patterns are combined with quantitative data on the incidence of congeneric deposits in various kinds of environments, the volume or area of favorable ground, and the extent to which it has been explored, they might yield more useful estimates of potential resources than are obtainable by any of the procedures so far applied. Thus estimates of total resources described in terms of their size- and grade-frequency distributions could be further analyzed in the light of economic criteria defining the size, grade, and accessibility of deposits workable at various costs, and then partitioned into feasibility-of-recovery and degree-of-certainty categories to provide targets for exploration and technologic development as well well as guidance for policy decisions.

Essential for such estimates, of course, is better knowledge than is now in hand for many minerals on the volume of ore per unit of favorable ground and on the characteristics of favorable ground itself. For petroleum the development of such knowledge is already well advanced. For example, whereas most estimates of resources have been based on an assumed average petroleum content of about 50,000 barrels per cubic mile of sediment, varied a little perhaps to reflect judgments of favorability, the range in various basins is from 10,000 to more than 2,000,000 barrels per cubic mile.

As shown by the recent analysis by Halbouty and his colleagues of the factors affecting the formation of giant fields, the geologic criteria are developing that make it possible to classify sedimentary basins in terms of their petroleum potential. Knowledge of the mode of occurrence and genesis of many metalliferous minerals and of the geology of the terranes in which they occur is not sufficient to support comprehensive estimates prepared in this way. But for many kinds of deposits enough is known to utilize this kind of approach on a district or regional basis, and I hope a start can soon be made in this direction.

NEED FOR REVIEW OF RESOURCE ADEQUACY

Let me return now to the question of whether or not resources are adequate to maintain our present level of living. This is not a new question by any means. In 1908 it was raised as a national policy issue at the famous Governors' Conference on Resources, and it has been the subject of rather extensive inquiry by several national and international bodies since then. In spite of some of the dire predictions about the future made by various people in the course of these inquiries, they did not lead to any major change in our full-speed-ahead policy of economic development. Some of these inquiries, in fact, led to immediate investigations that revealed a greater resource potential for certain minerals than had been thought to exist, and the net effect was to alleviate rather than heighten concern.

Now, however, concern about resource adequacy is mounting again. The overall tone of the recent National Academy of Sciences' report on Resources and Man was cautionary if not pessimistic about continued expansion in the production and use of mineral resources, and many scientists, including some eminent geologists, have expressed grave doubts about our ability to continue on our present course. The question is also being raised internationally, particularly in developing countries where concern is being expressed that our disproportionate use of minerals to support our high level of living may be depriving them of their own future.

Personally, I am confident that for millennia to come we can continue to develop the mineral supplies needed to maintain a high level of living for those who now enjoy it and to raise it for the impoverished people of our own country and the world. My reasons for thinking so are that there is a visible undeveloped potential of substantial proportions in each of the processes by which we create resources and that our experience justifies the belief that

these processes have dimensions beyond our knowledge and even beyond our imagination at any given time.

Setting aside the unimaginable, I will mention some examples of the believable. I am sure all geologists would agree that minable undiscovered deposits remain in explored as well as unexplored areas and that progress in our knowledge of regional geology and in exploration will lead to the discovery of many of them. With respect to unexplored areas, the mineral potential of the continental margins and ocean basins deserves particular emphasis, for the technology that will give us access to it is clearly now in sight. For many critical minerals, we already know of substantial paramarginal and submarginal resources that experience tells us should be brought within economic reach by technologic advance. The process of substituting an abundant for a scarce material has also been pursued successfully, thus far not out of need but out of economic opportunity, and plainly has much potential as a means of enlarging usable resources.

Extending our supplies by increasing the efficiency of recovery and use of raw materials has also been significant. For example, a unit weight of today's steel provides 43 percent more structural support than it did only ten years ago, reducing proportionately the amount required for a given purpose.

Similarly, we make as much electric power from one ton of coal now as we were able to make from seven tons around the turn of the century. Our rising awareness of pollution and its effects surely will force us to pay even more attention to increasing the efficiency of mineral recovery and use as a means of reducing the release of contaminants to the environment. For similar reasons, we are likely to pursue more diligently processes of recovery, re-use, and recycling of mineral materials than we have in the past.

Most important to secure our future is an abundant and cheap supply of energy, for if that is available we can obtain materials from low-quality sources, perhaps even country rocks, as Harrison Brown (1954, p. 174-175) has suggested. Again, I am personally optimistic on this matter, with respect to the fossil fuels and particularly to the nuclear fuels. Not only does the breeder reactor appear to be near enough to practical reality to justify the belief that it will permit the use of extremely low-grade sources of uranium and thorium that will carry us far into the future, but during the last couple of years there have been exciting new developments in the prospects for commercial energy from fusion. Gothermal energy has a large unexploited potential, and new concepts are also being developed to permit the commercial use of solar energy.

But many others do not share these views, and it seems likely that soon there will be a demand for a confrontation with the full-speed-ahead philosophy that will have to be answered by a deep review of resource adequacy. I myself think that such a review is necessary, simply because the stakes have become so high. Our own population, to say nothing of the world's, is already too large to exist without industrialized, high energy- and mineral-consuming agriculture, transportation, and manufacturing. If our supply of critical materials is enough to meet our needs for only a few decades, a mere tapering off in the rate of increase of their use, or even a modest cutback, would stretch out these supplies for only a trivial period. If resource adequacy cannot be assured into the far-distant future, a major reorientation of our philosophy, goals, and way of life will be necessary. And if we do need to revert to a low resource-consuming economy, we will have to begin the process as quickly as possible in order to avoid chaos and catastrophe.

Comprehensive resource estimates will be essential for this critical examination of resource adequacy, and they will have to be made by techniques of accepted reliability. The techniques I have described for making such estimates have thus far been applied to only a few minerals, and none of them have been developed to the point of general acceptance. Better methods need to be devised and applied more widely, and I hope that others can be enlisted in the effort necessary to do both.

SELECTED BIBLIOGRAPHY

Allais, M., 1957, Methods of appraising economic prospects of mining exploration over large territories: Management Sci., v. 2, p. 285-347.

Armstrong, F. C., 1970, Geologic factors controlling uranium resources in the Gas Hills District, Wyoming, in Wyoming Geol. Assoc. Guidebook 22d Ann. Field Conf.: p. 31-44.

Blondel, F., and Lasky, S. F., 1956, Mineral reserves and mineral resources: Econ. Geology, v. 60, p. 686-697.

Brown, Harrison, 1954, The challenge of Man's future: Viking Press, 290 p.

Bush, A. L., and Stager, H. K., 1956, Accuracy of ore-reserve estimates for uranium-vandium deposits on the Colorado Plateau: U.S. Geol. Survey Bull. 1030-D, p. 137.

Buyalov, N. I., Erofeev, N. S., Kalinen, N. A., Kleschev, A. I., Kudryashove, N. M., L'vov, M. S., Simakov, S. N., and Vasil'ev, V. G., 1964, Quantitative evaluation of predicted reserves of oil and gas: New York, Consultants Bur.; translation.

DeGeoffroy, J., and Wu, S. M., 1970, A statistical study of ore occurrences in the greenstone belts of the Canadian Shield: Econ. Geology, v. 65, no. 4, p. 496-504.

Elliott, M. A., and Linden, H. R., 1968, A new analysis of U.S. natural gas supplies: Jour. Petroleum Technology, v. 20, p. 135-141.

Griffiths, J. C., 1962, Frequency distributions of some natural resource materials, in Tech. Conf. on Petroleum Production, 23d: Mining Industry Expt. Sta. Circ. 63, p. 174-198.

---- 1964, Exploration for natural resources: Operations Research, v. 14, p. 189-209.

Halbouty, M. T., King, R. E., Klemme, H. D., Dott, R. H., Sr., and Meyerhoff, A. A., 1970, Factors affecting formation of giant oil and gas fields, and basin classification, pt. 2 of World's giant oil and gas fields, geologic factors affecting their formation, and basin classification, in Geology of giant petroleum fields

--a symposium: Am. Assoc. Petroleum Geologists Mem. 14, p. 528-
555.

Halbouty, M. T., Meyerhoff, A. A.,
King, R. E., Dott, R. H., Sr.,
Klemme, H. D., and Shabad,
Theodore, 1970, Giant oil and gas
fields, pt. 1 of World's giant oil
and gas fields, geologic factors
affecting their formation, and
basin classification, in Geology
of giant petroleum fields--a symposium: Am. Assoc. Petroleum
Geologists Mem. 14, p. 502-528.

Harris, D. P., and Euresty, D., 1969,
A preliminary model for the
economic appraisal of regional
resources and exploration based
upon geostatistical analyses and
computer simulation: Colorado
School Mines Quart., v. 64, p. 71-
98.

Haun, J. D., Barlow, J. A., Jr., and
Hallinger, D. E., 1970, Natural
gas resources, Rocky Mountain
region: Am. Assoc. Petroleum
Geologists Bull., v. 54, p.
1706-1708.

Hendricks, T. A., 1965, Resources of
oil, gas, and natural gas liquids
in the United States and the
world: U.S. Geol. Survey Circ.
522, 20 p.

Hendricks, T. A., and Schweinfurth, S.
P., 1969 (Unpub. memo., Sept. 14,
1966), cited in United States

petroleum through 1980: Washington, D.C., U.S. Dept. Interior,
Office of Oil and Gas, 92 p.

Hubbert, M. K., 1969, Energy resources,
chapter 8 in Resources and man, a
study and recommendations by the
Committee on Resources and Man of
the Division of Earth Sciences,
National Academy of Science-
National Research Council: San
Francisco, Calif., W. H. Freeman
and Co., p. 157-242.

Inglis, D. R., 1971, Nuclear energy and
the Malthusian dilemma: Atomic
Scientists Bull., v. 27, no. 2,
p. 14-18.

Kaufman, G. M., 1963, Statistical
decision and related techniques in
oil and gas exploration: Englewood, N.J., Prentice-Hall, 307 p.

Lasky, S. G., 1950, Mineral-resource
appraisal by the U.S. Geological
Survey: Colorado School Mines
Quart., v. 45, no. 1A, p. 1-27.

---- 1950, How tonnage and grade relations help predict ore reserves:
Eng. and Mining Jour., v. 151, no.
4, p. 81-85.

Lowell, J. D., 1970, Copper resources in 1970: Mining Eng.,
v. 22, p. 67-73.

McKelvey, V. E., 1960, Relations of
reserves of the elements to their
crustal abundance: Am. Jour.
Sci., v. 258-A (Bradley volume),
p. 234-241.

---- 1968, Contradictions in energy
resource estimates, in Energy--
Proc. 7th Bienn. Gas Dynamics
Symposium, Gas Dynamics Lab.,
Northwestern Univ.: Evanston,
Ill., Northwestern Univ. Press,
p. 18-26.

Moore, C. L., 1966, Projections of U.S.
petroleum supply to 1980:
Washington, D.C., U.S. Dept.
Interior, Office of Oil and Gas,
42 p.

National Academy of Sciences--National
Research Council, 1969, Resources
and man: San Francisco, Calif.,
W. H. Freeman and Co., 259 p.

National Petroleum Council, 1970,
Future petroleum provinces of the
United States--a summary: Wash-
ington, D.C., Natl. Petroleum
Council, 138 p.

Nolan, T. B., 1950, The search for
new mining districts: Econ.
Geology, v. 45, p. 601-608.

Oil and Gas Journal, 1969, Vast
Delaware-Val Verde reserve seen:
Oil and Gas Jour., v. 67, no. 16,
p. 44.

Olson, J. C., and Overstreet, W. C.,
1964, Geologic distribution and
resources of thorium: U.S. Geol.
Survey Bull. 1204, 61 p.

Orwell, George, 1937, The road to Wigan
Pier: London, Victor Gollancz,
Ltd., 264 p.; reprinted in 1958 by
Harcourt, Brace and World, New
York.

Potential Gas Committee, 1969, Guideline
lines for the estimation of
potential supply of natural gas in
the United States, in Potential
supply of natural gas in the
United States (as of December 31,
1968): Golden, Colo., Colorado
School Mines Found., Inc., p. 21-
30.

Pratt, W. E., 1950, The earth's petroleum resources, in Fanning, L. M.,
ed., Our oil resources (2d ed.):
New York, McGraw-Hill, p. 137-153.

Rodionov, D. A., 1965, Distribution
functions of the element and
mineral contents of igneous rocks
--A special research report: New
York, Consultants Bur., p. 28-29;
translation.

Sekine, Y., 1963, On the concept of
concentration of ore-forming
elements and the relationship of
their frequency in the earth's
crust: Internat. Geology Rev.,
v. 5, p. 505-515.

Slichter, L. B., 1960, The need of a
new philosophy of prospecting:
Mining Eng., v. 12, p. 570-576.

Slichter, L. B., and others, 1962,
Statistics as a guide to prospecting, in Math and computer applications in mining and exploration--
a symposium: Tucson, Arizona
Univ. Coll. Mines, Proc., p. F-
1-27.

U.S. Office of Science and Technology, 1965, Energy R. and D and national progress: Washington, D.C.: U.S. Govt. Printing Office.

Weeks, L. G., 1950, Concerning estimates of potential oil reserves: Am. Assoc. Petroleum Geologists Bull., v. 34, p. 1947-1953.

---- 1958, Fuel reserves of the future: Am. Assoc. Petroleum Geologists Bull., v. 42, p. 431-438.

---- 1965, World offshore petroleum resources: Am. Assoc. Petroleum Geologists Bull., v. 49, p. 1680-1693.

Zapp, A. D., 1962, Future petroleum producing capaicty of the United States: U.S. Geol. Survey Bull. 1142-H, 36 p.

# Human population and the global environment

## J.P. Holdren and P.R. Ehrlich (1974)

Three dangerous misconceptions appear to be widespread among decision-makers and others with responsibilities related to population growth, environmental deterioration, and resource depletion. The first is that the absolute size and rate of growth of the human population has little or no relationship to the rapidly escalating ecological problems facing mankind. The second is that environmental deterioration consists primarily of "pollution," which is perceived as a local and reversible phenomenon of concern mainly for its obvious and immediate effects on human health. The third misconception is that science and technology can make possible the long continuation of rapid growth in civilization's consumption of natural resources.

We and others have dealt at length with the third misconception elsewhere

Reprinted from American Scientist 62, 282-292 by permission of the authors. Copyright 1974 John P. Holdren and Paul R. Ehrlich.

(1). In this paper, we argue that environmental deterioration is a much more subtle, pervasive, and dangerous phenomenon than is implied by the narrow view of "pollution" alluded to above. We show further that population size and the rate of population growth, in rich countries as well as in poor ones, have been and continue to be important contributing factors in the generation of environmental disruption.

Environmental problems can be classified according to the nature of the damage to human beings:
1. Direct assaults on human welfare, including obvious damage to health (e.g. lead poisoning or aggravation of lung disease by air pollution), damage to goods and services (e.g. the corrosive effects of air pollution on buildings and crops), social disruption (e.g. displacement of people from their living areas by mining operations and hydroelectric projects), and other direct effects on what people perceive as the "quality of life" (e.g. congestion, noise, and litter).

2. Indirect effects on human welfare through interference with services provided for society by natural biological systems (e.g. diminution of ocean productivity by filling estuaries and polluting coastal waters, crop failure caused by pests whose natural enemies have been exterminated by civilization, and acceleration of erosion by logging or overgrazing).

Most of the attention devoted to environmental matters by scientists, politicians, and the public alike has been focused on the direst effects and, more particularly, on their acute rather than their chronic manifestations. This is only natural. It would be wrong, however, to interpret limited legislative and technical progress toward ameliorating the acute symptoms of environmental damage as evidence that society is on its way to an orderly resolution of its environmental problems. The difficulty is not merely that the discovery, implementation, and enforcement of treatment for the obvious symptoms is likely to be expensive and difficult, but also that the long-term human consequences of chronic exposure to low concentrations of environmental contaminants may be more serious--and the causes less amenable to detection and removal-- than the consequences of exposure to acute pollution as it is perceived today.

The most serious threats of all, however, may well prove to be the indirect ones generated by mankind's disruption of the functioning of the natural environment--the second category listed above, to which we will devote most of our attention here.

NATURAL SERVICES

The most obvious services provided for humanity by the natural environment have to do with food production. The fertility of the soil is maintained by the plants, animals, and microorganisms that participate in the great nutrient cycles--nitrogen, phosphorus, carbon, sulfur. Soil itself is produced from plant debris and weathered rock by the joint action of bacteria, fungi, worms, soil mites, and insects. The best protection against erosion of soil and flooding is natural vegetation.

At many stages of the natural processes comprising the nutrient cycles, organisms accomplish what humans have not yet learned to do-- the complete conversion of wastes into resources, with solar energy captured by photosynthesis as the driving energy source. Human society depends on these natural processes to recycle many of its own wastes, from sewage to detergents to industrial effluents (reflect on the term "biodegradable"). In the course of the same cycles, the environmental concentrations of ammonia, nitrites, and hydrogen sulfide--all poisonous--are biologically controlled (2,3).

Insects pollinate most vegetables, fruits, and berries. Most fish--the source of 10 to 15 percent of the animal protein consumed by mankind--are produced in the natural marine environment, unregulated by man. (As is well known, animal protein is the nutrient in shortest supply in a chronically malnourished world.) Most potential crop pests--one competent estimate is 99 percent--are held in check not by man but by their natural enemies and by characteristics of the physical environment such as temperature, moisture, and availability of breeding sites (3). Similarly, some agents of human disease are controlled principally not by medical technology but by environmental conditions, and some carriers of such agents are controlled by a combination of environmental conditions and natural enemies (4).

Finally, the natural environment in its diversity can be viewed as a unique library of genetic information. From this library can be drawn new food crops, new drugs and vaccines, new biological pest controls. The loss of a species, or even the loss of genetic diversity within a species, is the loss forever of a potential opportunity to improve human welfare.

These "public-service" functions of the global environment cannot be replaced by technology now or in the foreseeable future. This is so in some cases because the process by which the service is provided is not understood scientifically, in other cases because no technological equivalent for the natural process has yet been devised. But in the largest number of cases, the sheer size of the tasks simply dwarfs civilization's capacity to finance, produce, and deploy new technology.

The day is far away when food for billions is grown on synthetic nutrients in greenhouses free of pests and plant diseases, when the wastes of civilization are recycled entirely by technological means, and when all mankind lives in surroundings as sterile and as thoroughly managed as those of an Apollo space capsule. Until that improbable future arrives--and it may never come--the services provided by the orderly operation of natural biological processes will continue to be irreplaceable as well as indispensable.

SOME ELEMENTS OF ECOLOGY

How many of these natural services are actually threatened by human activities? Any of them? All of them? These questions call for a closer look at the operation of the biological systems that provide the services.

## Productivity

Plant communities are at the base of all food webs and are thus the basis of all life on earth. The fundamental measure of performance of a plant community is the rate at which solar energy is captured by photosynthesis to be stored in chemical bonds. In this context, gross primary productivity refers to the total rate of energy capture; net primary productivity is the total minus the rate at which captured energy is used to sustain the life processes of the plants themselves. Thus, net primary productivity measures the rate at which energy is made available to the remainder of the food web. Net community productivity is what remains after the other organisms in the biological community have used part of the net primary productivity to sustain their own life processes. The net community productivity may be exported (for example, in the form of grain from a wheat field) or it may remain in the community in the form of an enlarged standing crop of plants and animals. A community in balance may have no net community productivity at all--that is, the net primary productivity may be entirely burned up by the animals and microorganisms within the community. The productivities of various kinds of ecosystems are shown in Table 1 (5).

A critical point concerning energy flow in ecosystems is that each step in a food chain results in the eventual loss (as heat) of a substantial fraction of the energy transferred. A good rule of thumb for the loss is 90 percent. This means it takes 10,000 kilocalories of corn to produce 1,000 kcal of steer and, more generally, that available energy diminishes 10-fold at each higher trophic level. Thus, the food web is often described as an energy pyramid. Gains in production of animal protein come at high cost in primary calories, and the yield of prized food fishes such as cod and tuna

Table 1. Productivity of various ecosystems (in kilocalories of energy per square meter per year)

| Ecosystem | Net primary productivity | Net community productivity |
|---|---|---|
| Alfalfa field | 15,200 | 14,400 |
| Pine forest | 5,000 | 2,000 |
| Tropical rain forest | 13,000 | little or none |
| Long Island Sound | 2,500 | little or none |

Source:　Odum (5)

is limited by their position on the fourth or fifth trophic level of the oceanic food web.

## Complexity and Stability in Ecosystems

The intricate interlacing of most biological food webs provides a form of insurance against some kinds of disruptions. If one species of herbivore in a complex community is eradicated by disease or drought, the primary carnivores in the community may survive on other kinds of herbivores that are less susceptible to the disease. If a population of predators dwindles for one reason or another, an outbreak of the prey species is unlikely if there are other kinds of predators to fill the gap. Species diversity is one of a number of forms of biological complexity believed by many ecologists to impart stability to ecosystems.

Exactly what is meant by ecological stability? One definition is the ability of an ecosystem that has suffered an externally imposed disturbance to return to the conditions that preceded the disturbance. A more general meaning is that a stable ecosystem resists large, rapid changes in the sizes of its constituent populations. Such changes (called fluctuations or instabilities, depending on the circumstances) entail alteration of the orderly flow of energy and nutri-

ents in the ecosystem. Usually this will mean disruption of the "public-service" functions of the ecosystem, whether or not the instability is severe enough to cause any extinctions of species.

What kinds of complexity can influence stability, and how? Species diversity, already mentioned, presumably imparts stability by providing alternative pathways for the flow of nutrients and energy through the ecosystem. Another possible advantage of a large number of species in a community is that there will then be few empty niches--and thus few opportunities for invasion by a new species from outside the community, with possible disruptive effect. Sheer number of species is not the only determining factor in this type of complexity, however: a degree of balance in population sizes among the species is also required if the capacity of the alternative pathways is to be adequate and the niches solidly occupied. Measures of complexity exist at the population level as well as at that of the community. One is genetic variability, which provides the raw material for resistance against new threats. Another is physiological variability, in the form of a mixed age distribution. (Here the advantage of complexity manifests itself when threats appear that are specific to a particular stage in the organism's life cycle--say, a disease that strikes only juveniles.) There are other forms of complexity as well, including physical complexity of habitat and variety in the geographic distribution of a given species.

The causal links between complexity and stability in ecological systems are by no means firmly established or well understood, and exceptions do exist (6). The evidence of a general correlation between these properties is growing, however, and consists of theoretical considerations of the sort summarized above, general observations of actual ecosystems of widely varying complexity (the relatively simple eco-system of the boreal coniferous forest --the "north woods"--is observed to be less stable than the complex tropical rain forest), and a limited number of controlled laboratory and field experiments.

## Time Scales of Ecological Change

Ecological stability does not mean constancy or stagnation, and ecological change can take place over much longer time spans than the month-to-month or year-to-year time scale of fluctuations and instabilities. Ecological succes-sion refers to the orderly replacement of one community in an area with other communities over periods often measured in decades. Evolution refers to changes in the genetic characteristics of species, brought about by natural selection over time periods ranging from a few generations to hundreds of millions of years. Note that, in terms of human beings, evolution is not the solution to pollution. When significant evolutionary change does take place on the short time scale of a few generations, it is necessarily at the expense of the lives of a large fraction of the population.

HISTORY OF HUMAN ECOLOGICAL DISRUPTION

Ecological disruption on a large scale by human beings is not a new phenomenon. Even before the advent of agriculture, man as a hunter is thought to have contributed to a reduction in the number of species of large mammals inhabiting the earth (7). Much more significant, however, was the era of abuse of soils and habitat that was initiated by the agricultural revolu-tion about 10,000 years ago and has continued up to the present.

One of the best known early exam-ples is the conversion to desert of the lush Tigris and Euphrates valleys, through erosion and salt accumulation resulting from faulty irrigation practices (8). In essence, the down-fall of the great Mesopotamian civili-zation appears to have been the result of an "ecocatastrophe." Overgrazing and poor cultivation practices have contributed over the millennia to the expansion of the Sahara Desert, a process that continues today; and the

Rajasthan desert in India is also believed to be partly a product of human carelessness and population pressure (9).

Much of Europe and Asia were deforested by preindustrial men, beginning in the Stone Age; heavy erosion, recurrent flooding, and nearly permanent loss of a valuable resource were the result. Overgrazing by the sheep of Navajo herdsmen has destroyed large tracts of once prime pastureland in the American Southwest (10). Attempts to cultivate too intensively the fragile soils of tropical rain-forest areas are suspected of being at least in part responsible for the collapse of the Mayan civilization in Central America and that of the Khmers in what today is Cambodia (11). (The famous temples of Angkor Wat were built partly of laterite, the rock-like material that results when certain tropical soils are exposed to the air through cultivation.)

The practice of agriculture--even where quality of soils, erosion, or salt accumulation do not pose problems --may encounter ecological difficulties. The most basic is that agriculture is a simplifier of ecosystems, replacing complex natural biological communities with relatively simple man-made ones based on a few strains of crops. Being less complex, agricultural communities tend to be less stable

than their natural counterparts; they are vulnerable to invasions by weeds, insect pests, and plant diseases, and they are particularly sensitive to extremes of weather and variations in climate. Historically, civilization has attempted to defend its agricultural communities against the instabilities to which they are susceptible by means of vigilance and the application of "energy subsidies"--for example, hoeing weeds and, more recently, applying pesticides and fungicides. These attempts have not always been successful.

The Irish potato famine of the last century is perhaps the best-known example of the collapse of a simple agricultural ecosystem. The heavy reliance of the Irish population on a single, highly productive crop led to 1.5 million deaths when the potato monoculture fell victim to a fungus. To put it another way, the carrying capacity of Ireland was reduced, and the Irish population crashed.

## CONTEMPORARY MAN AS AN ECOLOGICAL FORCE

### Agriculture

Advances in agricultural technology in the last hundred years have not resolved the ecological dilemma of agriculture; they have aggravated it.

420 CONTROVERSIES IN THE EARTH SCIENCES

The dilemma can be summarized this way: civilization tries to manage ecosystems in such a way as to maximize productivity, "nature" manages ecosystems in such a way as to maximize stability, and the two goals are incompatible. Ecological succession proceeds in the directions of increasing complexity. Ecological research has shown that the most complex (and stable) natural ecosystems tend to have the smallest net community productivity; less complex, transitional ecosystems have higher net community productivity; and the highest net community productivities are achieved in the artificially simplified agricultural ecosystems of man (see Table 1). In short, productivity is achieved at the expense of stability.

Of course, mankind would have to practice agriculture to support even a fraction of the existing human population. A tendency toward instability in agricultural ecosystems must be accepted and, where possible, compensated for by technology. However, the trends in modern agriculture-- associated in part with the urgent need to cope with unprecedented population growth and in part with the desire to maximize yields per acre for strictly economic reasons--are especially worrisome ecologically. There are four major liabilities.

1. As larger and larger land areas are given over to farming, the unexploited tracts available to serve as reservoirs of species diversity and to carry out the "public service" functions of natural ecosystems become smaller and fewer (see Table 2) (12).

2. Pressure to expand the area under agriculture is leading to destructive attempts to cultivate land that is actually unsuitable for cultivation with the technologies at hand. For

Table 2.  World land use, 1966 (in millions of $km^2$)

|  | Total | Tilled | Pasture | Forest | Other* |
|---|---|---|---|---|---|
| Europe | 4.9 | 1.5 | 0.9 | 1.4 | 1.1 |
| U.S.S.R. | 22.4 | 2.3 | 3.7 | 9.1 | 7.3 |
| Asia | 27.8 | 4.5 | 4.5 | 5.2 | 13.7 |
| Africa | 30.2 | 2.3 | 7.0 | 6.0 | 15.0 |
| North America | 22.4 | 2.6 | 3.7 | 8.2 | 7.9 |
| South America | 17.8 | 0.8 | 4.1 | 9.4 | 3.5 |
| Oceania | 8.5 | 0.4 | 4.6 | 0.8 | 2.7 |
| Total† | 134.2 | 14.3 | 23.6 | 40.2 | 51.2 |
| Percentage | 100% | 10.6% | 21.3% | 29.9% | 38.2% |

*Deserts, wasteland, built-on land, glaciers, wetlands
†Less Antarctica

Source:  Borgstrom (12)

example, the expansion of agriculture to steep hillsides has led to serious erosion in Indonesia (13), the increasing pressure of slash-and-burn techniques is destroying tropical forests in the Philippines (14), and attempts to apply the techniques of temperate-zone agriculture to the tropical soils of Brazil and Southern Sudan have led to erosion, loss of nutrients, and laterization (15). Overlogging of tropical forests has had similar effects.

3. Even in parts of the world where land area under agriculture is constant or (for economic reasons) dwindling, attempts to maximize yields per acre have led to dramatic increases in the use of pesticides and inorganic fertilizers, which have far-reaching ecological consequences themselves (2).

4. The quest for high yields has led also to the replacement of a wide range of traditional crop varieties all over the world with a few, specially bred, high-yield strains. Unprecendented areas are now planted to a single variety of wheat or rice. This enormous expansion of monoculture has increased the probability and the potential magnitude of epidemic crop failure from insects or disease (16).

## Effects of Pollution on Ecosystems

The expansion and intensification of agriculture has been accompanied by a continuing industrial revolution that

has multiplied many times over both the magnitude and variety of the substances introduced into the biological environment by man. It is useful to classify these substances as qualitative pollutants (synthetic substances produced and released only by man) and quantitative pollutants (substances naturally present in the environment but released in significant additional amounts by man).

Well-known qualitative pollutants are the chlorinated hydrocarbon pesticides, such as DDT, the related class of industrial chemicals called PCB's (polychlorinated biphenyls), and some herbicides. These substances are biologically active in the sense of stimulating physiological changes, but since organisms have had no experience with them over evolutionary time the substances are usually not easily biodegradable. Thus, they may persist in the environment for years and even decades after being introduced and may be transported around the globe by wind and water (17). Their long-term effects will be discovered only by experience, but their potential for disruption of ecosystems is enormous.

Within the category of quantitative pollutants, there are three criteria by which a contribution made by mankind may be judged significant.

1. Man can perturb a natural cycle with a large amount of a substance

ordinarily considered innocuous, in several ways: by overloading part of the cycle (as we do to the denitrifying part of the nitrogen cycle when we overfertilize, leading to the accumulation of nitrates and nitrites in ground water) (18); by destabilizing a finely tuned balance (as we may do to the global atmospheric heat engine, which governs global climate, by adding $CO_2$ to the atmosphere via combustion of fossil fuels); or by swamping a natural cycle completely (as could happen to the climatic balance in the very long term from man's input of waste heat).

2. An amount of material negligible compared to natural global flows of the same substance can cause great damage if released in a sensitive spot, over a small area, or suddenly (for example, the destruction of coral reefs in Hawaii by silt washed from construction sites).

3. Any addition of a substance that can be harmful even at its naturally occurring concentrations must be considered significant. Some radioactive substances fall in this category, as does mercury.

The most general effect of pollution of all kinds of ecosystems is the loss of structure or complexity (19). Specifically, food chains are shortened by pollution via the selective loss of the predators at the top, because predators are more sensitive to envi-

ronmental stresses of all kinds--pesticides, industrial effluents, thermal stress, oxygen deficiency--than are herbivores. This increased sensitivity results from several mechanisms: the predator populations are usually smaller than those of the prey species, so the predator populations tend to have a smaller reservoir of genetic variability and, hence, less probability of evolving a resistant strain; top predators are often exposed to higher concentrations of toxic substances than organisms at lower trophic levels, owing to the phenomenon of biological concentration of pollutants as they move move up the food chain; and, finally, the direct effects of pollution on predators are compounded by the fact that pollutants may reduce the size of the prey population to the point where the predator population cannot be supported. Loss of structure may also occur at lower trophic levels when, for a variety of reasons, one species of herbivore or lower carnivore proves especially sensitive to a particular form of environmental stress. The food web does not have to be eradicated from top to bottom to show significant differential effects.

The adverse effects of loss of structure on the "public-service" functions performed by ecosystems are varied and serious. The vulnerable top predators in marine ecosystems are generally the food fishes most highly prized by man. The loss of predators on land releases checks on herbivorous pests that compete with man for his supply of staple crops. Damaging population outbreaks of these pests-- the classic "instability"--are the result. (A good example of the out- break phenomenon is the experience with pesticides and cotton pests in Peru's Canete Valley (20).) The loss of structure of ecosystems also increases the load on the aquatic food webs of decay, which are already heavily stressed by the burden of mankind's domestic and agricultural wastes. The resulting overload precipitates a vicious progression: oxygen depletion, a shift from aerobic to less efficient anaerobic bacterial metabolism, the accumulation of organic matter, and the release of methane and hydrogen sulfide gas (19).

## Vulnerability of the Sea

The ocean, presently indispensable as a source of animal protein, may be the most vulnerable ecosystem of all. Its vast bulk is deceiving. The great proportion of the ocean's productivity --over 99%--takes place beneath 10% of its surface area, and half of the productivity is concentrated in coastal upwellings amounting to only 0.1% of the surface area (21). The reason is

that productivity requires nutrients, which are most abundant near the bottom, and sunlight, available only near the top. Only in the coastal shelf areas and in upwellings are nutrients and sunlight both available in the same place.

The coastal regions, of course, also receive most of the impact of man's activities--oil spills, fallout from atmospheric pollutants generated on the adjacent land, and river outflow bearing pesticide and fertilizer residues, heavy metals, and industrial chemicals. Almost perversely, the most fertile and critical components in the ocean ecosystem are the estuaries into which the rivers empty; estuaries serve as residence, passage zone, or nursery for about 90% of commercially important fish (3). To compound the problem of pollution, the salt marshes that are an integral part of estuarine biological communities are being destroyed routinely by landfill operations.

Overfishing is almost certainly also taking a heavy toll in the ocean, although it is difficult to separate its effect from that of pollution and destruction of the estuarine breeding grounds and nurseries. The combined result of these factors is clear, however, even if the blame cannot be accurately apportioned. Since World War II, the catches of the East Asian sardine, the California sardine, the

Northwest Pacific salmon, the Scandinavian herring, and the Barents Sea cod
(among others) have entered declines
from which there has been no sign of
recovery (2).

The 1972 world fisheries production of somewhat over 60 million metric
tons was already more than half of the
100 million that some marine biologists
consider to be the maximum sustainable
yield (21). But recent interruptions
in the pattern of continuously increasing yields since World War II (22),
declining catches per unit effort, and
increasing international friction over
fishing rights make it seem unlikely
that theoretical maximum yields will
even be approached.

Flows of Material and Energy

Many people still imagine that
mankind is a puny force in the global
scale of things. They are persuaded,
perhaps by the vast empty spaces visible from any jet airliner in many parts
of the world, that talk of global
ecological disruption is a preposterous
exaggeration. The question of the
absolute scale of man's impact, however, is amenable to quantitative investigation. Natural global flows of
energy and materials can be reasonably
calculated or estimated, and these
provide an absolute yardstick against
which to measure the impact of human
activities.

The results are not reassuring.
As a global geological and biological
force, mankind is today becoming comparable to and even exceeding many
natural processes. Oil added to the
oceans in 1969 from tanker spills,
offshore production, routine shipping
operations, and refinery wastes exceeded the global input from natural seepage by an estimated 20-fold; the minimum estimate for 1980, assuming all
foreseeable precautions, is 30 times
natural seepage (23). Civilization is
now contributing half as much as nature
to the global atmospheric sulfur
burden, and will be contributing as
much as nature by the year 2000 (24).
In industrial areas, civilization's
input of sulfur (as sulfur dioxide) so
overwhelms natural removal processes
that increased atmospheric concentrations and acidic surface water are
found hundreds to thousands of kilometers downwind (25). Combustion of
fossil fuels has increased the global
atmospheric concentration of carbon
dioxide by 10% since the turn of the
century (26). Civilization's contribution to the global atmospheric burden
of particulate matter is uncertain:
estimates range from 5 to 45% of total
annual input (26). Roughly 5% of all
the energy captured by photosynthesis
on earth flows through the agricultural
ecosystems supporting the metabolic

Table 3. Mankind's mobilization of materials (in thousands of metric tons per year)

| Element | Geological rate (river flow) | Man's rate (mining and consumption) |
|---------|------------------------------|-------------------------------------|
| Iron | 25,000 | 319,000 |
| Nitrogen | 3,500 | 30,000 |
| Copper | 375 | 4,460 |
| Zinc | 370 | 3,930 |
| Nickel | 300 | 358 |
| Lead | 180 | 2,330 |
| Phosphorous | 180 | 6,500 |
| Mercury | 3 | 7 |
| Tin | 1.5 | 166 |

Source:  Institute of Ecology (2), SCEP (3)

consumption of human beings and their domestic animals--a few out of some millions of species (27). The rates at which mankind is mobilizing critical nutrients and many metals (including the most toxic ones) considerably exceeds the basic geological mobilization rates as estimated from river flows (see Table 3) (3). Such figures as these do not prove that disaster is upon us, but, combined with the ecological perspective summarized above, they are cause for uneasiness. In terms of the scale of its disruptions, civilization is for the first time operating on a level at which global balances could hinge on its mistakes.

Some of the forms of disruption just described are, of course, amenable in principle to elimination or drastic reduction through changes in tech-nology. Civilization's discharges of oil, sulfur dioxide, and carbon dioxide, for example, could be greatly reduced by switching to energy sources other than fossil fuels. In the case of these pollutants, then, the questions involve not whether the disruptions can be managed but whether they will be, whether the measures will come in time, and what social, economic, and new environmental penalties will accompany those measures. At least one environmental problem is intractable in a more absolute sense, however, and this is the discharge of waste heat accompanying all of civilization's use of energy. We refer here not simply to the well-publicized thermal pollution at the sites of electric generating plants, but to the fact that all the energy we use--as well as what we waste in generating electricity--ultimately arrives in the environment as waste heat. This phenomenon may be understood qualitatively by considering the heat from a light bulb, the heat from a running automobile engine and the heat in the exhaust, the heat from friction of tires against pavement and metal against air, or the heat from the oxidation of iron to rust--to name a few examples. Quantitatively, the ultimate conversion to heat of all the energy we use (most of which occurs near the point of use and almost immediately) is required by the laws of thermodynamics; the phenomenon cannot be averted by technological tricks.

The usual concern with local thermal polution at power plants is that the waste heat, which is usually discharged into water, will adversely affect aquatic life. Most of the waste heat from civilization's energy use as a whole, by contrast, is discharged directly into the atmosphere, and the concern is disruption of climate. Again, it is instructive to compare the scale of human activities with that of

the corresponding natural processes, in this case the natural energy flows that govern climate. One finds that the heat production resulting from (and numerically equal to) civilization's use of energy is not yet a significant fraction of the solar energy incident at the earth's surface on a global average basis (see Table 4); even if the present 5% per annum rate of increase of global energy use persists, it will take another century before civilization is discharging heat equivalent to 1% of incident solar energy at the surface worldwide (28).

Considerably sooner, however, as indicated in Table 4, mankind's heat production could become a significant fraction of smaller natural energy transfers that play a major role in the determination of regional and continental climate (e.g. the kinetic energy of winds and ocean currents and the poleward heat fluxes) (29). It is especially important in this connection that civilization's heat production is and will continue to be very unevenly distributed geographically. Human heat

Table 4.  Energy flows (in billion thermal kilowatts)

| | |
|---|---:|
| Civilization's 1970 rate of energy use | 7 |
| Global photosynthesis | 80 |
| 15 billion people at 10 thermal kilowatts/person | 150 |
| Winds and ocean currents | 370 |
| Poleward heat flux at 40° north latitude | 5,300 |
| Solar energy incident at earth's surface | 116,000 |

Sources:  Woodwell (27), Sellers (29), Hubbert (29)

production already exceeds 5% of incident solar radiation at the surface over local areas of tens of thousands of square kilometers, and will exceed this level over areas of millions of square kilometers by the year 2000 if present trends persist (26). Such figures could imply substantial climatic disruptions. In addition to the effects of its discharge of heat, civilization has the potential to disrupt climate through its additions of carbon dioxide and particular matter to the atmosphere, through large-scale alteration of the heat-transfer and moisture-transfer properties of the surface (e.g. agriculture, oil films on the ocean, urbanization), through cloud formation arising from aircraft contrails, and, of course, through the combined action of several or all of these disruptions.

Much uncertainty exists concerning the character and imminence of inadvertent climate modification through these various possibilities. It is known that a global warming of a few degrees centigrade would melt the icecaps and raise sea level by 80 meters, submerging coastal plains and cities. A few degrees in the opposite direction would initiate a new ice age. Although such global warming or cooling is certainly possible in principle, a more complicated alteration of climatic patterns seems a more probable and per-

haps more imminent consequence of the very unevenly distributed impacts of civilization's use of energy. It is particularly important to note that the consequences of climatic alteration reside not in any direct sensitivity of humans to moderate changes in temperature or moisture, but rather in the great sensitivity of food production to such changes (30) and, perhaps, in the possible climate-related spread of diseases into populations with no resistance against them (4).

The effect of climate on agriculture was once again dramatically demonstrated in early 1973. Because of "bad weather," famine was widespread in sub-Saharan Africa and was starting in India. Southeast Asia had small rice harvests, parts of Latin American were short of food, and crops were threatened in the United States and the Soviet Union. If there is another year of monsoon failure in the tropics and inclement weather in the temperate zones, the human death rate will climb precipitously. A telling symptom of overpopulation is mankind's inability to store sufficient carry-over food supplies in anticipation of the climatic variations that are a regular feature of the planet Earth.

ROLE OF POPULATION

It is beyond dispute that a population too large to be fed adequately in the prevailing technological and organizational framework, as is the case for the globe today, is particularly vulnerable to environmental disruptions that may reduce production even below normal levels. More controversial, however, are the roles of the size, growth rate, and geographic distribution of the human population in causing such environmental disruptions, and it is to this issue that we now turn.

## Multiplicative Effect

The most elementary relation between population and environmental deterioration is that population size acts as a multiplier of the activities, consumption, and attendant environmental damages associated with each individual in the population. The contributing factors in at least some kinds of environmental problems can be usefully studied by expressing the population/environment relation as an equation:

> environmental disruption=
> population x consumption per
> person x damage per unit of
> consumption

Needless to say, the numerical quantities that appear in such an equation will vary greatly depending on the problem under scrutiny. Different forms of consumption and

technology are relevant to each of the many forms of environmental disruption. The population factor may refer to the population of a city, a region, a country, or the world, depending on the problem being considered. (This point, of course, raises the issue of population distribution.) The equation, therefore, represents not one calculation but many.

For problems described by multiplicative relations like the one just given, no factor can be considered unimportant. The consequences of the growth of each factor are amplified in proportion to the size and the rate of growth of each of the others. Rising consumption per person has greater impact in a large population than in a small one-- and greater impact in a growing population than in a stationary one. A given environmentally disruptive technology, such as the gasoline-powered automobile, is more damaging in a large, rich population (many people own cars and drive them often) than in a small, poor one (few people own cars, and those who do drive them less). A given level of total consumption (population times consumption per person) is more damaging if it is provided by means of a disruptive technology, such as persistent pesticides, than if provided by means of a relatively nondisruptive one, such as integrated pest control.

The quantitative use of the population/environment equation is best illustrated by example. Suppose we take as an index of environmental impact the automotive emissions of lead in the United States since World War II. The appropriate measure of "consumption" is vehicle-miles per person, which increased twofold between 1946 and 1967. The impact per unit of consumption in this case is emissions of lead per vehicle-mile, which increased 83%, or 1.83-fold, in this period (31). Since the U.S. population increased 41%, or 1.41-fold, between 1946 and 1967, we have,

relative increase in emissions =

1.41 x 2.0 x 1.83 = 5.16 or 416%

Note that the dramatic increase in the total impact arose from rather moderate but simultaneous increases in the multiplicative contributing factors. None of the factors was unimportant--if population had not grown in this period, the total increase would have been 3.66-fold rather than 5.16-fold. (Contrast this result with the erroneous conclusion, arising from the assumption that the contributing factors are additive rather than multiplicative, that a 41% increase in population "explains" only one-tenth of 416% increase in emissions.)

Calculations such as the foregoing can be made for a wide variety of pollutants, although with frequent difficulty in uncovering the requisite data. Where data are available, the results show that the historical importance of population growth as a multiplicative contributor to widely recognized environmental problems has been substantial (32).

Between 1950 and 1970, for example, the world population increased by 46%. By regions, the figures were: Africa, 59%; North America, 38%; Latin America, 75%; Asia, 52%; Europe, 18%; Oceania, 54%; Soviet Union, 35% (33). On the assumption (which will be shown below to be too simplistic) that the patterns of technological change and rising consumption per capita that were experienced in this period would have been the same in the absence of population growth, one can conclude that the absolute magnitudes of damaging inputs to the environment in 1970 were greater by these same percentages than they would have been if population had remained at its 1950 value. Another way of saying this is that, under our simplistic assumption, the magnitude of damaging inputs to the global environment in 1970--a very large figure--would have been only 68% as large if population had not grown between 1950 and 1970. (This follows from the relation: 1970 inputs in absence of population growth equal actual 1970 inputs times 1950 population divided by 1970 population.)

Not only has population growth been important in absolute terms as a contributor of environmental damage, but it has been important relative to other sources of such damage. Perhaps the best way to illustrate this fact is with statistics for energy consumption per person, probably the best aggregate measure of both affluence and technological impact on the environment. One finds that energy consumption per person worldwide increased 57% between 1950 and 1970 (33,34). By this measure then, and under our simple assumption that population growth and trends in affluence and technology were independent, one finds that population growth in the period 1950-1970 was almost equal to the combined effect of

rising affluence and technological change as a contributor of damaging inputs to the environment. (The comparison of population growth and energy consumption broken down by major geographical regions is given in Table 5.) We shall argue, moreover, that the effect of the simplistic assumption of independence of population and other factors is more probably to underestimate the role of population than to overestimate it.

## Nonlinear Effects

While it is useful to understand what proportion of the historical increase in specific environmental problems has been directly attributable to the multiplier effect of population growth, there is a more difficult and perhaps more important question than this historical/arithmetical one. Specifically, under what circumstances may nonlinear effects cause a small increase in population to generate a disproportionately large increase in environmental disruption? These effects fall into two classes. First, population change may cause changes in consumption per person or in impact upon the environment per unit of consumption. Second, a small increase in impact upon the environment--generated in part by population change and in part by unrelated changes in the other

Table 5. Percentage increases in population and energy consumption per capita between 1950 and 1970

|  | Population (%) | Energy/person (%) |
|---|---|---|
| World | 46 | 57 |
| Africa | 59 | 73 |
| North America | 38 | 43 |
| Latin America | 75 | 122 |
| Asia | 52 | 197 |
| Europe | 18 | 96 |
| Oceania | 54 | 54 |

Source:   United Nations (33,34)

multiplicative factors--may stimulate a disproportionately large environmental change.

An obvious example in the first category is the growth of suburbs in the United States at the expense of central cities, which has had the effect of increasing the use of the automobile. Another is the heavy environmental costs incurred in the form of large water projects when demand (population times demand per person) exceeds easily exploited local supplies. Still another example is that of diminishing-returns phenomena in agriculture in which increases in yield needed to feed new mouths can be achieved only by disproportionate increases in inputs such as fertilizer and pesticides.

Many phenomena that have the effect of generating disproportionate consequences from a given change in demographic variables cannot easily be expressed in the framework of a single equation. One such class of problems involves technological change--the substitution of new materials or processes for old ones that provided the same types of material consumption. Obvious examples are the substitution of nylon and rayon for cotton and wool, of plastics for glass and wood and metals, of aluminum for steel and copper. Such substitutions may be necessitated by increasing total

demand, or they may be motivated by other factors such as durability and convenience. Substitutions or other technological changes that are motivated by the pressure of increased total demand, and that lead to increases in environmental impact per unit of consumption, should be considered as part of the environmental impact of population growth.

Environmental disruption is not, however, measured strictly by man's inputs to the environment--what we do to it. Equally important is how the environment responds to what we do to it. This response itself is often nonlinear: a small change in inputs may precipitate a dramatic response. One example is the existence of thresholds in the response of individual organisms to poisons and other forms of "stress." Fish may be able to tolerate a $10^{\circ}$ rise in water temperature without ill effect, whereas a $12^{\circ}$ rise would be fatal. Carbon monoxide is fatal to human beings at high concentrations but, as far as we know, causes only reversible effects at low concentrations. Algal blooms in overfertilized lakes and streams are examples of exceeding a threshold for the orderly cycling of nutrients in these biological systems.

Another nonlinear phenomenon on the response side of environmental problems involves the simultaneous action of two or more inputs. A disturbing example is the combined effect of DDT and oil spills in coastal waters. DDT is not very soluble in sea water, so the concentrations to which marine organisms are ordinarily exposed are small. However, DDT is very soluble in oil. Oil spills therefore have the effect of concentrating DDT in the surface layer of the ocean, where much of the oil remains and where many marine organisms spend part of their time (23). These organisms are thus exposed to far higher concentrations of DDT than would otherwise be possible, and as a result, the combined effect of oil and DDT probably far exceeds their individual effects. Many other synergisms in environmental systems are known or suspected: the interaction of sulfur dioxide and particulate matter in causing or aggravating lung disease; the interaction of radiation exposure and smoking in causing lung cancer; the enhanced toxicity of chlorinated hydrocarbon pesticides when plasticizers are present (35).

The exact role of population change varies considerably among the various forms of nonlinear behavior just described. A nonlinearity in the environment's response to growing total input--such as a threshold effect--increases the importance of all the multiplicative contributors to the input equally, whether or not population and the other contributors are causally related. Some other forms of nonlinearity, such as diminishing returns and certain substitutions, would occur eventually whether population or consumption per capita grew or not. For example, even a constant demand for copper that persisted for a long time would lead eventually to increasing expenditures of energy per pound of metal and to substitution of aluminum for copper in some applications. In such instances, the role of population growth--and that of rising consumption per capita--is simply to accelerate the onset of diminishing returns and the need for technological change, leaving less time to deal with the problems created and increasing the chances of mistakes. With respect to other phenomena, such as the effects of population concentration on certain forms of consumption and environmental impact, population change is clearly the sole and direct cause of the nonlinearity (e.g. additional transportation costs associated with suburbanization).

TIME FACTORS

The Pattern of Growth

All rational observers agree that no physical quantity can grow exponentially forever. This is true, for example, of population, the production of energy and other raw materials, and the generation of wastes. But is there anything about the 1970s--as opposed, say, to the 1920s or 1870s--that should make this the decade in which limits to growth become apparent? It should not be surprising that, when limits do appear, they will appear suddenly. Such behavior is typical of exponential growth. If twenty doublings are possible before a limit is reached in an exponentially growing process (characterized by a fixed doubling time if the growth rate is constant), then the system will be less than half "loaded" for the first nineteen doublings--or for 95% of the elapsed time between initiation of growth and exceeding the limit. Clearly, a long history of exponential growth does not imply a long future.

But where does mankind stand in its allotment of doublings? Are we notably closer to a limit now than we were 50 years ago? We are certainly moving faster. The number of people added to the world population each year in the 1970s has been about twice what it was in the 1920s. And according to one of the better indices of aggregate environmental disruption, total energy consumption, the annual increase in man's impact on the environment (in absolute magnitude, not percentage) is ten times larger now than then (33,36).

We have seen, moreover, that man is already a global ecological force, as measured against the yardstick of natural processes. While the human population grows at a rate that would double our numbers in 35 years, ecological impact is growing much faster. The 1970 M.I.T.-sponsored Study of Critical Environmental Problems estimated that civilization's demands upon the biological environment are increasing at about 5% per year, corresponding to a doubling time of 14 years (3). Continuation of this rate would imply a fourfold increase in demands on the environment between 1972 and the year 2000. It is difficult to view such a prospect with complacency.

## Momentum, Time Lags, and Irreversibility

The nature of exponential growth is such that limits can be approached with surprising suddenness. The likelihood of overshooting a limit is made even larger by the momentum of human population growth, by the time delays between cause and effect in many environmental systems, and by the fact that some kinds of damage are irreversible by the time they are visible.

The great momentum of human population growth has its origins in deep-seated attitudes toward reproduction and in the age composition of the world's population--37% is under 15 years of age. This means there are far more young people who will soon be reproducing--adding to the population-- than there are old people who will soon be dying--subtracting from it. Thus, even if the momentum in attitudes could miraculously be overcome overnight, so that every pair of parents in the world henceforth had only the number of children needed to replace themselves, the imbalance between young and old would cause population to grow for 50 to 70 years more before leveling off. The growth rate would be falling during this period, but population would still climb 30% or more during the transition to stability. Under extraordinarily optimistic assumptions about when re-placement fertility might really become the worldwide norm, one concludes that world population will not stabilize below 8 billion people (37).

The momentum of population growth manifests itself as a delay between the time when the need to stabilize popu-lation is perceived and the time when stabilization is actually accomplished. Forces that are perhaps even more firmly entrenched than those affecting population lend momentum to growth in per capita consumption of materials. These forces create time lags similar to that of population growth in the in-evitable transition to stabilized levels of consumption and technological

reform. Time delays between the initiation of environmental insults and the appearance of the symptoms compound the predicament because they postpone recognition of the need for any cor-rective action at all.

Such environmental time delays come about in a variety of ways. Some substances persist in dangerous form long after they have been introduced into the environment (mercury, lead, DDT and its relatives, and certain radioactive materials are obvious examples). They may be entering food webs from soil, water, and marine sedi-ments for years after being deposited there. The process of concentration from level to level in the food web takes more time. Increases in exposure to radiation may lead to increases in certain kinds of cancer only after decades and to genetic defects that first appear in later generations. The consequences of having simplified an environmental system by inadvertently wiping out predators or by planting large areas to a single high-yield grain may not show up until just the right pest or plant disease comes along a few years later.

Unfortunately, time lags of these sorts usually mean that, when the symp-toms finally appear, corrective action is ineffective or impossible. Species that have been eradicated cannot be restored. The radioactive debris of atmospheric bomb tests cannot be re-concentrated and isolated from the environment, nor can radiation exposure be undone. Soil that has been washed or blown away can be replaced by natural processes only on a time scale of centuries. If all use of persistent pesticides were stopped tomorrow, the concentrations of these substances in fish and fish-eating birds might continue to increase for some years to come.

## VIGOROUS ACTION NEEDED

The momentum of growth, the time delays between causes and effects, and the irreversibility of many kinds of damage all increase the chances that mankind may temporarily exceed the carrying capacity of the biological environment. Scientific knowledge is not yet adequate to the task of defining that carrying capacity unambiguously, nor can anyone say with assurance how the consequences of overshooting the carrying capacity will manifest themselves. Agricultural failures on a large scale, dramatic loss of fisheries productivity, and epidemic disease initiated by altered environmental conditions are among the possibilities. The evidence presented here concerning the present scale of man's ecological disruption and its rate of increase suggests that such

possibilities exist within a time frame measured in decades, rather than centuries.

All of this is not to suggest that the situation is hopeless. The point is rather that the potential for grave damage is real and that prompt and vigorous action to avert or minimize the damage is necessary. Such action should include measures to slow the growth of the global population to zero as rapidly as possible. Success in this endeavor is a necessary but not a sufficient condition for achieving a prosperous yet environmentally sustainable civilization. It will also be necessary to develop and implement programs to alleviate political tensions, render nuclear war impossible, divert flows of resources and energy from wasteful uses in rich countries to necessity-oriented uses in poor ones, reduce the environmental impact and increase the human benefits resulting from each pound of material and gallon of fuel, devise new energy sources, and, ultimately, stabilize civilization's annual throughput of materials and energy.

There are, in short, no easy single-faceted solutions, and no component of the problem can be safely ignored. There is a temptation to "go slow" on population limitation because this component is politically sensitive and operationally difficult, but the temptation must be resisted. The other approaches pose problems too, and the accomplishments of these approaches will be gradual at best. Ecological disaster will be difficult enough to avoid even if population limitation succeeds; if population growth proceeds unabated, the gains of improved technology and stabilized per capita consumption will be erased, and averting disaster will be impossible.

## REFERENCES

1. See, e.g., National Research Council/National Academy of Sciences. 1969. Resources and Man. San Francisco: W. H. Freeman and Co. Paul R. Ehrlich and John P. Holdren. 1969. Population and panaceas--a technological perspective. BioScience 12:1065-71.

2. Institute of Ecology. 1972. Man in the Living Environment. Madison: University of Wisconsin Press.

3. Report of the Study of Critical Environmental Problems (SCEP). 1970. Man's Impact on the Global Environment: Assessment and Recommendations for Action. Cambridge: M.I.T. Press.

4. Jacques M. May. 1972. Influence of environmental transformation in changing the map of disease. In The Careless Technology: Ecology and International Development, M. Taghi Farvar and John P. Milton, eds. Garden City, N.Y.: The Natural History Press.

5. Eugene P. Odum. 1971. Fundamentals of Ecology. 3rd ed. Philadelphia: Saunders, p. 46.

6. E. O. Wilson and W. A. Bossert. 1971. A Primer of Population Biology. Stamford, Conn: Sinauer Associates; and Brookhaven National Laboratory. 1969. Diversity and stability in ecological systems. Brookhaven Symposia in Biology, N. 22, BNL 50175 C-56. Upton, N.Y.: Brookhaven National Laboratory.

7. P. S. Martin and T. E. Wright, Jr., eds. 1957. Pleistocene Extinctions: The Search for a Cause. New Haven: Yale University Press.

8. Thorkild Jacobsen and Robert M. Adams. 1958. Salt and silt in ancient Mesopotamian agriculture. Science 128:1251-58.

9. M. Kassas. 1970. Desertification versus potential for recovery in circum-Saharan territories. In Arid Lands in Transition. Washington, D.C.: American Association for Advancement of Science. B. R. Seshachar. 1971. Problems of environment in India. In International Environmental Science. Proceedings of a joint

colloquium before the Committee on Commerce, U.S. Senate, and the Committee on Science and Astronautics, House of Representative, 92nd Congress. Washington, D.C.: U.S. Government Printing Office.

10. Carl O. Sauer. 1956. The agency of man on earth. In Man's Role in Changing the Face of the Earth, William L. Thomas, Jr., ed. Chicago: University of Chicago Press, p. 60.

11. Jeremy A. Sabloff. 1971. The collapse of classic Maya civilization. In Patient Earth, John Harte and Robert Socolow, eds. New York: Holt, Rinehart and Winston, p. 16.

12. Georg Borgstrom. 1969. Too Many. N.Y.: Macmillan.

13. Albert Ravenholt. 1974. Man-land-productivity microdynamics in rural Bali. In Population: Perspective, 1973, Harrison Brown, John Holdren, Alan Sweezy, and Barbara West, eds. San Francisco: Freeman-Cooper.

14. Albert Ravenholt. 1971. The Philippines. In Population: Perspective, 1971, Harrison Brown and Alan Sweezy, eds. San Francisco: Freeman-Cooper, pp. 247-66.

15. Mary McNeil. 1972. Lateritic soils in distinct tropical environments: Southern Sudan and Brazil. In The Careless Technology, op cit., pp. 591-608.

16. O. H. Frankel et al. 1969. Genetic dangers in the Green Revolution. Ceres 2(5):35-37 (Sept.-Oct.); and O. H. Frankel and E. Bennett, eds. 1970. Genetic Resources in Plants--Their Exploration and Conservation. Philadelphia: F. A. Davis Co.

17. See, e.g., R. W. Risebrough, R. J. Huggott, J. J. Griffin, and E. D. Goldberg. 1968. Pesticides: Transatlantic movements in the northeast trades. Science 159: 1233-36. G. M. Woodwell, P. P. Craig, and H. A. Johnson. 1971. DDT in the biosphere: Where does it go? Science 174:1101-07.

18. D. R. Keeney and W. R. Gardner. 1970. The dynamics of nitrogen transformations in the soil. In Global Effects of Environmental Pollution, S. F. Singer, ed. New York: Springer Verlag, pp. 96-103.

19. G. M. Woodwell. 1970. Effects of pollution on the structure and physiology of ecosystems. Science 168:429-33.

20. Teodoro Boza Barducci. 1972. Ecological consequences of pesticides used for the control of cotton insects in Canete Valley, Peru. In The Careless Technology, op. cit., pp. 423-38.

21. John H. Ryther. 1969. Photo-
    synthesis and fish production in
    the sea. Science 166:72-76.

22. FAO. 1972. State of Food and
    Agriculture 1972. Rome: Food
    and Agriculture Organization.

23. Roger Revelle, Edward Wenk,
    Bostwick Ketchum, and Edward R.
    Corino. 1971. Ocean pollution by
    petroleum hydrocarbons. In Man's
    Impact on Terrestrial and Oceanic
    Ecosystems, William H. Matthews,
    Frederick E. Smith, and Edward D.
    Goldberg, eds. Cambridge: M.I.T.
    Press, p. 297.

24. W. W. Kellogg, R. D. Cadle, E. R.
    Allen, A. L. Lazrus, and E. A.
    Martell. 1972. The sulfur cycle.
    Science 175:587.

25. Gene E. Likens, F. Herbert Bor-
    mann, and Noye M. Johnson. 1972.
    Acid rain. Environment 14(2):33.

26. Report of the Study of Man's
    Impact on Climate. 1971. Inad-
    vertent Climate Modification.
    Cambridge: M.I.T. Press, pp. 188-
    92.

27. George M. Woodwell. 1970. The
    energy cycle of the biosphere.
    Scientific American 223(3):64-74.

28. John P. Holdren. 1971. Global
    thermal pollution. In Global
    Ecology, J. P. Holdren and P. R.
    Ehrlich, eds. New York: Harcourt
    Brace Jovanovich.

29. William D. Sellers. 1965. Phys-
    ical Climatology. Chicago: Uni-
    versity of Chicago Press. M.
    King Hubbert. 1971. Energy re-
    sources. In Environment, William
    Murdoch, ed. Stamford, Conn.:
    Sinauer Associates.

30. Sherwood B. Idso. 1971. Poten-
    tial effects of global tempera-
    ture change on agriculture. In
    Man's Impact on Terrestrial and
    Oceanic Ecosystems, op. cit., p.
    184.

31. Barry Commoner. 1972. The envi-
    ronmental cost of economic growth.
    In Population Resources and the
    Environment, Vol. 3 of the Re-
    search Reports of the Commission
    on Population Growth and the
    American Future. Washington,
    D.C.: U.S. Government Printing
    Office, p. 339.

32. Paul R. Ehrlich and John P.
    Holdren. 1972. One-dimensional
    ecology. Science and Public
    Affairs: Bull. Atomic Sci. 28(5):
    16-27.

33. United Nations Statistical Office.
    1972. Statistical Yearbook, 1971.
    N.Y.: United Nations Publishing
    Service.

34. United Nations Statistical Office.
    1954. Statistical Yearbook, 1953.
    N.Y.: United Nations Publishing
    Service.

35. American Chemical Society. 1969.
    Cleaning Our Environment: The

Chemical Basis for Action. Washington, D.C.: American Chemical Society. U.S. Congress, Joint Committee on Atomic Energy. 1967. Hearings on Radiation Exposure of Uranium Miners, Parts 1 and 2. Washington, D.C.: U.S. Government Printing Office. E. P. Lichtenstein, K. R. Schulz, T. W. Fuhremann, and T. T. Liang. 1969.

Biological interaction between plasticizers and insecticides. J. Econ. Entom. 62(4):761-65.

36. Joel Darmstadter et al. 1971. Energy in the World Economy. Baltimore: Johns Hopkins Press.

37. Nathan Keyfitz. 1971. On the momentum of population growth. Demography 8(1):71-80.

†